MW00760923

Ordering and Phase Transitions in Charged Colloids

COMPLEX FLUIDS AND FLUID MICROSTRUCTURES

Series Editor

Raj Rajagopalan
Department of Chemical Engineering
and Physics
University of Houston
Houston, TX 72204-4792
USA

Advisory Board

Norio Ise
Central Laboratory
Rengo Co., Ltd.
186-1-4 Ohhiraki
Fukushima
Osaka 553
Japan

Jacob N. Israelachvili
Department of Chemical Engineering
University of California at Santa Barbara
Santa Barbara, CA 93106
USA

James Andrew McCammon
J.E. Mayer Chair in Theoretical Chemistry
Professor of Pharmacology
University of California at San Diego
La Jolla, CA 92093-0365
USA

Other Books in the Series

Kenneth S. Schmitz
Macroions in Solution and Colloidal Suspension

Ordering and Phase Transitions in Charged Colloids

Edited by

A. K. Arora
B. V. R. Tata

 WILEY-VCH

A. K. Arora
B. V. R. Tata
Materials Science Division
Indira Gandhi Centre for Atomic Research
Kalpakkam, Tamil Nadu 603 102
India

This book is printed on acid-free paper.

Library of Congress Cataloging-in-Publication Data

Ordering and phase transitions in charged colloids / edited by A.K. Arora and B.V.R. Tata
    p.   cm. — (Complex fluids and fluid microstructures)
   Includes bibliographical references (p. –) and index.

   1. Colloidal crystals–Structure. 2. Phase transformations
(Statistical physics)   I. Arora, A.K. (Akhilesh Kumar), 1952–
II. Tata, B.V.R. (Butchi Venkata Rao), 1956– . III. Series.
QD549.2.C64074 1995
541.3'45–dc20   95-30600
                                                               CIP

ISBN 1-47118-630-9 VCH Publishers, Inc.

Printing History:
10 9 8 7 6 5 4 3 2 1

Published jointly by

VCH Publishers, Inc.   VCH Verlagsgesellschaft mbH   VCH Publishers (UK) Ltd.
333 7th Ave.   P.O. Box 10 11 61   8 Wellington Court
New York, New York 10001   69451 Weinheim, Germany   Cambridge, CB1 1HZ
                                                         United Kingdom

Complex Fluids and Fluid Microstructures Series Preface

"Knowledge is one. Its division into subjects is a concession to human weakness."[1]

Bertrand Russell once wrote, with his characteristic dry humor, that

"Ordinary language is totally unsuited for expressing what physics really asserts, since the words of everyday life are not sufficiently abstract. Only mathematics and mathematical logic can say as little as the physicist means to say."[2]

In this sense, the term *complex fluids* is a happy compromise, for the words come from everyday language and yet convey, although not intentionally, very little to the uninitiated! Hiding behind these words, however, is a world of materials so rich in variety, phenomena, behavior, and practical utility that it has spawned an interdisciplinary adventure into its mysteries in the past decade in physics, chemistry, biology, and engineering.

---

[1]H.J. Mackinder, *Proc. R. Georg. Soc.* **9**, 141–60 (1887).

[2]B. Russell, *The Scientific Outlook*, Allen & Unwin, London, 1931.

Complex fluids (or, "soft matter," as they are also known) are materials composed of "supramolecular" species (e.g., polymers, surfactants, colloidal particles, etc.) whose collective behavior manifests itself in unusual ways structurally, mechanically, and chemically. Notable examples include liquid-crystalline materials, polymer solutions and gels, Coulombic Fluids, surfactant assemblies, vesicles and similar "fluid microstructures," self-assembled films and Langmuir–Blodgett layers, and colloidal dispersions. As noted by de Gennes[3] in his Nobel Lecture, the term *complex fluids* is meant to draw attention to two of the major features of this class of materials, viz., the *complexity* of the structure of these materials and their *flexibility* (which may, however, vanish or be modified drastically as a result of mild chemical or physical action). These materials are thus characterized by weak collective interactions among their supramolecular constituents, by mechanical softness, by sensitivity to the effects of thermal fluctuations, and, as a consequence, by a rich variety of physical and chemical behavior.

While the study of complex fluids and their behavior undoubtedly poses a significant challenge to modern science and promises, in return, rich intellectual rewards, the practical impact of such materials is not insignificant. The implications of being able to understand and manipulate complex fluids of biological significance alone are immense, be they biological membranes, pharmaceutical formulations, or drug delivery systems. Other applications are equally numerous; these include the many uses liquid-crystalline materials have found in display devices and the well-known and potential uses of polymers, surfactant solutions, microemulsions, and colloids in detergency, oil recovery, and advanced materials such as ceramics, electrophoretic imaging devices, and photonic "crystals."

The purpose of the *Complex Fluids and Fluid Microstructures* series is to highlight and consolidate the progress in the general area of complex fluids and soft condensed matter, both in pure sciences and in applied fields. In general, each volume will focus on a specific topic and develop it at a level suitable for beginning graduate students or advanced research scientists in chemistry, physics, biology, or engineering. Where appropriate, however, textbooks at senior undergraduate level in the above disciplines also will be included. It is our hope that in addition to documenting and synthesizing the emerging advances in this fascinating field, the Series will also serve as a stimulant for further progress.

Raj Rajagopalan
Series Editor

---

[3]P.G. de Gennes, *Science* **256**, 495–97 (1992).

Preface

Colloidal dispersions, apart from their numerous applications such as those in biotechnology, catalysis, rotary magnetic seals, and optical Bragg filters, have many interesting properties that almost mimic all the phases of condensed matter. Due to the extremely low value of elastic constants, these are also classified as soft condensed matter. The molar elastic constants and many thermodynamic parameters of colloidal crystals are found to be of the same order of magnitude as those in atomic solids, suggesting that the interparticle interaction strength in colloids is similar to that of atomic potentials. This has led many scientists to regard these as model systems that can be used to simulate the condensed matter. Several effects and phenomena are now being investigated with the help of colloidal dispersions, which cannot always be easily probed in atomic substances. For example, it has been possible to study the kinetics of nucleation and growth from undercooled melt, which occurs over a time scale of several seconds in the case of colloidal crystals, and confirm the predictions of the theoretical models. Similarly, shear stresses much in excess of the shear strength of colloidal crystals could be applied in the laboratory, and properties of exotic nonequilibrium phases have been investigated. It is not easy to accomplish this in atomic systems except under shock conditions. Another feature that makes charged colloids very attractive is the ease with which one can tailor the effective interparticle interaction. It is achieved by the variation of the particle concentration and ionic strength. This allows one to systematically study the structural transitions between different phases. The parameters that influence the stability and induce phase transitions are shear, ionic strength, photothermal effects, and polydispersity. Structural ordering in colloidal dispersions is conventionally investigated using light scattering. This is because the average interparticle separation is of the order of the wavelength of light. However, small-angle X-ray and neutron scattering and optical microscopy also have been extensively used to probe colloidal dispersions. During the last decade many new and important results have been obtained in the area of ordering and phase transitions in colloidal systems. Some of these include shear-induced melting and recrystallization, laser-induced freezing and melting, crystalline-to-amorphous transitions in polydisperse systems, the crossover behavior of diffusion mechanisms across the glass transition, and the observations of stable voids and the vapor–liquid condensation in charged colloids. These results have significantly contributed to the current understanding of the subject. This monograph reviews many of these most recent results.

This volume covers a wide range of experimental and theoretical investigations carried out to obtain a comprehensive understanding of ordering and phase transitions in charged colloids. The results obtained using the experimental techniques such as video microscopy, optical Bragg and Kossel diffraction, light scattering, and ultra-small-angle X-ray scattering are extensively discussed. Further, the role of the theoretical tools such as density functional theory, computer simulations, and the inversion methods in understanding various phase transitions and effective interaction potential is also discussed in detail. The contents of the book are organized in thirteen chapters. After the introductory chapter, four chapters deal with different complementary experimental investigations of ordering and colloidal crystal growth. The next five chapters discuss phase transitions and colloidal instabilities driven by ionic strength, polydisperisity, laser optical fields and other modulating fields. The last three chapters review the theory of colloidal interparticle interactions and inversion methods.

The situation with respect to the current understanding of interparticle interactions in charged colloids needs special mention. At present in the scientific community associated with the area of colloidal science, there appears to be two schools of thought. The first believes that the effective interparticle interaction is always repulsive at all interparticle distances and the traditional Derjaguin–Landau–Vervey–Overbeek (DLVO) potential is sufficient to explain the results. The other, which is recent, emphasizes the existence of attraction at large interparticle separations in addition to a repulsion that dominates at small distances. Attraction was initially proposed by Ise and co-workers. The recent theoretical formalism of interparticle interaction in charged colloids by Sogami could explain a large number of experimental results. However, scientists from each of these schools have made several claims for the correctness of their theoretical formalisms and pointed out the flaws in the arguments of the other. We have been fortunate in bringing together contributors from the two schools in this volume to present their latest views. Extra care has been taken to present their scientific contents with minimal editing and without diluting even the emotion. Thus one can find in this volume a given phenomenon being explained using the DLVO as well as the Sogami potential. This clearly shows that the debate is still on and only new, carefully devised experiments can cast further light on the subject.

This introduction to the book is given in Chapter 1. In this chapter Arora and Rajagopalan discuss the significance of research on colloidal dispersions with reference to its implications on condensed matter research. The new results on structural ordering, growth laws of colloidal crystals, and shear-induced ordering are briefly highlighted. A review of the current understanding of the colloidal interparticle interaction is presented in the context of the existing controversy.

In Chapter 2 Schätzel discusses the small-angle light scattering experiments carried out to investigate the growth of colloidal crystals. The early nucleation and growth and the subsequent coarsening or ripening regimes are identified. Diffusion and the interface-limited linear growth mechanisms are found to operate in dilute and concentrated suspensions, respectively. Growth and ripening kinetics at different concentrations are also reported. Yoshiyama and Sogami describe, in Chapter 3, optical Kossel diffraction from ordered colloids as an alternate technique to probe the structure and evolution of colloidal crystals. The three-point method and the two-ring method are reviewed in detail to analyze quantitatively the diffraction images. The complete sequence of evolution, from two-dimensional hexagonal closed-packed to three-dimensional body-centered cubic (BCC) structures, involving several intermediate stages, is characterized and discussed. The technique of video imaging of colloidal dispersions and the analysis of these images to obtain pair correlation functions are described in Chapter 4 by Grier and Murray. The interface between the face-centered cubic (FCC) and fluidlike ordered regions is found to be a single lattice spacing thick. Various freezing and melting rules, such as the Hansen–Verlet and the Lindemann criteria, are discussed in the contect of FCC-fluid transitions in colloidal systems. The results of nonequilibrium freezing from deeply supercooled fluid are also reported. Ise, Ito, Matsuoka, and Yoshida report, in Chapter 5, the experimental results on colloidal ordering by combining several techniques such as video and laser scanning microscopy and ultra-small-angle X-ray scattering (USAXS). The observation of the inhomogeneous nature of colloidal dispersions is pointed out, and the existence of large voids in the interior of the suspensions is reported. The non-space-filling nature of the suspension is also concluded from an analysis of the USAXS results. Possible implications of these results are discussed in the context of the existence of long-range attraction in charged colloids.

In Chapter 6 Tata and Arora review the recently reported vapor–liquid condensation in charged colloids and a transition to a reentrant homogeneous state, when deionized. Light scattering is used to obtain the structure factor of the concentrated and the dilute phases. The authors argue that this phenomenon demonstrates the existence of long-range attraction and the reentrant transition is unique to charged colloids. The estimated interparticle distance in the liquidlike phase is shown to agree with the prediction of the Sogami potential. A phase diagram calculated using this potential is found to be consistent with the Monte Carlo simulation results. An order–disorder transition in colloidal crystals driven by charge polydispersity is described in Chapter 7 by Arora and Tata. Monte Carlo simulations show that a colloidal crystal spontaneously evolves into an amorphous state if the polydispersity is beyond a critical value. An effective hard sphere model is also discussed that allows one to compare these results with those on size–polydisperse systems. Implications of this transition on the possible application of colloidal crystals as optical Bragg filters are also pointed out. In Chapter 8 Löwen discusses the results of Brownian-dynamics simulations carried out to investigate the freezing and glass transition in colloidal dispersion. The diffusion mechanism is found to change from hydrodynamic-like to jump diffusion across the glass transition. A new dynamic freezing rule, which is analogous to the Lindemann melting criterion, is also reported. Chakrabarti, Krishnamurthy, Sengupta, and Sood review, in Chapter 9, the density–functional theory, which describes the equilibrium BCC-FCC-liquid phase diagram. The phenomenon of laser-induced freezing is also briefly discussed. The theory is extended to incorporate the effect of an external periodic modulating potential. A first-order freezing transition at low magnitudes of the potential changes to a continuous one via a tricritical point. The stability of colloidal crystals against melting due to laser beam heating is considered by Kesavamoorthy in Chapter 10. The lattice parameter probed by the Kossel ring pattern of a weak probe beam is found to reduce upon illumination by an intense pump beam. A model to estimate the changes in the lattice parameter using an analytical temperature profile is discussed that takes into account the changes in the interparticle potential due to that of the dielectric constant. A novel technique to measure the collective diffusion constant from the relaxation of a modulated colloidal crystal is also described.

In Chapter 11 the theory of electrostatic interaction in charged colloids is reviewed by Jönsson, Åkesson, and Woodward. With the help of Monte Carlo simulations, it is shown that the DLVO theory fails in the case of divalent counterions as it neglects the ionic correlations. These are found to contribute an attractive component to the osmotic pressure. However, this attraction, which occurs within a few nanometers is different from the long-range attraction proposed by the Ise–Sogami school. The authors also present their views on the Sogami formalism. The subject of long-range attraction in charged colloids is addressed in Chapter 12 by Smalley. A theoretical formalism in the Gibbs ensemble shows that, similar to the charged spherical particles, charged plates in electrolytes also experience attraction. Experimental results on one-dimensional colloids that are consistent with the theory are also discussed. The opinions expressed by the DLVO school on the new theory are also reviewed. One can, in principle, obtain the interparticle potential from the measured static structure factor. This inverse problem is discussed by Rajagopalan in Chapter 13. An iterative predictor–corrector inversion method is described and tested on selected model potentials. The advantages and limitations of such procedures are also critically examined.

Thus this volume brings together state-of-the-art reviews in the field of ordering and phase transitions in colloidal dispersions and gives directions of future research. These have been written by some of the most active researchers in the field. It is hoped that this book will be of interest to graduate students and research scientists in the field of colloidal and interfacial science, condensed matter physicists, physical and colloidal chemists, biophysicists, and chemical engineers.

It is a pleasure to acknowledge Professor R. Rajagopalan, without whose encouragement and guidance at various stages this monograph would not have come to reality. We are indebted to Dr. Kanwar Krishan for many valuable suggestions and keen interest throughout the course of the evolution of this monograph. We thank our colleague Dr. R. Kesavamoorthy for several ideas. We are grateful to Dr. Baldev Raj for support and Dr. P. Rodriguez, Director, Indira Gandhi Centre for Atomic Research, for his kind permission to take up this project. We would also like to express our sincere gratitude to the contributors for their timely submission of the chapters.

Akhilesh K. Arora
B. V. R. Tata
Kalpakkam, T. N.
India

Contents

xiii

Contributors

TORBJÖRN ÅKESSON   Department of Physical Chemistry 2, Chemical Centre, P.O. Box 124, S-22100 Lund, Sweden.

AKHILESH K. ARORA   Materials Science Division, Indira Gandhi Centre for Atomic Research, Kalpakkam, Tamil Nadu, 603 102, India.

J. CHAKRABARTI   Department of Physics, Indian Institute of Science, Bangalore, 560 012, India.

DAVID G. GRIER   The James Frank Institute, The University of Chicago, 5640 S. Ellis Ave., Chicago, IL 60637 U.S.A.

KENSAKU ITO   Department of Chemical and Biochemical Engineering, Toyama University, Toyama 930, Japan.

NORIO ISE   Central Laboratory, Rengo Co. Ltd., 186-1-4 Ohhiraki, Fukushima, Osaka 553, Japan.

BO JÖNSSON   Department of Physical Chemistry 2, Chemical Centre, P.O. Box 124, S-22100 Lund, Sweden.

R. KESAVAMOORTHY   Materials Science Division, Indira Gandhi Centre for Atomic Research, Kalpakkan, Tamil Nadu, 603 102, India.

H. R. KRISHNAMURTHY   Department of Physics, Indian Institute of Science, Bangalore 560 012, India.

HARMUT LÖWEN   Sektion Physik der Universität München, Theresienstr. 37, D-80333 München, Germany.

HIDEKI MATSUOKA   Department of Polymer Chemistry, Kyoto University, Kyoto 606-01, Japan.

CHERRY A. MURRAY   AT&T Bell Laboratories, 600 Mountain Avenue, Murray Hill, NJ 07974, U.S.A.

RAJ RAJAGOLAPAN   Department of Chemical Engineering and Physics, University of Houston, Houston, TX 77204, U.S.A.

KLAUS SCHÄTZEL   Institut of Physik, Johannes Gutenberg Universität, D-55099 Mainz, Germany.

S. SENGUPTA   Materials Science Division, Indira Gandhi Centre for Atomic Research, Kalpakkam, Tamil Nadu, 603 102, India.

M. V. SMALLEY   Hashimoto Polymer Phasing Project, ERATO, JRDC, Keihanna Plaza, 1-7 Hikari-dai, Seika-cho, Kyoto 619-02, Japan.

IKUO S. SOGAMI   Department of Physics, Kyoto-Sangyo University, Kyoto 603, Japan.

A. K. SOOD   Department of Physics, Indian Institute of Science, Bangalore 560 012, India.

B. V. R. TATA   Materials Science Division, Indira Gandhi Centre for Atomic Research, Kalpakkam, Tamil Nadu, 603 102, India.

CLIFFORD E. WOODWARD   Department of Chemistry, University College ADFA, Campbell, ACT 2600, Australia.

HIROSHI YOSHIDA   Hashimoto Polymer Phasing Project, ERATO, JRDC, Keihanna Plaza, 1-7 Hikari-dai, Seika-cho, Kyoto 619-02, Japan.

TSUYOSHI YOSHIYAMA   Department of Physics, Kyoto-Sangyo University, Kyoto 603, Japan.

List of Common Symbols

a	Radius of a colloidal particle
A_H	Hamaker constant
B	Bulk modulus
b	Separation between charged walls in a planar double layer
$c^{(2)}(\mathbf{r}, \mathbf{r}_1)$	Second-order direct correlation function
$c^{(3)}(\mathbf{r}, \mathbf{r}_2, \mathbf{r}_3)$	Third-order direct correlation function
$C_\rho(t)$	Density autocorrelation function
d	Diameter of a particle
D_0	Free-particle diffusion constant
D_c	Collective diffusion constant
d_h	Effective hard-sphere diameter
D_L	Long-time self-diffusion constant
D_S	Short-time self-diffusion constant
d_{hkl}	Interplanar spacing
e	Electronic charge
f	Degree of dissociation
F	Free energy $(E - TS)$
$\mathbf{F}_i(t)$	Force on particle i
g	Acceleration due to gravity
G	Gibbs free energy $(E - TS + pV)$
$\mathbf{G}_{hkl}$	Reciprocal lattice vector
$g(r)$	Pair correlation function

$g_c(r)$	Coordinate-averaged pair correlation function
$G_s(\mathbf{r}, t), G_d(\mathbf{r}, t)$	Self- and disctinct van Hove correlation functions
g_{max}	Height of the first peak in $g(r)$
g_{min}	Value of the minimum after the first peak in $g(r)$
$h(r)$	Total liquid correlation function
hkl	Crystal plane index
$I_s(Q)$	Scattered intensity
k_B	Boltzmann constant
$\mathbf{K}_i$	Incident wave vector
$\mathbf{K}_s$	Scattered wave vector
K_T	Isothermal compressibility
ℓ	Natural length scale $(n_p^{-1/3})$
l_a	Lattice parameter
L_s	Length of simulation cell
m	Mass of a colloidal particle
m_r	Relative refractive index
N	No. of particles in volume V
N_A	Avogadro's number
n_i	Impurity ion concentration
n_p	Particle concentration (N/V)
p	Pressure/osmotic pressure
$P(Q)$	Particle form factor
$P(R)$	Distribution of nearest-neighbor distance
$P(Z)$	Charge distribution
$P_c(n)$	Cluster size distribution
$\mathbf{Q}$	Scattering vector
Q	Magnitude of scattering vector
Q_{max}	Position of the first peak in $S(Q)$
r	Interparticle separation
$\mathbf{r}_i$	Position of particle i
$< r^2(m) >$	Mean-square displacement after m Monte Carlo steps
$< r^2(t) >$	Mean-square displacement
R	Nearest-neighbor distance
R_0	Average nearest-neighbor distance
R^g	Gas constant
R_G	Radius of gyration
R_g	Structural parameter g_{min}/g_{max}
R_m	Position of the minimum of Sogami pair potential
R_{exp}	Measured interparticle spacing
S	Entropy
$S(Q)$	Structure factor
$S(Q, t)$	Intermediate scattering function (dynamic structure factor)
S_{max}	Height of first peak of $S(Q)$
t	Time
T	Absolute temperature
T^*	Relative or reduced temperature $(k_B T / U_0)$
T_{glass}^*	Glass transition temperature

$U(r)$	Interparticle potential
U_0	Pair interaction energy at $R[U_0 = U(R)]$
$U_A(r)$	van der Waals pair potential
U_m	Depth of the minimum of Sogami pair potential
$U_S(r)$	Sogami pair potential
U_T	Total interaction energy
$U_y(r)$	Yukawa pair potential
$U_{DLVO}(r)$	DLVO pair potential
$U_{scy}(r)$	Size-corrected Yukawa pair potential
V	Volume of suspension
V_e	Strength of the external modulation potential
$\mathbf{v}_i$	Velocity of particle i
v_p	Volume of the particle
V_s	Scattering volume
Ze	Charge of the particle
Z^*e	Effective or renormalized charge of the particle
Z_0	Average charge on the particle
z_i	Valency of ith type of impurity ion
β	$(k_B T)^{-1}$
χ	Dielectric susceptibility
Δ_s	Dimensionless supercooling parameter
ϵ	Dielectric constant
ϵ_0	Permitivity of free space
η	Viscosity
Γ	Dimensionless interaction strength $[U(R_0)/k_B T = U, '/k_B T)]$
κ	Inverse Debye screening length
λ	Wavelength of light
μ_i	Chemical potential of species i
μ_m	Refractive index of the medium
μ_p	Particle refractive index
μ_s	Refractive index of the suspension
ϕ	Volume fraction
ϕ_e	Effective volume fraction
ψ	Surface potential
$\psi(x)$	Electrostatic potential at distance x
$\Psi(x)$	Dimensionless electrostatic potential at x
$\Psi_6(r)$	six-fold bond-orientational order parameter
ρ	Bulk density of the particle
ρ_m	Bulk density of the medium
$\rho(\mathbf{r})$	Local density
ρ_0	Average density
ρ_e	Surface charge density on the particle
$\rho_\mathbf{Q}$	$\mathbf{Q}$ Fourier component of the density $\rho(\mathbf{r})$
θ	Scattering angle
ξ	Correlation length
$\xi(\mathbf{r})$	Molecular field

CHAPTER

1

Interactions, Ordering, and Phase Transitions in Charged Colloids: An Introduction

Akhilesh K. Arora and Raj Rajagopalan

Colloidal systems exhibit many of the phases of condensed matter. As a result and since the size of the particles, their charges, and the range and the strength of interparticle interaction forces can be controlled and varied systematically, colloidal dispersions offer themselves as very good model systems for the study of cooperative phenomena such as ordering, related phase transitions, and the stability of the resulting phases. The use of colloids in this fashion has led to a number of new results; in particular, several phenomena that are otherwise difficult to probe in atomic substances because of inaccessible time scales or other field variables are now being investigated with the help of colloidal dispersions (e.g., nucleation and growth from undercooled melt and shear-induced ordering and melting). Further, the crystalline and liquidlike structural ordering in charge-stabilized colloids have been described until recently using screened-Coulombic repulsion among the particles and the excluded-volume effect. However, there is a growing body of experimental evidence for the presence of long-range attraction between the particles. A long-range attraction has been predicted by Sogami in a new theoretical formalism, which differs from the Derjaguin-Landau-Verwey-Overbeek theory, and has led to a controversy over the nature of the pair - potential for some time. This chapter summarizes the experimental results that have direct implications for the current understanding of interparticle interactions and for the controversy relating to the theories of interaction forces. In addition, a brief discussion of the field variables that drive phase transitions and of the factors relevant to the stability of resulting homogeneous phases is also presented.

1

1.1 Colloidal Systems as Model Condensed Matter

Traditionally colloidal dispersions have been the domain of physical chemists and chemical engineers in view of the numerous industrial uses of colloids. Apart from the phenomena and processes of industrial interest, colloidal dispersions also display very rich phase behavior and exhibit structural ordering similar to that in crystalline atomic solids [1]-[3], liquids [4]-[6], gases [7], and even glasses [8]-[10]. The striking closeness of the magnitudes of the elastic constants [10]-[12] and latent heat of melting [13] of colloidal crystals (when expressed in the units of "per mole") with those of atomic solids allows one to regard colloidal dispersions as scaled-up versions of atomic solids, that is, macrosolids. A colloidal dispersion with a particle concentration $\sim 10^{13}$ cm^{-3} has elastic constants ~ 10 dyn/cm^2; on the other hand, atomic solids with atomic density $\sim 10^{22}$ cm^{-3} have elastic constants $\sim 10^{10}$ dyn/cm^2. The perfect scaling of the magnitudes of the physical constants with particle concentration suggests that the inter*particle* interaction energy U_0 in colloids must be of the same order as the inter*atomic* interaction in atomic systems. This fact has led many researchers to treat colloids as model condensed matter systems and investigate many effects and phenomena that are otherwise difficult to probe in atomic substances because of inaccessible time scales or other field variables. Although analogies can be made between charged colloids and metals or Wigner crystals, there are a few important differences that must be mentioned. First the delocalization of the small ions in charged colloids arises due to thermal energy, whereas that of electrons in metals is of quantum-mechanical origin. Second the damping (viscous) effects of the dispersing medium causes the particle motion to be Brownian. This leads the longitudinal phonons in colloidal crystals to be overdamped [14, 15] in contrast to the situation in atomic solids, where phonons can propagate.

A major advantage of using colloids as model systems is the ease with which the interparticle forces can be varied. In charged colloids the interaction between different species is electrostatic in origin. The presence of counterions and other "impurity" ions (free carriers) leads to the screening of the Coulombic interaction. The Debye screening length κ^{-1} depends on the concentration of the small ions; that is,

$$\kappa^2 = \frac{4\pi e^2}{\epsilon k_B T}(n_p Z + n_i z_i^2) \tag{1.1}$$

where n_p is the particle concentration, Z is the charge on the particle, n_i is the concentration of impurity ions of valency z_i, k_B is the Boltzmann constant, T is the temperature, e is the electronic charge, and ϵ is the dielectric constant of the medium. A remarkable feature of charged colloids is the ease with which one can vary the range and the strength of the interparticle interaction $U(r)$; that is, by controlling the ionic strength of the dispersion or, equivalently, the impurity ion concentration n_i, one can change κ^{-1} over a wide range.

In order to investigate phase transitions in atomic systems, one normally varies thermodynamic variables such as temperature or pressure. The ordering phenomena and related phase transitions basically occur due to a competition between the interaction energy U_0 and the thermal energy $k_B T$. Increasing the temperature lowers the ratio $U_0/k_B T$ below a critical value, and the ordered crystalline lattice melts into a disordered state. However, in colloidal systems the range of temperature variation is rather restricted because of the presence of the dispersing medium, and one must, therefore, identify other parameters equivalent to temperature and pressure. In charged colloids one can investigate the melting transition at a fixed temperature by increasing the ionic strength of the dispersion, thereby causing $U_0/k_B T$ to reduce below the critical value. Thus increasing n_i is equivalent to increasing the reduced temperature $T^* = k_B T/U_0$. The parameter equivalent to pressure in a colloidal dispersion is the osmotic pressure, which can be easily varied by changing the particle concentration n_p either by centrifugation or by dilution.

1.2 Condensed Matter Beyond Experimental Limits

As a result of the ratio of the dimension of colloidal particles to atomic dimensions (typically a factor of $\sim 10^3$), the time scales of particle diffusion and of all phenomena that are governed by diffusion are also scaled up appropriately. The relevant time scales in the case of colloids are therefore of the order of milliseconds. Moreover, the random force exerted by the solvent molecules on the particles (i.e., the Brownian force) also plays an important role in determining the *kinetics* of various processes (e.g., crystallization) in a dispersion. As a result, the use of colloids as model many-body systems allows one to probe in real time processes that are otherwise inaccessible in atomic systems. In what follows we outline two examples.

1.2.1 Nucleation and Growth

Major problems in atomic systems are the high speed of nucleation and crystal growth as well as the difficulty of preventing heterogeneous nucleation under the conditions of large supercooling [16], and these are circumvented when colloids are used as model systems. For instance, recent experiments on crystallization kinetics [17]-[23] in colloidal dispersions have led to a new understanding of nucleation and growth. Early experiments on the crystallization behaviour of shear-melted metastable liquid phase showed that the recrystallization occurred via nucleation and growth of single crystallites. The growth velocity of the rough interfaces of the crystallites was found to be determined by free diffusion [17]. In the early stage, the growth velocity is $\sim D_0/\xi$, where D_0 is the free-particle diffusion coefficient and ξ is a characteristic length comparable to the interparticle spacing. The crystallization kinetics has also been found to depend on the range of interaction, which is altered by the addition of a free polymer. Rapid crystallization occurred in dispersions with long-range interaction [19]. Time-resolved small-angle light scattering experiments have been used to investigate the power-law behavior of the ripening process at large times [20]. A $t^{1/3}$-Lifshitz-Slyozov behavior is found at low volume fractions ϕ, whereas at large ϕ the dispersions show $t^{1/2}$-Lifshitz-Allen-Cahn behavior due to the coupling of the conserved density parameter with the nonconserved crystal-order parameter [21]. A change of crystallization mechanism in nearly hard-sphere colloids from homogeneous isometric crystallization to highly asymmetric needle- and platelike crystal formation across the glass transition at an effective volume fraction of $\phi_e \sim$ 0.58 has also been recently reported [23].

1.2.2 Shear-Induced Melting and Ordering

Another important consequence of the scaling up of the particle size is the reduction of particle concentration with respect to atomic density by a factor $\sim 10^{10}$. This causes the shear and the bulk moduli of colloidal crystals to be lowered by a similar magnitude, thus making the colloidal crystals extremely fragile. However, this fact has been exploited to investigate the consequences of external shear on crystalline order, since the external shear can now be several orders of magnitude higher than its shear modulus [24]-[32]. Obviously, one cannot imagine conducting such experiments on atomic solids except using shock waves.

Shear-induced melting of colloidal crystals has been known for some time [24, 25]. At low shear the structure is identified as a "flowing crystal," which exhibits a transition from three- to two-dimensional order with increasing shear. Further transitions to "sliding-layer structure" and to "one-dimensionally-ordered strings of particles" have also been reported [26]. The shear-melted dispersion is an anisotropic fluid. Accurate measurements of shear viscosity are possible in concentrated colloidal dispersions subjected to shear. The viscosity of colloidal dispersions shows a decrease first (shear thinning), then an increase, and a subsequent decrease at high shear rates [28, 29]. Since the high turbidity causes multiple scattering of light in concentrated colloids, neutron scattering experiments have been performed to probe the structure under steady-state shear. At large shear rates a shear-induced amorphous state has also been reported [29]. On the other hand, oscillatory shear, when applied on liquidlike ordered dispersions, has been found to induce crystalline order [30] in nearly hard-sphere systems. In order to obtain a complete understanding of shear-induced effects, Brownian- [31] and molecular- [32] dynamics simulations have also been carried out. Brownian-dynamics simulations qualitatively reproduce the experimental results of Ackerson and Pusey [30] and also predict shear thinning. The nonequilibrium phase diagram obtained using molecular-dynamics simulations show that shear-melted dispersions can recrystallize to a new structure at high shear rates [32]. Figure 1.1 shows a phase diagram obtained under shear. Note that at low

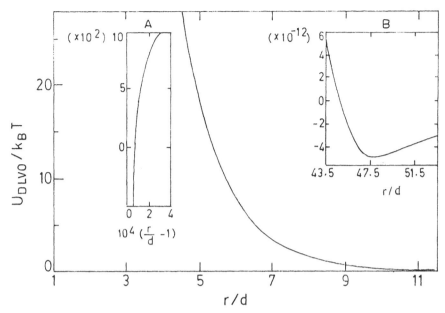

Figure 1.2 DLVO potential calculated for a charged colloidal dispersion of polystyrene particles of diameter 109 nm and charge $600e$ at a concentration of 1.33×10^{12} cm^{-3}. The impurity ion concentration is set to 3.35×10^{15} cm^{-3} ($4.2 n_p Z$). The Hamaker constant is taken as 10^{-13} ergs. Inset (A) shows the primary minimum and the Coulomb barrier very close to the particle. Inset (B) shows the secondary minimum of extremely small depth, which occurs at a distance of $\sim 47d$.

1.4 Interactions in Charged Colloids: Possibilities of Long-Range Attraction

1.4.1 Experimental Evidence

Although for a long time ordering in charged colloidal dispersions has been believed, and also explained, to arise as a result of screened-Coulombic repulsion, there is now a growing body of experimental evidence that suggests the presence of long-range attraction between similarly charged particles. The effects are most dramatic in dilute dispersions and at moderate ionic strengths. A system of particles interacting via a purely repulsive potential is expected to remain homogeneous except for fluid-solid coexistence. However, in a number of cases the dispersions have been found to be inhomogeneous under appropriate conditions [56]-[59]. For example:

1. The particle concentration in dispersions with liquidlike order estimated using light scattering and that in colloidal crystals using Bragg diffraction [60]-[63] have been reported to be significantly higher than that expected

for a homogeneous dispersion. This disagreement between the estimated and actual particle concentration has been attributed either to the error in supplier's statement about the volume fraction [62] or to the hypothesis of coexistence of crystalline and liquidlike orders with widely different particle concentrations [63]. Subsequent careful experiments on coexisting crystalline and liquidlike orders have proved the above hypothesis to be incorrect, and an estimation of exact volume fraction by the evaporation method has shown the observed disagreement to be real [64]. This suggests that the crystalline and the liquidlike phases need not occupy the full volume of the dispersions.

2. Optical-microscopic investigations show that the measured nearest-neighbor distance R_{exp} in ordered regions is lower than that expected for homogeneous dispersions R_0 [65] and that the ordered regions are found to coexist with disordered regions (labeled the "two-state structure") with widely different particle concentrations [66, 67].

3. The observation of stable voids in the interior of the dispersion [57, 68] suggests the possible existence of long-range attractive interaction in the dispersion.

4. A vapor-liquid-type condensation reported recently in dilute aqueous colloidal dispersions of polystyrene particles [59, 69, 70] can occur only if the particles get trapped into a potential minimum occurring at a distance shorter than the average interparticle separation $l \sim n_p^{-1/3}$. This minimum in the potential can occur if there exists an attraction that dominates at large r. It may be mentioned that a "vapor-liquid" transition has also been predicted for DLVO potential, due to the trapping of particles in the secondary minimum at high ionic strengths [71]; however, the observed dependence of the particle concentration in the liquid phase on ionic strength and magnitudes of other parameters rule out the possibility that the vapor-liquid transition [69] is caused by the secondary minimum in the DLVO potential.

5. The report of isolated bound pairs (with large interparticle separations) executing coupled motion in dilute dispersions [58] also suggests trapping of particles into a potential minimum.

6. Recent estimation of the effective pair potential by Kepler and Fraden [72] from experimental pair correlation functions in dispersions confined between two glass plates indeed shows an attraction at large r, with the dependence of the position of the potential minimum on the ionic strength being qualitiatively similar to that observed in the vapor-liquid condensation [69].

1.4.2 Theoretical Understanding

Thus there are a large number of experimental results that suggest attraction between similarly charged particles at large r, with repulsion continuing to dominate at small r. One must therefore look for the possible origin of such an attraction. Ise and Okubo [73] have proposed that the observed attraction arises due to presence of counterions, that is, "counterion-mediated attraction." In an attempt to examine this further, the theory of electrostatic interaction in dispersions has been reformulated in the Gibbs ensemble by Sogami [74] and by Sogami and Ise [75], and their results show that the effective pair potential contains, in addition to the screened-Coulombic repulsive term, an attractive term that dominates at large r.

The Sogami pair potential $U_s(r)$ is given by

$$U_s(r) = \frac{Z^2 e^2}{2\epsilon} \left(\frac{\sinh(\kappa a)}{\kappa a} \right)^2 \left(\frac{A}{r} - \kappa \right) e^{-\kappa r} \tag{1.6}$$

where $A = 2[1 + \kappa a \coth(\kappa a)]$. This potential has a minimum at $R_m = \{A + [A(A+4)]^{1/2}\}/2\kappa$. Figure 1.3 shows $U_s(r)$ at several ionic strengths [48], and Fig. 1.4 shows the dependence of R_m and the potential minimum U_m on κ.

One can see from Fig. 1.4 that the position and the depth of the potential minimum strongly depend on κ or, equivalently, on the ionic strength. The basic assumptions of the formalism are the following: (1) the motion of the particles is adiabatically cut off from those of small ions; (2) the Debye charging process is used to charge the ions; (3) the linearized Poisson-Boltzmann equation is used to calculate the total electrostatic energy; (4) the primitive model is used with the solvent treated as a dielectric continuum; and (5) the volume occupied by the

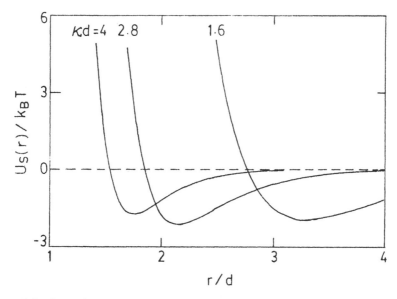

Figure 1.3 Sogami potential calculated for particles of diameter 109 nm and charge 600e at several values of κ. Note the strong dependence of the position and the depth of the potential minimum on κ.

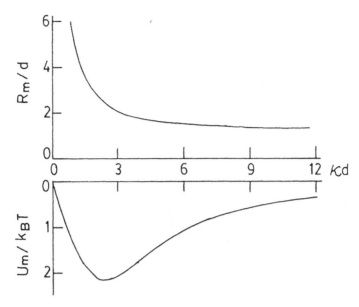

Figure 1.4   Dependence of the position (R_m) and the depth (U_m) of the potential minimum on κ for Sogami potential. Other parameters are same as those of Fig. 1.3.

dispersion need not be the same as the total volume of the solution. Assumptions (1) - (4) are the same as those of the DLVO formalism; however, assumption (5) allows the dispersion volume to be treated as a thermodynamic variable, and hence the treatment of the problem in the Gibbs ensemble is considered reasonable. Even though the Sogami potential could explain most of the experimental results of Ise and co-workers [57, 65] and several other results [58, 59] that otherwise could not be understood on the basis of the DLVO potential, the new formalism [76]-[78] has come under some criticisms, which have been subsequently shown to be incorrect and invalid [79]-[84]. In particular, Overbeek has commented [76] that the contribution of the solvent to electrical free energy has been missed in the new formalism and has shown that when it is incorporated the attractive term disappears. However, Ise et al. [79] have pointed out that Overbeek, while considering the solvent contribution, has misused the Gibbs-Dühem equation by omitting the contribution of the macroions. The Gibbs-Dühem relation states that the solvent is not an independent component when considering the thermodynamic properties of the dispersion. Subsequently Smalley [80, 81] further pointed out the error in the treatment of Overbeek and showed that it leads to incorrect conclusion that "there is no energy associated with the electrical double layers." Woodward [77, 78] objected to the use of the Gibbs ensemble on the basis that the total volume of the solution [solvent plus solute] must be considered in the treatment because the counterions sample all the available solution volume no matter what configurations of macroions exist at any time. This objection has been countered by the argument that the solution volume available to the counterions is restricted to the vicinity of the region of macroions due to the condition of electroneutrality [81, 84]. It is further pointed out that an enormous electrical restoring force would arise if any significant proportion of the counterions were to simultaneously sample a region of the solution not bounded by the macroions (see [84] for a more detailed review of the controversies). Macroionic solutions having two types of regions and screening lengths have been considered by other investigators as well [85, 86].

It is worth pointing out that there have been some attempts to see whether one can use $U_s(r)$ to explain other experimental results that were earlier understood on the basis of the DLVO or screened-Coulombic repulsion. Ito et al. [87] have obtained the same values of the elastic constants that had been obtained earlier using U_{scy} and a renormalized charge [10]. It has been shown that $U_s(r)$ can also predict a liquid-solid (disorder-order) transition [88]. Tata et al. [89] have shown using computer simulations that both $U_s(r)$ and $U_{scy}(r)$ fit the structure factor of a homogeneous liquidlike ordered dispersion equally well. The photothermal comparison of colloidal crystals [90, 91], which was earlier explained using the Yukawa potential, is now shown to be also understandable on the basis of $U_s(r)$ [92].

At the outset it may appear surprising that two widely different interaction potentials can explain the same experimental result. This apparent difficulty can be understood if one considers the position of the potential minimum R_m of $U_s(r)$ with respect to the average interparticle separation $l \sim n_p^{-1/3}$. Figure 1.5 compares $U_{scy}(r)$ and $U_s(r)$ at different ionic strengths. At low ionic strengths if $R_m > l$ the particle experiences essentially the repulsive part of the potential and hence the structure and other properties calculated using either of the two potentials can explain the data [87–89, 92]. One must also recognize the fact that, although one can measure the actual charge on the particle using a combination of techniques [93], in most cases only an effective or renormalized charge Z^* has been used in the theoretical calculations [10, 40]. Further, the exact determination of the residual impurity ion concentration even in the best deionized dispersion is also difficult. In view of these uncertainties, Z and n_i are treated as parameters in most of the cases. Subsequent to the fitting of the experimental data to appropriate theoretical models, one can only argue about the physical reasonability of the parameters used. On the other hand, at high ionic strengths, if $R_m < l$, the predictions based on the two potentials are very different. The size-corrected Yukawa potential $U_{scy}(r)$ can under no circumstances predict an inhomogeneous dispersion. Hence appropriate experiments at suitable particle concentrations and ionic strengths become the testing grounds for the applicability of the effective pair potentials and the validity of corresponding formalisms.

In view of the preceding questions concerning the nature of interaction forces in charged systems, there have also been attempts to obtain the effective interparticle potential from experimental structure factors. The determination of pair

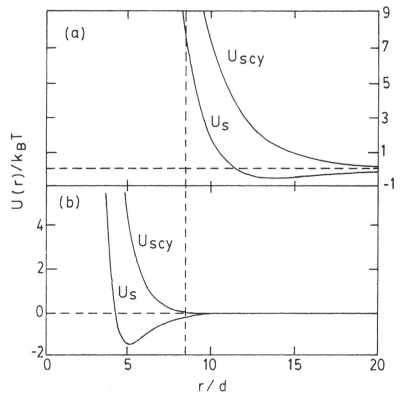

Figure 1.5   Comparison of the size-corrected Yukawa (DLVO) and the Sogami potentials at low and high ionic strengths. (a) $n_i = 4 \times 10^{14}$ cm^{-3} $(0.5 n_p Z)$ and (b) $n_i = 8 \times 10^{15}$ cm^{-3} $(10 n_p Z)$. Other parameters are: $Z = 600$, $n_p = 1.33 \times 10^{12}$ cm^{-3}. The vertical dashed line corresponds to average interparticle separation $l = n_p^{-1/3}$. In the case of Sogami potential at low n_i, $R_m > l$, and hence the repulsive part dominates, whereas at high n_i, $R_m < l$ and consequently the attractive term dominates. On the other hand, U_{scy} remains repulsive for all ionic strengths.

potentials from structure factors [94, 95] or pair correlation functions [72] is, of course, of general interest, and the most successful methods employ predictor-corrector iterative techniques, in which the "predictor" is usually based on a theoretical formalism (such as an integral-equation method), and the "corrector" (computer simulations, for best results) serves to improve the result for subsequent iterations. The predictor-corrector methods developed recently have shown rapid convergence and also good reproducibility of test potentials [95, 96]. However, the success of such inversion schemes depends on the quality of the experimental data and a detailed knowledge of polydispersity (if appropriate). An extended discussion of these is presented in Chapter 13 in this volume [96].

1.5 Phase Transitions and Colloidal Stability

Apart from the appearance of long-range order, a variety of other very interesting phase transitions and instabilities in colloidal dispersions have also been reported in the literature. As mentioned earlier, the relevant thermodynamic parameters in the case of dispersions are the particle concentrations and the ionic strength. The field variables other than shear rate are the laser optical field [97] and the electrical or magnetic fields [98, 99]. The polydispersities of size and charge, inherent to the colloidal systems, also influence the structure [100] and ordering [101]. Careful investigations of the evolution of structural order using Kossel line analysis show the existence of a number of intermediate structures during crystallization [102]; on the other hand, single layers of colloidal crystals are found to melt through two second-order phase transitions [103]. The investigation of the dynamics of charged colloids across freezing has yielded a dynamic scaling law [104] analogous to the Lindemann melting criterion. Another aspect of the phase transition is the stability of the homogeneous phase. Investigations of phase transitions driven by various thermodynamic or field variables allow one to identify the regions of stable homogeneous phase and those of two-phase coexistence in the phase diagram [70, 50]. Stability considerations are also important from the point of view of applications such as optical Bragg filters [105].

The discussions in this chapter are merely meant to highlight the advantages of the use of charged colloids as model systems to study the nature of interaction forces and order/disorder phenomana in condensed matter. Some of the topics mentioned are taken up in more detail in the various chapters of this volume.

Acknowledgements

AKA wishes to thank Dr. B. V. R. Tata for stimulating discussions and Dr. Kanwar Krishan for encouragement.

References

1. Hiltner, P. A., and Krieger, I. M., *J. Phys. Chem.* **73**, 2686 (1969).
2. Williams, R., and Crandall, R. S., *Phys. Lett.* **48A**, 225 (1974).
3. Pieranski, P., *Contemp. Phys.* **24**, 25 (1983).
4. Pusey, P. N., *Phil. Trans. R. Soc.* **A293**, 429 (1979).
5. Arora, A. K., *J. Phys. E: Sci. Instr.* **17**, 1119 (1984).
6. Kesavamoorthy, R., Tata, B. V. R., Arora, A. K., and Sood, A. K., *Phys. Lett.* **138A**, 208 (1989).
7. Brown, J. C., Pusey, P. N., Goodwin, J. W., and Ottewill, R. H., *J. Phys. A: Math. Gen.* **8**, 664 (1975).
8. Pusey, P. N., and van Megen, W., *Nature* **320**, 340 (1986); *Phys. Rev. Lett.* **59**, 2083 (1987).
9. Kesavamoorthy, R., Sood, A. K., Tata, B. V. R., and Arora, A. K., *J. Phys. C: Solid State Phys.* **21**, 4737 (1988).
10. Lindsay, H. M., and Chaikin, P. M., *J. Chem. Phys.* **76**, 3774 (1982).
11. Crandall, R. S., and Williams, R., *Science* **198**, 293 (1977).

12. Kesavamoorthy, R., and Arora, A. K., *J. Phys. A: Math. Gen.* **18**, 3389 (1985).
13. Williams, R., Crandall, R. S., and Wojtowicz, P. J., *Phys. Rev. lett.* **37**, 348 (1976).
14. Hurd, A. J., Clark, N. A., Mockler, R. C., and O'Sullivan, W. J., *Phys. Rev.* **A 26**, 2869 (1982).
15. Cotter, L. K., and Clark, N. A., *J. Chem. Phys.* **86**, 6616 (1987).
16. Turnbull, D., *J. Chem. Phys.* **20**, 411 (1952).
17. Aastuen, D. J. W., Clark, N. A., Cotter, L. K., and Ackerson, B. J., *Phys. Rev. Lett.* **57**, 1733 (1986).
18. Okubo, T., *J. Chem. Phys.* **87**, 3022 (1987).
19. Smits, C., van Duijneveldt, J. S., Dhont, J. K. G., Lekkerkerker, H.N.W., and Briels, W. J., *Phase Transitions* **21**, 157 (1990).
20. Schätzel, K., and Ackerson, B. J., *Phys. Rev. Lett.* **68**, 337 (1992).
21. Schätzel, K. Chapter 2, this volume.
22. Gast, A. P., *Nature* **351**, 535 (1991).
23. van Megen, W., and Underwood, S. M., *Nature* **351**, 616 (1993).
24. Clark, N. A., and Ackerson, B. J., *Phys. Rev. Lett.* **44**, 1005 (1980).
25. Ackerson, B. J., and Clark, N. A., *Phys. Rev. Lett.* **46**, 123 (1981); *Physica A* **118**, 221 (1983).
26. Ackerson, B. J., and Clark, N. A., *Phys. Rev.* **A 30**, 906 (1984).
27. Weitz, D. A., Dozier, W. D., and Chaikin, P. M., *J. de Physique* **46**, C3-257 (1985).
28. di Meglio, J. M., Weitz, D. A., and Chaikin, P. M., *Phys. Rev. Lett.* **58**, 137 (1987).
29. Ackerson, B. J., Hayter, J. B., Clark, N. A., and Cotter, L., *J. Chem. Phys.* **84**, 2344 (1986).
30. Ackerson, B. J., and Pusey, P. N., *Phys. Rev. Lett.* **61**, 1033 (1988).
31. Xue, W., and Grest, G. S., *Phys. Rev. Lett.* **64**, 419 (1990).
32. Stevens, M. J., Robbins, M. O., and Belak, J. F., *Phys. Rev. Lett.* **66**, 3004 (1991).
33. Hachisu, S., Kobayashi, Y., and Kose, A., *J. Coll. Interface Sci.* **42**, 342 (1973).
34. Hachisu, S., and Kobayashi, Y., *J. Coll. Interface Sci.* **46**, 470 (1974).
35. Fujita, H., and Ametani, K., *Jap. J. Appl. Phys.* **16**, 1091 (1977).
36. Monovoukas, Y., and Gast, A. P., *J. Coll. Interface Sci.* **128**, 533 (1989).
37. Sirota, E. B., Ou-Yang, H. D., Sinha, S. K., Chaikin, P. M., Axe, J. D., and Fujji, Y., *Phys. Rev. Lett.* **62**, 1528 (1989).
38. Shih, W. H., and Stroud, D., *J. Chem. Phys.* **79**, 6254 (1982).
39. Hone, D., Alexander, S., Chaikin, P. M., and Pincus, P., *J. Chem. Phys.* **79**, 1474 (1983).
40. Alexander, S., Chaikin, P. M., Grant, P., Morales, G. J., and Pincus, P., *J. Chem. Phys.* **80**, 5776 (1984).
41. Smith, B. B., Chan, D. Y. C., and Mitchell, D. J., *J. Coll. Interface Sci.* **105**, 216 (1985).
42. Härtl, W., and Versmold, H., *J. Chem. Phys.* **88**, 7157 (1988).
43. Salgi, P., and Rajagopalan, R., *Langmuir* **7**, 1383 (1991).
44. Sengupta, S., and Sood, A. K., *Phys. Rev.* **A 44**, 1233 (1991)
45. Kremer, K., Robbins, M. O., and Grest, G. S., *Phys. Rev. Lett.* **57**, 2694 (1986).
46. Rosenberg, R. O., and Thirumalai, D., *Phys. Rev.* **A 36**, 5690 (1987).
47. Verwey, E. J. W., and Overbeek, J. Th. G., *Theory of the Stability of Lyophobic Colloids* (Elsevier, New York, 1948).
48. Tata, B. V. R., Ph. D. Thesis (1992) University of Madras (Unpublished).
49. Hamaker, H. C., *Physica* **4**, 1058 (1937).
50. Chakrabarti, J., et al., Chapter 9, this volume.
51. Robbins, M. O., Kremer, K., and Grest, G. S., *J. Chem. Phys.* **88**, 3286 (1988).
52. Belloni, L., *J. Chem. Phys.* **85**, 519 (1986).
53. Patey, G. N., *J. Chem. Phys.* **72**, 5763 (1980).
54. Wennerström, H., Jönsson, B., and Linse, P., *J. Chem. Phys.* **76**, 4665 (1982).
55. Jönsson, B., et al., Chapter 11, this volume.
56. Ise, N., Okubo, T., Sugimura, M., Ito, K., and Nolte, H. J., *J. Chem. Phys.* **78**, 536 (1983).
57. Ise, N., Ito, K., and Yoshida, H., *Polymer Preprints* **33**, 769 (1992).
58. Yoshino, S., in *Ordering and Organisation in Ionic Solutions*, Eds. N. Ise and I. S. Sogami (World Scientific, Singapore, 1988) p. 449.

59. Arora, A. K., Tata, B. V. R., Sood, A. K., and Kesavamoorthy, R., *Phys. Rev. Lett.* **60**, 2438 (1988).
60. Pusey, P. N., *J. Phys. A: Math. Gen.* **12**, 1805 (1979).
61. Daly, J. G., and Hasting, R., *J. Phys. Chem.* **85**, 294 (1981).
62. Grüner, F., and Lehmann, W. P., *J. Phys. A: Math. Gen.* **15**, 2847 (1982).
63. Udo, M. K., and de Souza, M. F., *Solid State Commun.* **35**, 907 (1980).
64. Arora, A. K., and Kesavamoorthy, R., *Solid State Commun.* **54**, 1047 (1985).
65. Ito, K., Nakamura, H., and Ise, N., *J. Chem. Phys.* **85**, 6136 (1986).
66. Ito, K., Nakamura, H., Yoshida, H., and Ise, N., *J. Am. Chem. Soc.* **110**, 6955 (1988).
67. Ise, N., et al., Chapter 5, this volume.
68. Kesavamoorthy, R., Rajalakshmi, M., and Baburao, C., *J. Phys.: Condens. Matter* **1**, 7149 (1989).
69. Tata, B. V. R., Rajalakshmi, M., and Arora, A. K., *Phys. Rev. Lett.* **69**, 3778 (1992).
70. Tata, B. V. R., and Arora, A. K., Chapter 6, this volume.
71. Hansen, J. P., *J. de Physique* **46**, C3-9 (1985).
72. Kepler, G. M., and Fraden, S., *Phys. Rev. Lett.* **73**, 356 (1994).
73. Ise, N., and Okubo, T., *J. Phys. Chem.* **70**, 2400 (1966).
74. Sogami, I., *Phys. Lett.* **96A**, 199 (1983).
75. Sogami, I., and Ise, N., *J. Chem. Phys.* **81**, 6320 (1984).
76. Overbeek, J. Th. G., *J. Chem. Phys.* **87**, 4406 (1987).
77. Woodward, C. E., *J. Chem. Phys.* **89**, 5140 (1988).
78. Jönsson, B., et al. Chapter 11, this volume.
79. Ise, N., Matsuoka, H., Ito, K., Yoshida, H., and Yamanaka, J., *Langmuir* **6**, 296 (1990) (see footnote 32).
80. Smalley, M. V., *Molec. Phys.* **71**, 1251 (1990).
81. Smalley, M. V., Chapter 12, this volume.
82. Dosho, S., et al., *Langmuir* **9**, 394 (1993).
83. Ise, N., and Matsuoka, H., Macromolecules **27**, 5218 (1994).
84. Schmitz, K. S., *Macroions in Solution and Colloidal Dispersions* (VCH Publishers, New York, 1993), p. 122.
85. Alexandrowicz, Z., and Katchalsky, A., *J. Polym. Sci. Part A* **1**, 3231 (1963).
86. Oosawa, F., *J. Polym. Sci.* **23**, 421 (1957).
87. Ito, K., Sumaru, K., and Ise, N., *Phys. Rev. B* **46**, 3105 (1992).
88. Matsumoto, M., and Kataoka, Y., in *Ordering and Organisation in Ionic Solutions*, Eds. N. Ise and I. Sogami (World Scientific, Singapore, 1988) p. 574.
89. Tata, B. V. R., Sood, A. K., and Kesavamoorthy, R., *Pramana J. Phys.* **34**, 23 (1990).
90. Rundquist, P. A., Jagannathan, S., Kesavamoorthy, R., Brnadic, C., Xu, S., and Asher, S. A., *J. Chem. Phys.* **94**, 711 (1991)
91. Kesavamoorthy, R., Chapter 10, this volume.
92. Ise, N., and Smalley, M. V., *Phys. Rev. B*, **50**, 16722 (1994).
93. Ito, K., Ise, N., and Okubo, T., *J. Chem. Phys.* **82**, 5732 (1985).
94. Reatto, L., Levesque, D., and Weis, J. J., *Phys. Rev. A* **33**, 3451 (1986).
95. Rajagopalan, R., *Langmuir* **8**, 2898 (1992).
96. Rajagopalan, R., Chapter 13, this volume.
97. Chowdhury, A., Ackerson, B. J., and Clark, N. A., *Phys. Rev. Lett.* **55**, 833 (1985).
98. Klingenberg, D. J., van Swol, F., and Zukoski, C. F., *J. Chem. Phys.* **91**, 7888 (1989).
99. Skjeltrop, A. T., *J. Appl. Phys.* **55**, 2587 (1984).
100. Salgi, P., and Rajagopalan, R., *Adv. in Colloid Interface Sci.* **43**, 163 (1993).
101. Arora, A. K., and Tata, B. V. R., Chapter 7, this volume.
102. Yoshiyama, T., and Sogami, I. S., Chapter 3, this volume.
103. Grier, D. G., and Murray, C. A., Chapter 4, this volume.
104. Löwen, H., Chapter 8, this volume.
105. Asher, S. A., Flaugh, P. L., and Washinger, G., *Spectroscopy* **1**, 26 (1986).

2

Kinetics of Colloidal Crystallization

Klaus Schätzel [1]

The crystallization of colloids is determined by two fields: the nonconserved crystalline order and the conserved particle density. The spatial fluctuations of both fields were simultaneously recorded as a function of time for hard-sphere-type colloidal particles during their crystallization from an initial shear molten state. Small-angle laser light scattering provided information about spatial density fluctuations, that is, the morphology of growing crystal grains, typically in the form of a bright ring pattern that grows in intensity and shrinks in diameter as the crystallites grow. General features like the existence of two distinct growth regimes — early nucleation and growth, late coarsening or ripening — were clearly detectable in both scattering experiments. But there were also surprising differences, such as an apparent slowing of crystal growth or even shrinking of crystals prior to the onset of coarsening that could only be detected in small-angle scattering. Quantitative data were obtained on growth laws in both regimes. Depending on the initial sample density as a control parameter, both classical

---

[1] It is with deep regret that we inform the readers of the untimely death of Klaus Schätzel. Sections 2.1 through 2.5 were written by Klaus and appear with minor editing. Sections 2.6 and 2.7 were written by Andreas Heymann,who collected the data as part of his doctoral thesis work, and by Bruce J. Ackerson with whom Klaus had an ongoing collaboration focused on the crystallization process in colloidal suspensions. A. H. and B. J. A. accept responsibilty for any errors or misrepresentation of viewpoint in these sections.

limiting growth behaviors were evidenced: diffusion-limited growth, with crystal diameters increasing like the square root of time as well as interface-limited linear growth. The ripening of the samples below melting density followed the Lifshitz-Slyozov model, which is typical for phase transitions with conserved order parameter. Samples with higher mean densities indicated larger growth exponents during ripening, more like the Lifshitz-Allen-Cahn law commonly found for transitions with nonconserved order parameter.

2.1 Introduction

Among first-order phase transitions, crystallization and melting processes immediately come to everyone's mind, and certainly these have a long and respectable history of research in the past. But due to particular difficulties of the subject, we cannot yet claim a thorough and complete understanding either of the macroscopic kinetics of the transition or of the underlying microscopic processes. What are these particular difficulties presented by the liquid-to-solid phase transformation?

Theoretically, one may like to treat the transition with the proven methods of field theory. But in addition to the nonconserved field of the order parameter, one also has to consider five more conserved fields, that is, particle density, energy, and three components of momentum. The nonlinear coupling of all these fields presents an extremely challenging task. Most work along these lines has hence been restricted to a small subset of all the relevant fields [1].

Experimentally, high-resolution techniques like X-ray and neutron scattering or raster tunnel microscopy have made it possible to study the equilibrium structures of liquids and solids in great detail. A huge body of empirical data about slow crystal growth is also available due to the technological importance of the controlled growth of single crystals. However, under conditions of large undercooling, crystallization becomes an extremely fast process, and kinetic data in this regime are rare and difficult to obtain experimentally. Techniques that offer sufficient temporal resolution generally lack microscopic spatial resolution.

In view of substantial theoretical difficulties and limited experimental access, our mental picture of the crystallization transition is still largely shaped by rather simple empirical model concepts that were developed during the first half of this century [2]. According to these concepts, the transition starts through the formation of small nuclei that then grow into individual crystals. The growth process continues until thermal equilibrium is closely approached. At even later times, the polycrystalline material typically shows coarsening or ripening, that is, the slow growth of large crystallites at the expense of smaller ones. All these concepts originated from indirect, often macroscopic observations like the large supersaturations neccessary to drive homogeneous nucleation or measurements of grain sizes by scattering or microscopy, which were performed after completion of the transition. The direct experimental study of these models on the microscopic level remains a challenging task.

This is where colloids come into the game. As compared to atomic or molecular systems, where the relevant spatial and temporal scales (0.1 nm and 1 ps, typically) are extremely difficult to resolve in a single experiment,[2] colloids offer comparatively easy real-time experimental access with microscopic spatial resolution, just because of the less extreme length and time scales involved (100 nm and 1 ms, typically). With today's availability of pure and rather monodisperse model suspensions and a suitable range of laser-based optical investigation tools, the crystallization of colloids looks like an extremely rewarding topic and has indeed received a lot of recent attention [3–13].

However, one word of caution is necessary at this stage. While structural equilibrium properties completely agree between atoms and colloids, their dynamics are known to be fundamentally different [14]. On the time scale where atoms experience their mutual interactions, their motion is ballistic, governed by a set of Newton's equations. Colloidal particles, on the other hand, exchange energy and momenta with the suspending liquid very rapidly — in fact much faster than the time they need in order to move a significant fraction of their diameter. On these time scales, colloids show their characteristic Brownian random motion, governed by a many-particle Smoluchowski equation. In spite of this fundamental difference, studies of atomic and colloidal liquids have still shown similarities of their respective dynamics [14]. Whether or not such similarities also exist for the crystallization kinetics of both systems has yet to be determined. The first experimental evidence, including this work, is rather encouraging.

The difference in the dynamic behavior of atoms and colloids contains an important implication for the theorist. Four of the five conserved fields, which were mentioned previously, are irrelevant for colloidal systems! Energy and momentum are freely (and rapidly) exchanged with the suspending liquid molecules and particle density remains the only conserved quantity. This constitutes a considerable simplification, which should encourage theoretical research on colloidal crystallization.

This work focuses on experimental data of hard-sphere-type colloids during their process of crystallization. The experiments were time-resolved static light scattering in two different angular ranges, a technique described in Section 2.1. Section 2.2 contains details of sample preparation and characterization. Section 2.3 explains the data processing. Experimental results are shown in Section 2.4 and discussed in Section 2.5. The emphasis in the discussion is laid upon the identification of the steps in our classical picture of crystallization: nucleation/growth and coarsening.

---

[2]Dynamic neutron scattering appears to do the trick, but such experiments require lengthy averaging times — ranging from minutes to hours, whereas crystallization kinetics often proceed within a few nanoseconds.

2.2 Time-Resolved Static Light Scattering

The spatial dimensions of many colloidal particles are in the range of the wavelength of visible light, and laser light scattering techniques are suitable and readily available tools for the investigation of these systems. There are three basic observational techniques: static and dynamic light scattering [15–17] plus (video-enhanced) microscopy. Static scattering accesses the equilibrium structure in terms of the product of the single-particle form factor $P(Q)$ and static structure factor $S(Q)$. Q denotes the magnitude of the scattering vector. Dynamic light scattering further includes information about particle motion, the dynamic structure factor $S(Q, \tau)$, which is the spatial Fourier transform of the van Hove space-time correlation function [18, 14]. Both scattering techniques yield data that are averaged over many particles. The ergodicity of liquid samples allows any level of statistical accuracy to be obtained by suitable time averaging.

The major drawbacks of scattering techniques, limitation of access to reciprocal space and two-particle correlations, may both be overcome by video-enhanced microscopy. Ordinary microscopes limited first applications to quasi-two-dimensional samples [19], but this restriction may now be lifted by real-time confocal microscopes. Time-consuming image analysis procedures still limit the statistical accuracy of quantitative results, but the immediate access to time-resolved images in real space carries a lot of appeal for the qualitative analysis of microscopic processes.

All three techniques have already been successfully applied to colloidal crystals as well as to liquids [20, 21]. A common difficulty arises due to the high particle densities that typically are required at the phase transition. Many samples already produce a lot of multiple scattering at these densities unless they are used in extremely thin layers. A second method to reduce multiple scattering involves careful matching of the index of refraction of particles and suspending liquid, usually by suitable mixing of two different liquids. This technique is used in the present work. Within dynamic light scattering, multiple-scattering artifacts can also be eliminated by dual-beam and dual-detector cross-correlation schemes [22–24]. But these new techniques have not been applied to crystalline samples yet.

In order to obtain accurate kinetic information about a colloidal sample while it undergoes a crystallization transition, two of the three basic techniques mentioned seem of limited use only. Dynamic light scattering typically requires extended measurement times of the order of several 10^4 fluctuation times [25] and limits the access to one scattering vector at a time, at least with present-day commercial equipment.[3] Video microscopy typically observes only a limited number of particles (or crystals) at a time. Hence both techniques cannot easily combine high statistical accuracy with real-time access.

---

[3]Real-time processing of multiangle data detected by a video camera may become a useful technique for dynamic measurements on samples, where kinetic effects are sufficiently slow [26].

There remains the classical approach of static light scattering, the visible light analogue of X-ray diffraction. Yet there are still problems with the use of commercially available equipment. Quite clearly, goniometric devices, which measure at different scattering vectors Q one after the other, have a severe speed handicap. The few available multidetector setups were all designed for polymer analysis and do not provide sufficient angular resolution. Furthermore, all these devices only detect light that is scattered into a single plane. While this procedure is adequate for ergodic liquid samples, given sufficient time for the averaging of speckle noise, it again severely limits the statistical accuracy of data from nonergodic solid samples. Instead, our ideal device would detect all the light scattered into a cone in 3D space for a particular scattering vector Q. Spatial averaging around the cone would significantly improve the statistical accuracy and allow for rather short measurement times.

Given the desire for a new apparatus, there remains one important question: Which scattering vectors should be accessible? At first sight, the optimum choice seems to be the same as for the study of dense colloidal liquids, say $Ql = 2$– 10, the range of the lowest-order peaks in the static structure factor $S(Q)$, if l denotes a typical distance of nearest-neighbor particles. Here the experiment accesses the local ordering of the colloidal particles. As crystallization proceeds, one observes a gradual change from a liquidlike $S(Q)$ obtained from the initial metastable colloidal liquid to a final much more spiked Debye-Scherrer pattern from the colloidal polycrystal. Experiments performed within the coexistence range of the phase diagram show a finite liquidlike contribution to $S(Q)$ throughout the experiment. The nontrivial problem of separating liquid and crystalline contributions will be addressed in Section 2.3. If this separation can be done, three pieces of valuable information are available: the local structure of liquid and solid as well as their relative volumes in the sample.

Provided the experimental resolution in scattering vector is sufficient, the width of Debye-Scherrer peaks may be analyzed in order to access the typical dimensions of growing crystals. If l_c denotes such a dimension, the resolution should be better than π/l_c. This requirement immediately implies a number of data points in Q space that exceeds the number of close-packed colloidal particles in a string of length l_c, possibly a number that becomes quite large for big crystals.[4] If one wants all the data at the same time, the device needs a large number of detectors or a flexible combination of low-resolution detectors, which cover the whole liquid structure factor, with a limited range of high-resolution detectors centered at a suitable Debye-Scherrer peak.

So far attention has been limited to access of local order with the scattering experiment yielding (averaged) information about the order parameter field. It is, however, quite feasible to access the second relevant field, particle density, as well! Once again, the appropriate procedure may be borrowed from X-ray

---

[4]This holds for hard-sphere-type particles. Particles with long-distance repulsion should be arranged at their mean next-neighbor separations.

scattering. Fluctuations of the coarse-grained particle density (coarse compared to l) lead to scattering at very small angles or scattering vectors. In the present context, the most prominent feature of the sample will be the density difference between the growing crystal grains and the surrounding liquid. Strong small-angle scattering may be expected and was indeed observed [25, 27] at scattering vectors with $Ql_c = 2$–20, in particular for hard-sphere-type systems where the density difference between coexisting liquid and solid is known to be large. A major practical problem with small-angle scattering is the sufficient reduction of flare from optical surfaces and cuvette walls.

The attempt to combine all these requirements resulted in the design and construction of the dual-range static light scattering apparatus of Fig. 2.1. The illuminating beam is generated by a small HeNe laser (optical power 5 mW, linearly polarized, wavelength 633 nm), spatially filtered (microscope lens with 10x magnification, pin hole with 0.04 mm diameter), and slightly focused using a single best form lens ($f = 100$ mm). This arrangement produces a clean beam with approximately Gaussian intensity profile ($1/e^2$ radius of 3 mm) within the sample cell. The beam leaves the sample through a flat surface that was ground perpendicular to the axis of a special, almost hemispherical lens (radius 50 mm, index of refraction 1.5, resulting in a focal length of 100 mm measured from the exit surface). Scattered light may leave the sample in two different ways: If scattering angles are small, say below 5 degrees, all (singly) scattered radiation will pass through a flat surface ground onto the hemispherical side of the lens. For a given angle or scattering vector this radiation will hence come to a focus in the same plane perpendicular to the beam axis as the illuminating laser beam.[5] A linear array of photodiode detectors senses the small-angle scattering in this plane. If scattering angles are larger, say between 15 and 75 degrees, the scattered radiation passes through the spherical section of the special lens

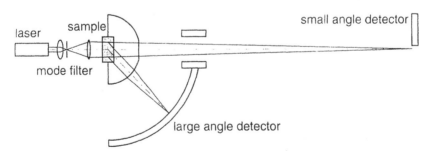

Figure 2.1 Schematic experimental setup for simultaneous access to small-angle and Debye-Scherrer light scattering.

---

[5]This small-angle scattering design was based on earlier work, where a screen in the focal plane was employed to detect scattering patterns [25]. Again the attempt was made to keep the number of necessary optical surfaces as small as possible in order to reduce stray light.

surface.[6] Ignoring, for the moment, the slight convergence of the light scattered at one particular scattering vector, this light will come to a focus on a spherical surface centered around the special lens. The distance of this surface from the lens is given by the focal length of the lens (100 mm). Several short linear arrays of photodiodes are arranged close to the focal surface in order to detect Debye-Scherrer scattering or liquid structure factors.

Both sets of detectors are mechanically coupled and can be rotated around the beam axis through an angle of 360 degrees, driven by a computer-controlled stepping motor. Thus data at all the accessible scattering angles can be captured at once. If the scattering is restricted to a single plane, the achievable time resolution is a few milliseconds. For a full scan of both detectors around 360 degrees, the present version of the apparatus takes about 25 seconds.

A common electrooptic module was designed for use within both detectors. It is based upon a 16-element linear photodiode array (spacing and width of elements approximately 1.5 mm). Sixteen individual current-to-voltage converters are surface mounted close to the array in order to provide low-noise amplification (4.7 V/μA) and slight low-pass filtering (time constant of 0.5 ms). Two 8-to-1 multiplexers provide a selectable pair of output signals. These signals are wired to a fast analog-to-digital conversion board with 24 multiplexed balanced inputs, which provides optional additional signal amplification (factor 16) and completes one 16-bit conversion every 0.01 ms. The converter accepts signals between 0 and 2.5 V corresponding to intensities up to 590 nW/mm^2 at the detectors, or 37 nW/mm^2 with the optional amplification. For comparison, scattered light from typical samples very rarely exceeded 500 nW/mm^2, even at small angles, with typical levels more like 10 nW/mm^2. The noise level of individual photodiodes as well as the current noise of the input amplifiers is typically below 0.0005 nW/mm^2, roughly the magnitude of a single conversion step of the A/D converter.

Digitized data are stored in a personal computer. The combination of individual rather slow amplifiers and a fast multiplexed converter achieves optimum noise reduction, as long as conversion cycles are repeated on a time scale similar to the time constant of the amplifiers. With about 100 individual detectors, both times are of the order of 1 ms — fast enough to combine with the quasicontinuous motion of the stepping motor, which rotates both detectors around the beam axis. A full rotation cycle corresponds to 6272 motor steps. Hence data are typically averaged over 6272 samples.

Determined by the spatial dimensions of the optical setup and the detector modules, the small-angle detector covers a range of scattering vectors between 0.02 and 0.4/μm (with 2 modules) or 0.77/μm (with 4 modules) at a resolution better than 0.012/μm. The Debye-Scherrer detector covers scattering vectors between 4 and 18/μm (with 6 modules) or any subset of this range (with less

---

[6]This design is a 3D version of a rather common 2D setup based on the refraction of light by a cylindrical index matching bath [28, 29].

modules) at a resolution of approximately $0.15/\mu m$. Both resolutions are essentially determined by the rather coarse dimensions of the individual photodiodes. The optical diffraction limit for small-angle detection lies roughly one order of magnitude below the stated resolution limit, the smearing of Debye-Scherrer data due to convergent beam illumination is less than half of a detector element, and spherical abberration effects are more than one order of magnitude smaller.

In order to reduce flare from outer walls of the cuvette and to provide a controlled temperature environment for the sample, the sample cell is placed within an index-matching fluid bath. The bath is enclosed between a flat high-quality window for beam entrance, a surrounding body of solid aluminum, and the special almost hemispherical lens shown in Fig. 2.1. All outer optical surfaces are antireflection coated for 633 nm. Thermal control is provided through Peltier elements coupled to the aluminum body. An analog control unit allows heating as well as cooling of the sample cell with a long-term accuracy of the cell temperature better than 0.1 K. Index matching is realized both ways, with respect to the sample (by using the same or a similar solvent or solvent mixture) and to the special lens (by having it made out of suitable glass). Thus one does not have to consider any refraction distortions due to sample cell geometry or bath, as long as the index of refraction of the sample remains sufficiently close to 1.5, the index of refraction of polymethylmethacrylate (PMMA).

2.3 Samples

Hard-sphere particles with no other interaction than mutual volume exclusion are among the simplest model systems considered by theoreticians. Such systems have also been investigated on the computer, mainly by way of simulations [30]. These calculations first demonstrated the existence of a liquid-solid transition in a system without attractive interparticle forces, that is, solely driven by entropic effects [31]. Further simulations contributed equation-of-state data for metastable liquids as well as for crystalline solids, estimates of interfacial energies, and phase diagrams [30].

Unfortunately, interaction forces cannot typically be ignored in atomic systems, and hence theoretical or computer results obtained for hard-sphere systems could not be tested against experiments. This situation changed with the development of carefully prepared colloidal model suspensions. If such suspensions are prepared in nonpolar liquids, one can avoid the buildup of surface charges and essentially eliminate electrostatic repulsion forces. Careful matching of the (frequency-dependent) dielectric properties between particles and solvent furthermore reduces dispersion forces and hence van der Waals attraction between the particles. A perfect match is, however, difficult to achieve, and residual, extremely short-range attractive forces will eventually lead to aggregation of the sample. These forces can be overcome by a process technically known as steric stabilization [32], the coating of the colloidal particles with a short-chain-length polymer, for which the suspension liquid is a good solvent. Such a polymer coating provides a steep repulsive potential with a range not exceeding the chain length, which restores perfect colloidal stability. Hence particles with radii well in excess of the chain length of the coating polymer are rather close to perfect hard spheres.

Within this work we use particles made of PMMA and coated with hydroxy-stearic acid. This particle type was first developed by Ottewill's group at Bristol [33]. The particle radius is $a = 500$ nm, and the chain length of the coating is 10 nm. The particles are suspended in a mixture of decalin and tetralin, as suggested by Ackerson [34], which is chosen to match the index of refraction of the particles to better than 0.001. This accuracy of the match is necessary in order to keep the scattering by concentrated dispersions of the particles with diameters larger than the wavelength of light sufficiently small to neglect multiple scattering.[7] A perfect optical match cannot be achieved due to the presence of the coating layer. Another practical difficulty arises due to differences in thermal expansion of particles and suspending liquid. The close optical match can only be stably maintained with proper thermal control of the sample, say to 0.1 K. Such a level of thermal control was not available during the first low-angle scattering experiments on colloidal hard spheres [25, 35, 27] and resulted in poor quality of long-time data.

Luckily, optical matching is also typically an efficient way of establishing a sufficient reduction of dispersion forces [14]. Hence by tuning the mixture of solvents to achieve a sufficiently small level of scattering, one simultaneously switches off van der Waals attraction and obtains rather pure and short-range steric repulsion between the particles, close to the behavior of ideal hard spheres. Experimental proof of this hard sphere character are provided by liquid structure factor data close to Percus-Yevick calculations [14], the expected viscosity dependence on particle density [36], the expected sedimentation velocity dependence on particle density [34], and measurements of the correct densities at freezing and melting (volume fractions of 0.494 and 0.545, respectively) [9].

Other previous work on PMMA hard-sphere colloids included the dynamics of their glass transition at volume fractions around 0.58 [37] and a structure investigation of polycrystalline solids [38], which identified them to be mostly random stacks of hexagonal layers, rather than perfect FCC or hcp crystals. A first series of kinetic measurements on PMMA hard spheres by small angle light scattering [25, 35, 27] were already mentioned. These data will be included in the discussion in Section 2.5. A recent paper [37] also addressed the crystallization kinetics of PMMA hard-sphere systems (particle radius 201 nm), starting from metastable liquids or glasses. Time-resolved Debye-Scherrer light scattering indicated a qualitative change of the growth process at the glass transition. Nucleation rate estimates showed a peak around the melting density and fell essentially to zero upon approaching the glass transition, while finite velocity growth (or ripening?) was demonstrated to occur even above the glass transition.

---

[7]Total scattering of the sample is less than 20% of the incident laser light.

An important peculiarity of colloidal particles, as compared to atoms, is the unavoidable size differences between individual particles, known as polydispersity. It affects both, the scattering properties as well as the structure of the samples [14, 39]. While both effects are reasonably understood for liquid systems, it remains a challenging task to determine polydispersity influences on crystallization or the formation of glasses. Obviously, increased levels of polydispersity (i.e., wider particle size distributions) make crystalline order more difficult to establish and favor glass formation. A relative standard deviation of 10% in particle size is typically considered as the upper limit for colloidal crystallization [14]. The particle systems used in this work have much lower polydispersities, well below 5%.

The accurate measurement of such small polydispersities calls for special techniques. Besides the tedious use of electron microscopy, there are two light scattering approaches. Conventional dynamic light scattering is inherently limited in terms of particle size resolution and does not allow accurate estimates of polydispersities below 10% [40]. However, mean particle diameters may be determined to within the 1% level, and a clever scheme based on a series of dynamic measurements performed at scattering angles close to a minimum in the single-particle form factor $P(Q)$ was suggested by Pusey et al. [41]. Essentially the dramatic changes of the form factor result in different weightings of the (narrow) particle size distribution, and a kink observed in measured particle diameter as a function of scattering vector may be analyzed to yield reliable estimates of small polydispersities. In practice, the weak scattering close to the form factor minimum makes the experiment extremely susceptible to multiple scattering artefacts [42]. Successful experiments were, however, recently reported, when cross correlation was utilized to eliminate multiple scattering [43].

As an alternative to the dynamic measurements, a careful study of particle form factors by static scattering may also provide accurate polydispersity information. A finite width of the particle size distribution leads to some smearing of the minima in the form factor. Again, multiple scattering leads to very similar smearing effects and typically becomes a severe practical problem. Accurate corrections are necessary and may be achieved by the use of dynamic cross-correlation experiments, which allow one to determine the fraction of multiple scattering from measured contrast factors. Such experiments have been performed, and a preliminary analysis indicates a relative standard deviation of particle size close to 3% for the particles used in this investigation. A more thorough analysis of static data with samples, where the index of refraction of the suspension liquid was varied around the match point, is under progress and should also yield information about the core-shell structure of the particles.

Sample preparation included careful cleaning of the particles as well as the suspending liquid mixture by centrifugation and filtering. Unfortunately, some of these cleaning steps and the small amount of available particles limited the accuracy with which the final volume fraction of the samples could be determined. As a preliminary measure, the time-dependent scattering data of the new samples were compared with earlier small-angle light scattering experiments [27], where almost identical particle systems with carefully determined volume fractions had been used. More accurate means of sample density characterization are under development. Between successive experiments with the same sample, there is typically some small loss of solvents due to imperfect sealing. If this loss is monitored by accurate weighing of the samples, one can track changes in sample density.

As hard-sphere systems cannot be directly controlled by temperature, their crystallization was not initiated by temperature jumps. The only accessible control parameter is particle density or volume fraction. However, no quick methods to change this parameter are yet known for colloidal systems. Hence one has to resort to a slightly unusual melting mechanism: Samples are molten by the application of shear forces. Typical crystals were easily molten by tumbling of the cuvettes ($5 \times 10 \times 40$ mm^3, filled to a height of about 15 mm), which contained the samples, at a moderate rate of one revolution in a few seconds. Please note the importance of the presence of air in the sample. The rapid passage of an air bubble through the suspension during the rotation was responsible for a sufficiently large level of turbulent random shear. Rigid sediments, which tend to build up from the bottom of the sample cells, if they were kept in an upright position for long times, needed more vigorous shearing, which could be achieved by a mechanical stirrer.

In a typical experiment, care was taken to melt the whole volume of the sample. The shear molten sample was kept under continuous rotation and in an environment with a temperature close to that within the index matching bath of the scattering device. In order to start a crystallization experiment, the rotation was stopped and the sample was manually transferred to the scattering experiment. This process took a few seconds, a negligible time as compared to the fastest observed times scales of the crystallization process ($\sim$ 10 min). The time needed for viscous damping of existing flows within the cell was of the order of 1 s, again a negligible magnitude. During the actual experiment, the sample was left at rest inside the thermostatted index-matching bath.

Typical durations of experiments ranged from 16 h to more than 2 d. While scattering data were recorded quite rapidly at the beginning of an experiment, a geometric grid of sampling times was employed at later times. Thus the total data volume was kept manageable, and an adequate time resolution was achieved throughout the experiments.

2.4 Data Processing

Raw data consist of Debye-Scherrer and small-angle scattering data, which were preaveraged over a full rotation of both detectors around the beam axis. Each data point corresponds to an individual photodiode detector and hence needs individual corrections for dark current and sensitivity. Dark current estimates are easily obtained by switching the laser off momentarily. Sensitivity factors can be estimated by using dilute samples of particles with known, smooth form factors $P(Q)$.

Data at very small angles often showed residual stray light, such as that due to flare from imperfections of, or dirt on, the walls of the sample cell. Using the reasonable assumption that this kind of scattering does not change much during the experiment, we generally subtracted an early-time small-angle data set from all other small-angle data sets of the same series. This is the same procedure as was used in earlier measurements [27]. Please note that small-angle data do not need form factor corrections. The covered range of scattering vectors typically extends out to values much less than 2π/particle diameter, and $P(Q)$ is essentially flat and close to 1 in the small-angle region.[8]

Further analysis of the small-angle scattering data is based on an important general feature of these data, the occurrence of a single peak at some finite scattering vector. The presence of this peak indicates the onset of crystal formation, which is absent in purely liquid initial samples [25]. While the peak height I_{SA} can typically be determined with good accuracy by fitting a parabola to a range of data points including the absolute maximum of the data set, the simultaneous estimates of peak position Q_m show significant scatter. More stable characteristic scattering vectors are obtained by searching for the scattering vector Q_{SA},[9] where the scattered intensity falls to $I_{SA}/2$ with $Q_{SA} > Q_m$.

The analysis of Debye-Scherrer data is considerably more difficult due to the more complicated nature of scattering at large angles. In principle, form factor corrections should be applied first. However, this form factor $P(Q)$ is often not known with sufficient accuracy, particularly around the important first minimum, which typically falls in the range at or near the lowest-order peaks in the structure factor for dense hard-sphere systems, if one works close to index-matching conditions and considers the optical core-shell structure of the particles due the presence of the polymer coating. Together, these conditions do not just change the depth of the minima in $P(Q)$, but also shift these positions significantly [44]. Additional measurements on dilute samples appear to be necessary in order to obtain sufficiently reliable form factor data and are planned for future measurements.

Even with perfect form factor correction, the separation of liquid and solid structure factors during crystallization is not straightforward. Simple analytic approximations are available for the liquid contribution only (e.g., Percus-Yevick). The expected crystal structures contain variable degrees of random stacking, resulting in structure factors that depend upon the initial sample density control parameter in addition to the time-dependent size of individual crystals [38]. Random stacking smears one of the major low-order FCC reflections (200) into a broad feature that is not easily separated from the first peak in the liquid structure factor. The other low-order reflections, however, remain sharp. This includes the

---

[8]Some noticeable deviations may, however, occur very close to perfect index-matching conditions due to the core-shell structure of the particles [44].

[9]$q_{1/2}$ in the notation of [27].

(111), (220), and (311) peaks in the usual FCC indexing.[10] Further complications include the presence of small but noticeable changes in the density of both liquid and solid, as the crystallization proceeds, as indicated theoretically and in crystal growth calculations [46]. Solid density changes were also found experimentally [37]. Early crystals are formed with densities above the initial liquid density. Consequently, the mean density of the liquid falls. As crystallization proceeds, the mean crystal density falls as well, in fact down to the initial liquid density for samples that undergo full conversion. Some of the Debye-Scherrer data to be discussed clearly prove the existence of this effect for the solid. Finally, these density changes are really not just time dependent, but also vary with position in space, as a function of distance from a liquid-crystal interface.

The resulting picture is far too complicated to be modeled faithfully yet. More restricted attempts to extract limited information from the Debye-Scherrer scattering must be applied instead. A possible approach that was recently suggested [37] is the analysis of a single crystalline peak. The suggested procedure uses the lowest-order sharp reflection, the scattering by a set of (111) planes in the usual FCC indexing.[11] The integral over the peak area is known to be proportional to the crystallized volume, and would thus constitute valuable information about the progress of the crystallization process. However, the (111) peak is located close to the first peak in the initial liquid structure factor and hence rides on a rather time-dependent background, which complicates the separation. Furthermore, the (111) peak falls within the range of the broadened (200) structure. As a simple practical procedure Harland et al. [37] used the integrated intensity over the full length of their detector array, which covered only part of the first peak in the liquid structure factor, and normalized these data by division with its initial value and subtraction of 1. The resulting number $X(t)$ rises from 0 and constitutes a coarse measure of the degree of crystallization, albeit with a rather arbitrary and slightly sample-dependent normalization.

The availability of data at higher-order reflections allowed a somewhat less ambiguous data analysis to be used in this work. The (220) peak happens to fall close to the third crossing of the initial liquid structure factor with its large Q unit asymptote and is also reasonably well separated from all other crystalline peaks.[12] Figure 2.2 shows corresponding data. In this range of scattering vectors, the background contribution of liquid may be faithfully modeled by a simple straight line. Data at a sufficient distance from the peak position are used in order to estimate slope and intercept of this line, which is then subtracted from the raw data for further analysis of the (220) peak. This analysis typically includes the determination of the area under the peak I_{DS} as a measure proportional

---

[10]The hexagonal indexing used in [38] of these peaks is (001), (110), and (111), respectively. Within the conventional hcp basis [45] the reflections are from (002), (110), and (112) planes.

[11](001) in the notation of [38].

[12]There occurs slight overlap with the third peak at very early times, when both peaks are rather broad.

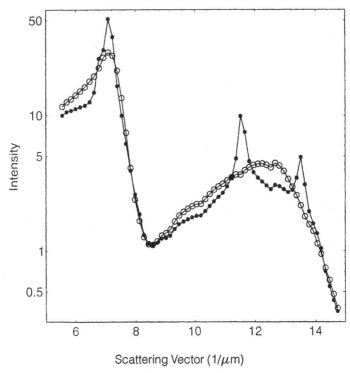

Figure 2.2 Debye-Scherrer data obtained on a hard-sphere sample prior to crystallization (metastable fluid open circles) and after significant progress of the crystallization
process (closed symbols). Neither curve has been corrected for the form factor or gain
variations of individual detector elements.

to the total crystal volume, and the peak width (standard deviation ΔQ_{DS}) as
an inverse measure of crystal size. In order to check the accuracy of the peak
analysis, Debye-Scherrer data were obtained from different samples, which all
underwent full conversion from liquid to solid (initial densities between melting
and glass formation). These data indeed gave I_{DS} values that approached almost
the same asymptotic value at long times. Normalization of I_{DS} by division with
this asymptote would hence appear to be a plausible and accurate measure of the
crystalline volume fraction as a function of time. This is in marked contrast to
the behavior of $X(t)$ data presented by Harland et al., where asymptotic values
scatter over almost an order of magnitude with the smallest-density sample (well
below melting) coming out highest [37].

Please note some distinct differences between the small-angle characteristic
scattering vector Q_{SA} and the width of the (220) peak ΔQ_{DS}. Although both
values should be inversely proportional to the diameter of a single scattering
crystal, they will generally be different from each other not just by a trivial
scale factor. ΔQ_{DS} responds to the size (and shape) of single crystals. Light

scattered by one such single crystal (or a single crystalline domain of a larger polycrystal) into the Debye-Scherrer range of scattering vectors will typically contain a random phase factor, which changes strongly with crystal position in space.[13] Hence interference effects between light scattered by different single crystals (or domains) can be ignored, and the total scattered intensity is just a sum of individual single crystal contributions. ΔQ_{DS} will reflect the mean size of single crystals or domains, weighted by their volumes. For a spherical single crystal of diameter l_c, there results $\Delta Q_{DS} \approx \pi / l_c$.

In contrast, the small-angle scattering senses the density structure of the whole sample, but is uninfluenced by features of crystalline order like the orientation of some lattice planes. Hence a single crystal will scatter at small angles exactly like a polycrystal of same size and density. Furthermore, any particular density structure within the surrounding liquid also contributes to the small-angle scattering. In particular depletion zones, which develop around a growing crystal, are responsible for the typical ring-shaped scattering patterns [25], because owing to density conservation, the higher than average density within the crystal is exactly balanced by the density deficit in the surrounding liquid, which leads to a cancellation of density fluctuations on very large length scales, corresponding to very small scattering angles. For spherical crystals of diameter l_c with a depletion layer,[14] Q_{SA} can take on values ranging from about $4/l_c$ for a rather wide and shallow depletion layer to much larger values for a narrow and deep depletion layer, for example, to about $8/l_c$ for an exponential depletion layer with a $1/e$ width of $l_c/20$. Finally, the fact that there are no strongly position-dependent (and hence random) phase factors in near forward scattering leads to a coherent superposition of light that has been scattered by individual single or polycrystals. Essentially the crystals may be looked upon as scattering particles, and the small-angle scattering will not just reflect their form factors (including depletion zones), but also contain a structure factor contribution with information about spatial ordering of crystals. Early on in the crystallization process, when crystals are typically well separated and at random positions, such structure factor effects should be small, and a pure form factor analysis appears appropriate.

A brief comment on the scattering intensities I_{DS} and I_{SA} will close this section. In both scattering regimes we obtain coherent superposition of light scattered from colloidal particles within one single crystal (or polycrystal plus depletion layer in the case of small-angle scattering). Coherent superposition implies that the scattered amplitude is proportional to the volume of the scattering crystal, V_c, or the scattered intensity is proportional to V_c^2. Indeed this can be observed for I_{SA} in small-angle scattering [27]. But a particular domain will contribute to the Debye-Scherrer signal only if it is properly oriented in space. The required accuracy of orientation rises with increasing volume V_c. The angular

---

[13]The phase changes by 2π for a crystal translation by one lattice constant in the direction of any basis vector that is not perpendicular to $\vec{Q}$.

[14]Located at $r > l_c/2$.

width in one dimension scales as $V_c^{-1/3}$. As long as measurements are carried out on sufficiently large samples and using the possibility of detector rotation around the sample, a two-dimensional angular average over domain orientation is being performed. The explicit integration over the width of the peak extends averaging to all three dimensions and results in a Debye-Scherrer signal I_{DS} that is proportional to $V_c^2(V_c^{-1/3})^3 = V_c$.

2.5 Experimental Results

Results of a first series of simultaneous scattering measurements at small angles and in the Debye-Scherrer range are displayed in Figs. 2.3 to 2.6. Initial sample densities were effective volume fractions of 0.54 (open circles), 0.55 (full circles), and 0.555 (open triangles). As indicated, all densities were estimated by comparison of the small-angle scattering data with earlier measurements performed on samples of identical preparation [27]. Figures 2.3 and 2.4 show small-angle peak intensities I_{SA} and characteristic wave vectors Q_{SA}, respectively.

Debye-Scherrer integrated peak intensities I_{DS} and widths ΔQ_{DS} are displayed in Figs. 2.5 and 2.6. All plots use log-log scales in order to cover the large ranges in time and intensities as well as to ease the fitting of power laws.

The highest-density sample (volume fraction 0.555) falls rather close to the sample P6 used in [27] (volume fraction 0.556). The measured characteristic wavevectors Q_{SA} (Fig. 2.4) indeed follow the earlier small-angle measurements closely. This includes the peculiar upturn of the data at times exceeding 500 min, which is also accompanied by corresponding downturns in maximum intensity I_{SA} in both data sets (Figs. 3 and 5 of [27]).

The middle-density sample (volume fraction 0.55) fills a gap in the earlier set (between samples at 0.545 and 0.556). It provides valuable information about the behavior of samples with initial densities well above freezing.

The lowest-density sample provided a true surprise. Although its volume fraction of 0.54 falls just below a range of volume fractions studied earlier (samples P3 and P5 of [27] with volume fractions of 0.542 and 0.545, close to melting), and the long-time behavior of all three data sets is very similar, the new sample shows a much steeper rise in maximum intensity I_{SA} (Fig. 2.3) and a correspondingly steeper fall in characteristic wave vector Q_{SA} (Fig. 2.4) than the two other sets of data.

2.6 Discussion

The nucleation, growth, and ripening of colloidal crystals have been studied by scattering techniques. The density fluctuations produced by this process have been monitored by small-angle light scattering [25, 27, 47], while the crystal

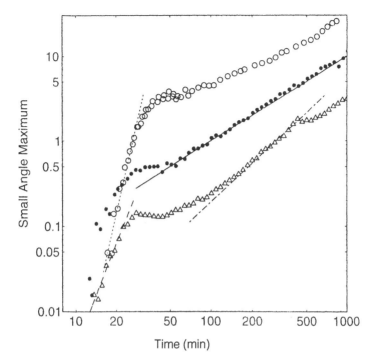

Figure 2.3 Intensity maximum I_{SA} in small-angle scattering as a function of time since the start of crystallization. Initial volume fractions were 0.54 (open circles), 0.55 (full circles), and 0.555 (open triangles). Possible power laws are indicated by dotted (t^7), dashed (t^4), dash-dotted ($t^{3/2}$), and full (t^1) lines.

order parameter has been monitored by Debye-Scherrer light scattering [37]. Both techniques show qualitatively similar behavior, having two distinct growth regimes. However, there are also differences such as the apparent shrinking of crystal growth prior to the onset of coarsening seen in SA but not in DS scattering. Because SA experiments were carried out independently and on different samples from the DS experiments, it is not clear if the observed growth regions correspond between the two experiments. It is the purpose of the experiment reported here to provide a direct comparison of the two measurements made simultaneously using a single sample.

A comparison of I_{SA} in Fig. 2.3 with I_{DS} in Fig. 2.5 shows that the nucleation and growth region observed in SA for times less than 50 min corresponds to the initial growth of the DS signal in the same time range. For times greater than 50 min, a qualitatively different growth behavior is observed in both DS and SA. The two signals are correlated!

A simple model for the nucleation and growth region (for times less than 50 min) further correlates the behavior observed in DS and SA. This model was

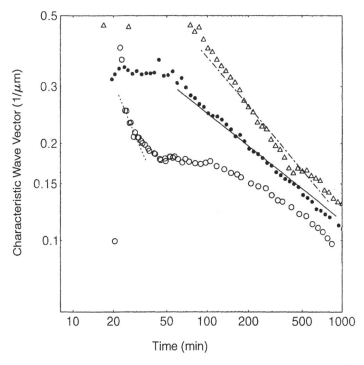

Figure 2.4 Characteristic wave vector Q_{SA} in small-angle scattering as a function of time since the start of crystallization. Initial volume fractions and symbols as in Fig. 2.3. Possible power laws are indicated by dotted (t^{-1}), dashdotted ($t^{-1/2}$), and full ($t^{-1/3}$) lines.

presented earlier to understand the SA scattering [27] and posits the small-angle scattered intensity to be proportional to the number of crystallites times the square of the volume of a single crystal, $\sim N_c r_c^6$. The DS scattering is known to be proportional to the total crystal volume and is given by $\sim N_c r_c^3$. Here the number of crystallites is given by N_c, and r_c is the radius of one of the assumed spherical crystals. If the nucleation rate density is constant, then N_c increases proportional to the time after shear melting, t. If the crystal radius increases in time as $r_c \sim t^\eta$, then $I_{SA} \sim t^{6\eta+1}$ and $I_{DS} \sim t^{3\eta+1}$. In Fig. 2.3 the growth exponent for the $\phi = 0.54$ sample is 7 and is 4 in Fig. 2.5. This is consistent with $\eta = 1$, which is shown as a dotted line in Fig. 2.4. For the $\phi = 0.555$ sample the growth exponent in Fig. 2.3 is 4 and is $\frac{5}{2}$ in Fig. 2.5. This is consistent with $\eta = \frac{1}{2}$. While this exponent is not evident in the data presented in Fig. 2.4, we note that the angular range of these measurements was too restricted to provide accurate measurements of Q_{SA} in this range. However, this exponent was observed at the melting transition for the same sample system in the earlier SA work [27]. These results for the two values of η are summarized in Table 2.1.

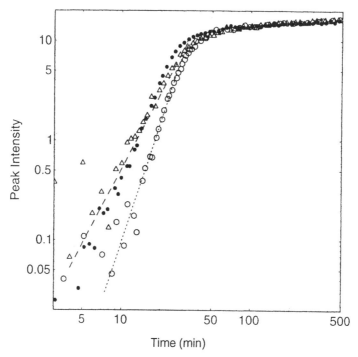

Figure 2.5 Integrated peak intensity I_{DS} in Debye-Scherrer scattering as a function of time since the start of crystallization. Initial volume fractions and symbols as in Fig. 2.3. Possible power laws are indicated by dotted (t^4), dashed ($t^{5/2}$) lines.

Table 2.1 Growth laws for experimental parameters.

	Nucleation and reaction-limited growth	Nucleation and diffusion-limited growth	Lifshitz-Allen-Cahn ripening	Lifshitz-Slyozov ripening
	$r_c \sim t$ $N_c \sim t$	$r_c \sim t^{1/2}$ $N_c \sim t$	$r_c \sim t^{1/2}$ $N_c r_c^3 = \text{const}$	$r_c \sim t^{1/3}$ $N_c r_c^3 = \text{const}$
$I_{SA} \sim N_c r_c^6$	t^7	t^4	$t^{3/2}$	t
$I_{DS} \sim N_c r_c^3$	t^4	$t^{5/2}$	const	const
$Q_{SA} \sim 1/r_c$ $\Delta Q_{DS} \sim 1/r_c$	t^{-1} t^{-1}	$t^{-1/2}$ $t^{-1/2}$	$t^{-1/2}$ $t^{-1/2}$	$t^{-1/3}$ $t^{-1/3}$

The growth exponents for η are suggestive of certain limiting growth conditions. Diffusion-limited growth, which requires the diffusion of heat or impurities away from an interface or in our case diffusion of particles to the interface, gives

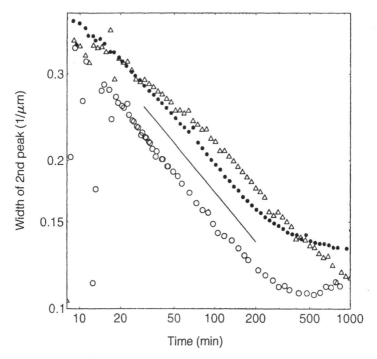

Figure 2.6   Peak width ΔQ_{DS} in Debye-Scherrer scattering as a function of time since start of crystallization. Initial volume fractions and symbols as in Fig. 2.3.

a growth exponent $\eta = \frac{1}{2}$ [48]. On the other hand, interface-limited growth gives a growth exponent $\eta = 1$ [49]. While it is tempting to identify these two exponents with these growth processes, the diffusion-limited growth occurs above the melting point and interface-limited growth in the coexistence region. This is the reverse of what is expected intuitively. Above the melting volume fraction, the metastable fluid and the equilibrium crystal density are the same. Direct conversion of fluid to crystal should be possible without any diffusion limitation. On the other hand, in the coexistence region the equilibrium crystal density is larger than the metastable fluid density. Thus a growing crystal requires more particles than can be supplied locally by the metastable fluid, and a depletion layer results. Particles diffuse from ever-increasing distances in response to the particle gradient as growth proceeds. While the quality of the Q_{SA} data may lead one to question these limiting values for η, they have been confirmed in other SA studies using "hard" colloidal particles [27, 37]. The Wilson-Frenkel model has recently been evaluated for the growth of hard-sphere colloidal crystals [46], with the particles assumed to diffuse through the suspending solvent and with a mechanical equilibrium (pressure balance) enforced at the crystal-liquid interface. Depending on the growth conditions, such as the rate of incorporation of particles into the interface, different growth "exponents" are possible. High incorporation rates give

$\eta = \frac{1}{2}$ in the coexistence region and $\eta = 1$ above the melting volume fraction. Low incorporation rates can increase η in the coexistence region and decrease η above the melting point, consistent with the results presented here. However, for low incorporation rates the growth process is transient, and the exponents are really pseudoexponents. A detailed comparison between theory and experiment has not yet been done.

In the ripening region for times greater than 50 min, I_{DS} has saturated to nearly the same value for all samples. Since the integrated intensity is proportional to the total crystal volume, these data indicate that the crystal volume has saturated in this time regime. Thus the DS results support the claim that SA scattering monitors the ripening process where larger crystals grow at the expense of smaller ones without changing the total fraction of crystal volume in the sample. As in earlier SA studies of the ripening process [27], the growth exponent for the characteristic length is compared in Fig. 2.4 to $\frac{1}{3}$ in the coexistence region (near melting) and to $\frac{1}{2}$ in the fully crystalline region. The comparison is reasonable and correlates with the Lifshitz-Slyozov growth in the coexistence region and Lifshitz-Allen-Cahn growth in the fully crystalline region. Lifshitz-Slyozov growth describes growth with a conserved order parameter and is appropriate in the coexistence region, with particle density being the conserved and controlling parameter. Lifshitz-Allen-Cahn growth describes ripening involving a nonconserved order parameter. In this case the ripening evolves via domain wall motion. With the total crystal volume fixed, the crystal number is given by $N_c \sim r_c^{-3}$, and the DS and SA intensities are determined as shown in Table 2.1. The exponents for I_{SA} give a reasonable representation of the data presented in Fig. 2.3. We have already seen that I_{DS} has a fixed value.

Figure 2.6 shows the evolution of the width of the DS peaks for the same volume fractions presented in the previous figures. The analysis procedure uses data in a fixed angular range, which is less than the angular distance between the centers of peaks on either side of the measured peak. At early times less than ~ 30 min the measured peak is broad and overlaps with other neighboring peaks. Thus some distortion of the DS peak width measurement is expected in this time range. At times greater than 200 min the peak width is of the order of the detector element separation; so the observed saturation of the peak width is an instrumental artifact. However, the solid line with slope $\frac{1}{3}$ gives a good representation of the data at large times when ignoring the instrumentally induced saturation, and the data points also fit this line in the intermediate region! Due to the above-mentioned early-time distortion, we do not yet know if the DS peak width will exhibit the same growth characteristics seen by SA scattering in the nucleation and growth regime.

It is interesting to compare these peak width data with those given in Fig. 2.4. The lower-volume-fraction samples, $\phi = 0.54$ and 0.55, match the same $\frac{1}{3}$ power-law behavior at large times in the ripening region. In contrast to the DS peak width, the SA rate slows to a stop in the intermediate region before entering the ripening region. We do not fully understand this behavior at this time. We only note that the DS measurements are sensitive to those crystals or to that portion of a crystal aggregate oriented to scatter to the detector. The SA scattering is sensitive to the full size of a crystal aggregate, including any depletion zone extending into the metastable fluid. Thus SA scattering may give larger growth rates due to depletion zones or crystal aggregates in the nucleation and growth region. In ripening the depletion zones are gone, and the crystal aggregates may be stressed and break into pieces. While SA scattering would be sensitive to this, DS scattering would not be.

2.7 Conclusions

In this chapter combined SA and DS measurements have been presented for the first time for the study of the nucleation, growth, and ripening of colloidal crystals. These measurements have provided conclusive evidence that the two different growth regions observed in separate SA and DS measurements are indeed related. A simple model correlates measured intensities of both SA and DS with the growth rates of the SA length scale as shown in Table 2.1. In the nucleation and growth region, the measured length scales evolve quite differently in SA and DS measurements. More detailed theoretical models of the growth process necessarily will be constrained by these data.

Acknowledgements

W. van Megen, S. M. Underwood, and B. J. Ackerson are thanked for giving samples and for continuing discussions. The North Atlantic Treaty Organization is thanked for a travel grant to promote collaboration with B. J. Ackerson. The experiment was done with financial support of the Deutsche Forschungsgemeinschaft.

References

1. Oxtoby, D. W., and Harrowell, P. R., *J. Chem. Phys.* **96**, 3834 (1992).
2. Frenkel, J., *Kinetic Theory of Liquids* (Oxford University Press, Oxford, 1946).
3. Hiltner, P. A., and Krieger, I. M., *J. Chem. Phys.* **73**, 2386 (1969).
4. Pieransky, P., Dubois-Violette, E., Rothen, F., and Strzelecki, L., *J. Phys.* **42**, 53 (1981).
5. Ackerson, B. J., and Clark, N. A., *Phys. Rev. Lett.* **46**, 123 (1981).
6. Lindsay, H. M., and Chaikin, P. M., *J. Chem. Phys.* **76**, 3774 (1982).
7. Yoshiyama, T., Sogami, I., and Ise, N., *Phys. Rev. Lett.* **53**, 2153 (1984).
8. Aastuen, D. J. W., Clark, N. A., Cotter, L. K., and Ackerson, B. J., *Phys. Rev. Lett.* **57**, 1733 (1986); Errata *Phys. Rev. Lett.*, **57**, 2772 (1986).
9. Pusey, P. N., and van Megen, W., *Nature* **320**, 340 (1986).
10. Monovoukas, Y., and Gast, A. P., *J. Colloid Interface Sci.* **128**, 533 (1989).
11. Ackerson, B.J., editor. *Phase Transitions* **21**, (1990).
12. Löwen, H., Palberg, T., and Simon, R., *Phys. Rev. Lett.* **70**, 1557 (1993).
13. Palberg, T., and Streicher, K., *Nature* **367**, 51 (1994).
14. Pusey, P. N., in *Liquids, Freezing and the Glass Transition*, Eds. D. Levesque, J. P. Hansen, and J. Zinn-Justin (Elsevier, Amsterdam, 1990).
15. Berne, B. J., and Pecora, R., *Dynamic Light Scattering* (Wiley, New York, 1976).

16. Brown, W., editor, *Dynamic Light Scattering* (Clarendon Press, Oxford, 1992).
17. Schätzel, K., *Adv. Colloid Interface Sci.* **46**, 309 (1993).
18. Hansen, J. P., and McDonald, I. R., *Theory of Simple Liquids* (Academic, New York, 1986).
19. Murray, C. A., Sprenger, W. O., and Wenk, R. A., *Phys. Rev.* B **42**, 688 (1990).
20. Smits, C., van Duijneveldt, J. S., Dhont, J. K. G., Lekkerkerker, H. N. W., and Briels, W. J., *Phase Transitions* **21**, 157 (1990).
21. Aastuen, D. J. W., Clark, N. A., Swindal, J. C., and Muzny, C. D., *Phase Transitions* **21**, 139 (1990).
22. Phillies, G. D. J., *Phys. Rev.* **24**, 1939 (1981).
23. Drewel, M., Ahrens, J., and Podschus, U., *J. Opt. Soc. Am. A* **7**, 206 (1990).
24. Schätzel, K., *J. Modern Optics* **38**, 1849 (1991).
25. Schätzel, K., and Ackerson, B. J., *Phys. Rev. Lett.* **68**, 337 (1992).
26. Grier, D. G., and Murray, C. A., *J. Chem. Phys.* **100**, 9088 (1994).
27. Schätzel, K., and Ackerson, B. J., *Phys. Rev.* E **48**, 3766 (1993).
28. Härtl, W., Klemp, R., and Versmold, H., *Phase Transitions* **21**, 229 (1990).
29. Dhont, J. K. G., Smits, C., and Lekkerkerker, H. N. W., *J. Colloid Interface Sci.* **152**, 386 (1992).
30. Hoover, W. G., and Ree, F. H., *J. Chem. Phys.* **49**, 3609 (1968).
31. Alder, B. J., and Wainwright, T. E., *Phys. Rev.* **127**, 359 (1962).
32. Hunter, R. J., *Foundations of Colloid Science* (Clarendon, Oxford, 1987).
33. Antl, L., Goodwin, J. W., Hill, R. D., Ottewill, R. H., Owens, S. M., Papworth, S., and Waters, J. A., *Colloids Surfaces* **17**, 67 (1986).
34. Paulin, S. E., and Ackerson, B. J., *Phys. Rev. Lett.* **64**, 2663 (1990).
35. Ackerson, B. J., and Schätzel, K., in *Sitges Conference*, Ed. L. Garrido (Springer, Heidelberg, 1992).
36. Mewis, J., Frith, W. J., Strivens, T. A., and Russel, W. B., *AIChE J.* **35**, 415 (1989).
37. Harland, J. L., Henderson, S. I., Underwood, S. M., and van Megen, W., Preprint (1995).
38. Pusey, P. N., van Megen, W., Bartlett, P., Ackerson, B. J., Rarity, J. G., and Underwood, S. M., *Phys. Rev. Lett.* **63**, 2153 (1989).
39. D'Aguanno, B., Klein, R., Méndez-Alcaraz, J. M., and Nägele, G., in *Complex Fluids*, Ed. L. Garrido (Springer, New York, 1993).
40. Stock, R. S., and Ray, W. H., *J. Polym. Sci.: Polym. Phys. Ed.* **23**, 1393 (1985).
41. Pusey, P. N., and van Megen W., *J. Chem. Phys.* **80**, 3513 (1984).
42. Nobbmann, U., Private communication (1994).
43. Segrè, P. N., van Megen, W., Pusey, P. N., Schätzel, K., and Peters, W., *J. Mod. Optics* **42**, 1929 (1995).
44. Philipse, A. P., Smits, C., and Vrij, A., *J. Colloid Interface Sci.* **129**, 335 (1989).
45. Kittel, C., *Introduction to Solid State Physics* (John Wiley, New York, 1976).
46. Ackerson, B. J., and Schätzel, K., *Phys. Rev.* E **52**, 6448 (1995).
47. He, Y., Ackerson, B. J., van Megen, W., Underwood, S. M., and Schätzel, K., Unpublished (1994).
48. Frank, F. C., *Proc. Roy. Soc. (London)* **A201**, 586 (1950).
49. Langer, J. S., *Rev. Mod. Phys.* **52**, 1 (1980).

3

Crystal Growth in Colloidal Suspensions: Kossel Line Analysis

Tsuyoshi Yoshiyama and Ikuo S. Sogami

The laser diffraction technique is effective for structure analysis of colloidal crystals. In particular, Kossel diffraction images faithfully reflect three-dimensional information on the symmetry of colloidal crystals in suspensions. After a brief description on the mechanism of Kossel diffraction, a detailed review is given on methods of quantitative analyses of diffraction images, namely, a three-point method and a two-ring method. These enable us to determine the structures, lattice constants, and orientations of colloidal crystals. Using photographic data obtained by real-time observations on semidilute suspensions of highly charged latex particles, an investigation was made into the time evolution of crystal structures. The results show that the crystallization takes place by way of the following intermediate processes: (1) two-dimensional hcp structure, (2) random layer structure, (3) layer structure with one sliding degree of freedom, (4) stacking disorder structure, (5) stacking structure with multivariant periodicities, (6) FCC twin structure, (7) normal FCC structure, (8) BCC twin structure, and (9) normal BCC structure. The early several stages of these processes (the era of layer structures) are characterized by strong anisotropy, originating in the wall effect. Thermal agitation and interaction among colloidal particles gradually rectify the anisotropy and advance the formation of crystals with cubic symmetry (the era of cubic structures).

3.1 Introduction

When divergent monochromatic beams from a point source are diffracted by lattice planes of a single crystal, a characteristic diffraction pattern disclosing the reciprocal lattice structure of the crystal is produced. The phenomenon was first discovered in 1934 by W. Kossel et al. [1] in a diffraction experiment on a Cu single crystal when diverging X rays emanating from the Cu atoms were excited by an electron beam incident on the crystal. Diffraction images (patterns, lines, and rings) of electromagnetic radiation so obtained in similar experiments are generically called Kossel images (patterns, lines, and rings) irrespective of the wavelength involved.

The Kossel pattern of laser beams turns out to be a convenient messenger for conveying accurate three-dimensional information on symmetries of crystalline arrays formed in salt-free aqueous suspensions of colloidal particles. In particular, the Kossel diffraction method enables us to record and analyze the evolution of ordered structures in suspensions. The purpose of this article is to review the methods of analysis of photographic diffraction data in detail and to investigate crystal growth in colloidal suspensions.

Aqueous suspensions of latex particles that have a diameter of order $0.1 \sim 1.0 \mu m$ and large surface-charge density provide ideal thermodynamic systems for studying ordering processes. It is possible to control the properties of these systems over a wide range by adjusting the radius, the surface-charge density and the concentration ϕ of the particles, the salt concentrations, and the temperature. Crystalline orderings appear and gradually develop in semidilute suspensions of monodisperse latex particles at low ionic strength, where excess electrolytes were removed by anion and cation ion-exchange resins. For quantitative analysis of laser diffraction it is convenient to use rectangular quartz cuvettes as containers for the suspensions.

Crystallization proceeds so slowly, over a time scale ranging from days to several months, that it is possible to make real-time observations. Note that, since the colloidal crystals grow in a fluid and have abnormally small elastic constants [2–4], it is indispensable to observe the colloidal crystals in a static state. The Kossel diffraction of an optical laser is suitable for such a purpose [5–10]. The Kossel images photographed on sheets of films record precise information on the structural development of colloidal crystals.

Through our experiences of diffraction studies on colloidal crystals grown in semidilute suspensions ($\phi < 10$ vol%) in cuvettes, we have found several empirical laws that afford bases for the Kossel line analyses of the systems as follows [7, 8, 10]:

Law 1. Colloidal crystals grow with a surface parallel to the cuvette surface.

Law 2. Almost without exception, the crystal surface is the most dense plane, that is, (111) for the FCC structure and (110) for the BCC structure.

Law 3. Crystallization proceeds to the FCC structure for comparatively concentrated suspensions ($3 < \phi < 10$ vol%), and to the BCC structure via the FCC structure for dilute suspensions ($\phi < 2$ vol%) of small latex particles.

The first two laws are a unique characteristic of colloidal crystals that exist only in suspensions in containers. Namely, they reflect, in a global sense, the so-called *wall effects*. The first law allows us to fix the orientations of the crystals and, as a result, to determine the axes of the Kossel cones. It is the second law that has enough information to provide quick criteria for deciding on the crystal structure. With the aid of the third law, we are able to make it a rule to assign a crystallographic index to the Kossel line observed in the early stages of the ordering formation, in which it is hard to classify the crystal system, by making reference to the FCC structure.

Analyses of more than 1000 photographic records of Kossel images that were taken over several years showed that the ordering formation proceeded, through intermediate multistages of phase transitions, to the normal FCC and BCC structures in suspensions with high and low particle concentrations, respectively. This result suggests that crystallization in systems of atoms and molecules also evolves along such a multistage course of phase transitions, although it would be extremely hard to observe such processes because they occur too fast.

In Section 3.2, after brief remarks on the preparation of specimens and photographing of diffraction images, we explain procedures for refractive effect corrections and derive relations among physical quantities used in our analysis. Two methods of Kossel line analysis, the three-point and two-ring methods, are explained in Section 3.3. Complimentary use of these methods allows us to extract sufficient information on the Kossel cones in the colloidal crystal to determine the precise value of the lattice constant and the crystal symmetry from measured quantities. Simple criteria for inferring the crystal structure and indexing the Kossel lines are also given in this section. Dividing the process of crystallization into an earlier era of layer structures and a later era of cubic structures, we analyze the Kossel diffraction images for crystal growth in Section 3.4. The results show that crystallization develops, through multistages of phase transitions, into the final cubic structure. A discussion is presented in Section 3.5.

3.2 Experimental Methods and Corrections to Observed Quantities

3.2.1 Experimental Method

The development of techniques in polymer chemistry has opened the way to obtaining a variety of polymer colloids with high monodispersity. Among them, polystyrene latices are the most popular colloids. To avoid sample dependence and to achieve universality, we used different kinds of polystyrene latices whose properties are listed in Table 3.1.

Table 3.1 Specimens of polystyrene latices used in this study

Specimen	Diameter 2a(nm)	Surf. charge $-Q(\times 10^4 e)$	Charge density $\sigma(e/nm^2)$	Maker
X-1	77			Cosmo Bio.
1B22	109	0.07	0.02	Dow Chem.
SSH6	120	1.2	0.27	Kyoto Univ.
N100	120	1.8	0.40	Sekisui Co.
N150	150	1.7	0.24	Sekisui Co.
SS32	156	3.0	0.39	Kyoto Univ.
IM08-05	193	5.4	0.46	JSR Co.
N200	220			Sekisui Co.
N300	320	27.7	0.86	Sekisui Co.
N400	400	32.0	0.64	Sekisui Co.
G5401	419	25.0	0.45	JSR Co.
N601	600	49.0	0.43	Sekisui Co.

Impurity ions contained in the original suspension were reduced to a minimum by repeating ultrafiltration in an Amicon 8050 cell until the conductivity of the suspension came down to less than 10^{-6} S/m. The purified suspensions thus obtained were diluted by pure water or pure heavy water to get specimen suspensions with a definite concentration. The small amount of impurities still remaining in the suspension were removed with a mixed ion exchange resin (AG 501-X8(D)) in a Teflon test tube. The specimen suspensions that began to show iridescence were introduced into rectangular quartz cuvettes of dimension $1 \times 10 \times 45$ mn.[3]

The colloidal crystals were examined by diffraction methods using HeNe and Ar lasers mainly because their wavelengths (λ =632.8 nm for HeNe, 488.0 nm for Ar) are comparable to interparticle distances in the crystals. An ultraviolet HeCd laser (λ =325.0 nm) was also used for crystals with shorter interparticle distances. Note that high transparency is reached only at the last stage of crystallization in very dilute specimens. Therefore, backward Kossel images were photographed for opaque specimens at various stages of the ordering formation in suspensions of different particle concentrations.

To examine as wide a range of diffraction angles as possible, it was necessary to use a rotating crystal camera. In our experiment we used a modified rotating camera of a cylindrical film holder with diameter $2R_{cam} = 57.3$ mm, a collimation slit with diameter 1 mm, and a goniometer head on which the specimen cuvette was mounted with its wide surface held vertically. Backward Kossel images were photographed on Fuji FG films for electron microscopy, being sensitive to lasers of various wavelengths. Fine parallel laser beams were incident on the specimen through a hole made in the film. The diffraction experiments had to be done in a dark room under a safety (Chloro-Bromide) lamp to prevent the films from exposure to white light. The specimen and experimental apparatus used in our experiments are shown in Fig. 3.1(a) and (b).

(a)

(b)

Figure 3.1 (a) Specimens in the quartz cuvettes and (b) the experimental apparatus for photographing the backward Kossel diffraction images.

3.2.2 Corrections for Refractive Effects

Refraction at the boundary between the colloidal crystal surface and a vacuum (air) changes the direction of the diffracted beams reaching the film. Although the refraction occurs twice, at the interface between the colloidal crystal surface and the cuvette's inner surface and at the interface between the cuvette's outer surface and the vacuum, it is sufficient to take only the effect of refraction for the colloidal crystal surface – vacuum interface into consideration, because the existence of a thin parallel quartz plate merely causes a little parallel displacement of the beam. Therefore, for practical analyses of diffraction, it is allowable to assume that the cuvette's quartz plate is infinitely thin.

The angles of incidence and refraction, i and r, are subject to Snell's law

$$\frac{\sin i}{\sin r} = \frac{1}{n} \tag{3.1}$$

where n is the refractive index of the colloidal crystal. The refractive index n of the colloidal suspension varies depending on the particle concentration ϕ. Measurement of n by an Abbe refractometer over a semidilute concentration range ($\phi < 10$ vol%) showed that n is linearly proportional to the concentration ϕ in accordance with Hiltner et al. [11]. In this analysis we adopt the following linear formula as

$$n(\phi) = 1.332 + 0.268 \times 10^{-2}\phi \tag{3.2}$$

3.2.3 Measured Quantities

The divergent beams emanating from a point source within the crystal are reflected by the lattice planes only at angles satisfying Bragg's law. The reflected beams from a set of planes with the index hkl will then generate the surface of a cone (Kossel cone) whose central axis is parallel to the reciprocal lattice vector $\mathbf{G}_{hkl}$. The magnitude G_{hkl} of the reciprocal lattice vector and the semiapex angle α_{hkl} of the cone are related by the equation

$$G_{hkl} = \frac{2n}{\lambda} \cos \alpha_{hkl} \tag{3.3}$$

where λ/n is the wavelength of the beam within the crystal and n is the refractive index of the crystal. Aspects of the Kossel diffraction in real and reciprocal spaces are shown, respectively, in Fig. 3.2(a) and (b).

Diffracted beams in the crystal are refracted at the crystal surface. Owing to the empirical law (Law 1) stated in the introduction, it is legitimate to presume that the crystal maintains its surface parallel to the wide plane of the cuvette, which is held vertically during observation. Let us consider an incident beam on a Kossel cone in the crystal. On refraction the incident beam, the refractive beam, and the surface normal are in a plane. We call such an auxiliary plane introduced for each incident beam an R-plane. Angles of incidence and refraction, i and r,

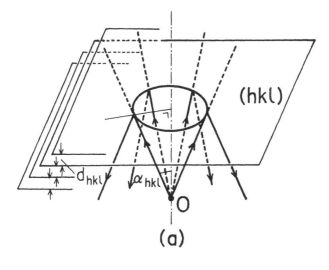

(a)

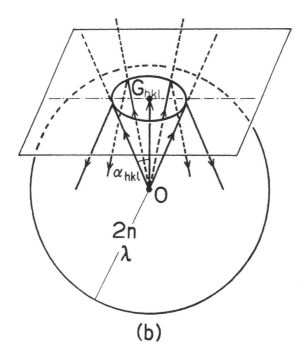

(b)

Figure 3.2 Diagram representation of Kossel diffraction in (a) real space and (b) reciprocal space. The axis of the Kossel cone is parallel to the reciprocal lattice vector G_{hkl} and the normal of the lattice plane (hkl), and d_{hkl} and α_{hkl} are, respectively, the interplanar distance and the semiapex angle of the cone.

and related quantities are shown schematically in Fig. 3.3, where the effects of the cuvette's plate are safely ignored, as discussed in Subsection 3.2.2.

To relate the angles of incidence and refraction i and r with the measured quantities, it is convenient to decompose these angles into their horizontal components i_h and r_h and vertical components i_v and r_v. The incident angle and its components satisfy

$$\tan i_h = \tan i \cos \omega, \quad \tan i_v = \sin i_h \tan \omega \tag{3.4}$$

and

$$\cos i = \cos i_h \cos i_v \tag{3.5}$$

where ω is the angle between the R-plane and a horizontal plane. Likewise, the refractive angle and its components satisfy

$$\tan r_h = \tan r \cos \omega, \quad \tan r_v = \sin r_h \tan \omega \tag{3.6}$$

and

$$\cos r = \cos r_h \cos r_v \tag{3.7}$$

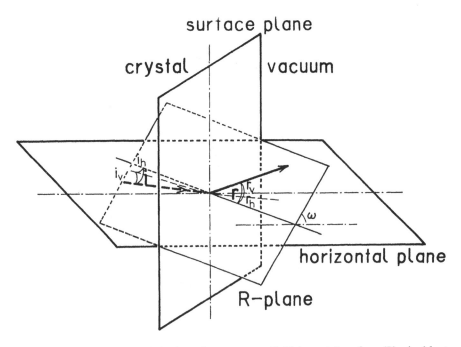

Figure 3.3   Refraction of the laser beams at a colloidal crystal surface. The incident beam, the refractive beam, and the surface normal lie in the R-plane, which has the angle of inclination ω from the horizontal plane. Angles of incidence and refraction, i and r, are decomposed into their horizontal components i_h and r_h and vertical components i_v and r_v.

These relations hold for each beam.

For each Kossel pattern photographed on a film by the cylindrical camera with camera radius R_{cam}, there is a point of symmetry O of the pattern of the diffraction image on the film. On the film we take the coordinate system with a horizontal x axis and a vertical y axis and the point of symmetry O as its origin and fix the position of a point on a Kossel line on the film by its coordinates (x, y). Then the components of the refractive angles r_h and r_v of a beam are derived from the measured coordinates (x, y) of the point as follows:

$$r_h = \frac{x}{R_{cam}}, \quad r_v = \tan^{-1} \frac{y}{R_{cam}} \tag{3.8}$$

which determine the value of the refractive angle r through Eq. (3.7).

3.2.4 Relations among Components of Incident and Refractive Angles

For the structure analysis to be discussed in the next section, it is necessary to extract the values of the incident angle i and its associated angle ω from the measured quantities. Equations (3.4) and (3.6) lead to

$$\tan \omega = \frac{\tan i_v}{\sin i_h} = \frac{\tan r_v}{\sin r_h} \tag{3.9}$$

through which ω is fixed by r_v and r_h. Equations (3.4), (3.5), and (3.9) prove that

$$1 + \tan^2 i_v = 1 + \sin^2 i_h \left(\frac{\tan r_v}{\sin r_h} \right)^2 = \frac{1}{\cos^2 i_v} = \left(\frac{\cos i_h}{\cos i} \right)^2 \tag{3.10}$$

which results readily in

$$\sin^2 i_h \left[\left(\frac{\tan r_v}{\sin r_h} \right)^2 + \frac{1}{\cos^2 i} \right] = \frac{1}{\cos^2 i} - 1 = \tan^2 i \tag{3.11}$$

Using Snell's law, we get the equation for i_h as

$$\sin^2 i_h = \frac{\sin^2 i \sin^2 r_h}{\tan^2 r_v(1 - \sin^2 i) + \sin^2 r_h}$$

$$= \frac{\sin^2 r \sin^2 r_h}{\tan^2 r_v(n^2 - \sin^2 r) + n^2 \sin^2 r_h} \tag{3.12}$$

Then, with the aid of Eqs. (3.5) and (3.9), the values of i_v and i are also expressed in terms of r_v and r_h. In this way, all components of the incident and refractive angles of the beam are determined by measuring the coordinates x and y of a point on the Kossel line.

3.3 Methods for Kossel Line Analyses

3.3.1 Three-Point Method

As shown in the previous section, the measurement of the coordinates of a point on a Kossel line at which a beam impinges determines the angles ω and i of the beam. Coordinates of three points on a Kossel line are sufficient to define the geometry of the Kossel cone within the colloidal crystal. Therefore, by measuring the coordinates of three points on a Kossel line, we are able in principle to determine the semiapex angle α of the Kossel cone and the direction of its axis. We call this procedure the three-point method and adopt it as the standard for the Kossel line analysis.

Corresponding to the three points with coordinates (x_a, y_a) $(a = 1, 2, 3)$ on a Kossel line, there exist three unit vectors $\mathbf{r}_a$ $(a = 1, 2, 3)$ along generating lines of the Kossel cone in the crystal. As shown in Fig. 3.4, the polar and azimuthal angles (θ_a, ϕ_a) of the unit vector $\mathbf{r}_a$ are determined by the relations

$$\theta_a = i_a, \quad \phi_a = \omega_a \tag{3.13}$$

where the incident angle i_a and the angle ω_a between the R-plane and the horizontal plane are estimated by following the procedure in subsections 3.2.3 and 3.2.4.

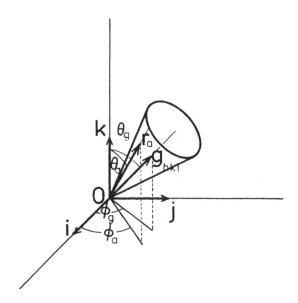

Figure 3.4 A beam on the Kossel cone. The polar and azimuthal angles of a beam $\mathbf{r}_a$ on the Kossel cone and the cone axis $\mathbf{g}_{hkl}$ are expressed, respectively, by (θ_a, ϕ_a) and (θ_g, ϕ_g).

The axis of the Kossel cone is parallel to the reciprocal lattice vector $\mathbf{G}_{hkl}$. Setting the polar and azimuthal angles of the unit vector $\mathbf{g}_{hkl}$ of $\mathbf{G}_{hkl}$ to be (θ_g, ϕ_g), the formulae for the semiapex angle α of the Kossel cone in the crystal are given by

$$\cos \alpha = \mathbf{g}_{hkl} \cdot \mathbf{r}_a \tag{3.14}$$

$$= \sin \theta_g \sin \theta_a \cos(\phi_g - \phi_a) + \cos \theta_g \cos \theta_a \tag{3.15}$$

for $a = 1, 2,$ and 3. From these formulae, we get the equations

$$\tan \theta_g = \frac{\cos \theta_2 - \cos \theta_1}{\sin \theta_1 \cos(\phi_g - \phi_1) - \sin \theta_2 \cos(\phi_g - \phi_2)} \tag{3.16}$$

$$= \frac{\cos \theta_3 - \cos \theta_2}{\sin \theta_2 \cos(\phi_g - \phi_2) - \sin \theta_3 \cos(\phi_g - \phi_3)} \tag{3.17}$$

which lead readily to

$$\tan \phi_g = \frac{C_{12}(\cos \theta_3 - \cos \theta_2) - C_{23}(\cos \theta_2 - \cos \theta_1)}{S_{23}(\cos \theta_2 - \cos \theta_1) - S_{12}(\cos \theta_3 - \cos \theta_2)} \tag{3.18}$$

where

$$C_{ab} = \sin \theta_a \cos \phi_a - \sin \theta_b \cos \phi_b, \quad S_{ab} = \sin \theta_a \sin \phi_a - \sin \theta_b \sin \phi_b \tag{3.19}$$

In this way the three sets of estimated angles (θ_a, ϕ_a) $(a = 1, 2, 3)$ determine the orientation of the Kossel cone axis, that is, the angle ϕ_g through Eq. (3.18) and then the angle θ_g through Eq. (3.17).

Once the orientation of the Kossel cone axis is fixed, Eq. (3.15) determines the semiapex angle α of the cone. Then the magnitude of the reciprocal lattice vector $\mathbf{G}_{hkl}$ is obtained from Eq. (3.3), since the wavelength λ of the beam in vacuum and the refractive index n of the crystal are known beforehand. For crystals with cubic structure the interplanar spacing d_{hkl}, which is the inverse of the magnitude G_{hkl} of the reciprocal lattice vector, is given by the lattice constant l_a as

$$d_{hkl} = \frac{1}{G_{hkl}} = \frac{l_a}{\sqrt{h^2 + k^2 + l^2}} \tag{3.20}$$

We apply this procedure as many times as possible to various lines included in the Kossel pattern. If the crystal structure and the indices for the lines are properly chosen, we get, within the limits of experimental error, a unique value for the lattice constant l_a from all lines.

In principle three points can be chosen arbitrarily on a Kossel line, that is, a part of a Kossel ring, for this method. However, the errors involved can be considerably reduced in practice if one of the points, say (x_1, y_1), is selected so that the condition $\phi_1 = \phi_g$ is fulfilled. It is evident that the axis of the Kossel cone with the azimuthal angle ϕ_g is in the R-plane associated with the beam, which impinges on the film at (x_1, y_1). In actual analysis this point is found at the point on the ring that has the minimum distance from the point of symmetry O of the whole Kossel pattern.

3.3.2 Two-Ring Method

The three-point method applied to each Kossel ring forms a useful standard for the Kossel diffraction analysis. However, it may lose its accuracy for Kossel lines that extend like straight lines, since small errors in the coordinate measurements lead to large fluctuations in reconstructing Kossel cones that have large values for the semiapex angle α. In such cases it is possible to extract information on the reciprocal lattice vector space by investigating relations among different Kossel rings on a film. The simplest way that is used in a complementary manner to the three-point method in our analysis is to measure the relative orientation of two Kossel rings. We call this method the two-ring method.

Let us consider two Kossel cones with indices m ($= hkl$) and m' ($= h'k'l'$). The plane spanned by the axes of the cones cuts each cone at two generating lines. As shown in Fig. 3.5, the relative orientation of two cones is specified by the angle Θ between the cone axes and the angle β between the closest generating lines cut out of both cones by the plane spanned by the axes. These angles are related in terms of the semiapex angles α_m and $\alpha_{m'}$ as

$$\Theta = \alpha_m + \alpha_{m'} \pm \beta \tag{3.21}$$

where the $-$ or $+$ sign must be taken depending on whether the two Kossel cones cross or do not cross each other.

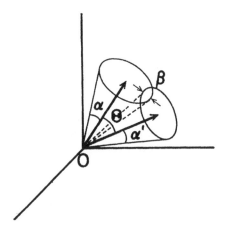

Figure 3.5   The angle Θ between the axes of two Kossel cones and the angle β between two generating lines of the cones in the plane spanned by the axes.

Bragg's condition tells us that

$$\cos \alpha_m = \frac{\lambda}{2nd_m}, \quad \cos \alpha_{m'} = \frac{\lambda}{2nd_{m'}} \tag{3.22}$$

where, for a cubic crystal system, the interplanar distances d_m and $d_{m'}$ are given by

$$d_m = \frac{l_a}{|m|}, \quad d_{m'} = \frac{l_a}{|m'|} \tag{3.23}$$

with the abbreviation

$$|m| = \sqrt{h^2 + k^2 + l^2}, \quad |m'| = \sqrt{h'^2 + k'^2 + l'^2} \tag{3.24}$$

Eliminating α_m and $\alpha_{m'}$ from Eqs. (3.21)–(3.24), we get

$$l_a = \frac{\lambda}{2n \sin(\Theta \pm \beta)} \sqrt{|m|^2 + |m'|^2 - 2|m||m'| \cos(\Theta \pm \beta)} \tag{3.25}$$

for the lattice constant l_a.

Once the indices m and m' of the cubic crystal structure are fixed for the Kossel lines, the angle Θ takes a definite value, irrespective of the lattice constant, for the cones associated with reciprocal vectors $\mathbf{G}_m$ and $\mathbf{G}_{m'}$. On the other hand, the procedure in Subsections 3.2.3 and 3.2.4 determines the value of the angle β between the closest generating lines characterized by the polar and azimuthal angles (θ, ϕ) and (θ', ϕ'). The angle β is given by $\cos \beta = \sin \theta \sin \theta' \cos(\phi - \phi') + \cos \theta \cos \theta'$.

Note that the line connecting the points of symmetry on each Kossel ring intersects the rings at the points (x, y) and (x', y'), where the beams along those two generating lines impinge on the film. The coordinates of two points (x, y) and (x', y') determine the angles, respectively, as $(\theta, \phi) = (i, \omega)$ and $(\theta', \phi') = (i', \omega')$.

Evidently it is convenient to choose a pair of points on each Kossel cone that share a common R-plane. In such a case the angle β is given by the difference $\theta - \theta' = i - i'$, since $\phi = \phi'$. In actual analysis such a pair is given by the points on the Kossel lines each of which has the minimum distance from the point of symmetry O of the Kossel pattern.

3.3.3 Criteria for Determining Crystal Structures and Line Indices

Along with the quantitative methods for precise data analysis, it is useful to have a quick rule to infer crystal structures and indices from the Kossel images. The empirical law (Law 2) gives us such a convenient criterion for determining crystal structures. Conditions for persistent multiple diffraction and properties of the structure factor of diffracted beams provide useful rules for proper indexing of the Kossel lines.

In a cuvette the colloidal crystal grows maintaining its most dense lattice plane, such as the (111) plane for the FCC structure and (110) plane for the BCC structure, parallel to the wide cuvette surface. Therefore, while the Kossel diffraction pattern from the colloidal crystal with FCC structure (FCC twin structure) has threefold (sixfold) rotation symmetry, the Kossel pattern from the crystal with BCC (BCC twin structure) has twofold (doubly twofold) symmetry. We then have the following useful rule to determine the crystal structure:

3-fold symmetry → FCC structure,
6-fold symmetry → FCC twin structure,
2-fold symmetry → BCC structure,
doubly 2-fold symmetry → BCC twin structure

As an interesting characteristic of the Kossel image, it is observed that three or more specified Kossel lines meet at a point irrespective of the values of the lattice constant and the wavelength of the laser beams used. This phenomenon, called persistent multiple diffraction, happens when three or more reciprocal lattice vectors (reciprocal lattice points) lie in a plane intersecting the Ewald sphere, and all the reciprocal lattice points are circumscribed by a circle (a reflection circle) whose radius r_0 is smaller than that of the Ewald sphere (see Fig. 3.6).

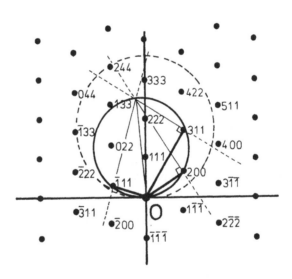

Figure 3.6 The reflection circle (solid line) in the plane intersecting the Ewald sphere (dashed line). Persistent multiple diffractions occur from the lattice plane with indices 311, 200, and $\bar{1}$11, since corresponding reciprocal lattice points are on the circumference of the reflection circle.

Let us consider the case where the Kossel lines with indices m ($= hkl$), m' ($= h'k'l'$), m'' ($= h''k''l''$), ... meet at a point on the film. The condition for the reciprocal lattice vectors $\mathbf{G}_m$, $\mathbf{G}_{m'}$, $\mathbf{G}_{m''}$, ... to be coplanar is

$$(\mathbf{G}_m \times \mathbf{G}_{m'}) \cdot \mathbf{G}_{m''} = 0 \tag{3.26}$$

for $m \neq m' \neq m'' \neq m$, and the radius r_0 of the reflection circle is determined by [12]

$$r_0 = \frac{|\mathbf{G}_m||\mathbf{G}_{m'}||\mathbf{G}_m - \mathbf{G}_{m'}|}{2|\mathbf{G}_m \times \mathbf{G}_{m'}|} \tag{3.27}$$

The Kossel lines with the indices $m, m', m'', \ldots$ satisfying these conditions for persistent multiple diffraction meet at a point. This is a strong criterion for correct indexing of the Kossel lines. A breakdown of this criterion suggests that some lattice defects exist in the crystal.

When the diffraction condition is satisfied, the scattering amplitude of diffracted beams from the lattice plane (hkl) is expressed by the product of the number of unit cells and the structure factor, $S(hkl)$, which takes the form [13, 14][1]

$$S(hkl) \propto 1 + e^{-i\pi(k+l)} + e^{-i\pi(l+h)} + e^{-i\pi(h+k)} \tag{3.28}$$

for the FCC structure and

$$S(hkl) \propto 1 + e^{-i\pi(h+k+l)} \tag{3.29}$$

for the BCC structure. The strength of diffraction becomes a maximum for crystals of FCC structure when all indices are even or odd integers, and for crystals of BCC structure when the sum of all indices is even. Therefore, the indices must be assigned to the Kossel lines so that all indices are even or odd integers for crystals of FCC structure and the sum of all indices is even for crystals of BCC structure.

As an illustrative example of the structure analysis, we have investigated the Kossel image in Fig. 3.7. The threefold symmetry of the image tells us that the colloidal crystal has the normal FCC structure. All indices of the Kossel lines were chosen to be either the sets of even integers or those of odd integers. Tables 3.2 and 3.3 show, respectively, results of the analyses by the three-point method and the two-ring method.

---

[1]For crystals of hexagonal close-packed strucure (hcp), the interplanar spacing d_{hkl} is related to the lattice constant l_a as

$$d_{hkl} = \frac{l_a}{\sqrt{\frac{4}{3}(h^2 + hk + k^2) + \frac{8}{3}l^2}}$$

and the structure factor is given by

$$S(hkl) \propto 1 + e^{-i2\pi(\frac{2}{3}h + \frac{1}{3}k + \frac{1}{2}l)}.$$

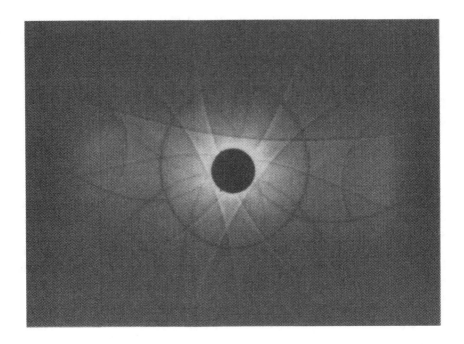

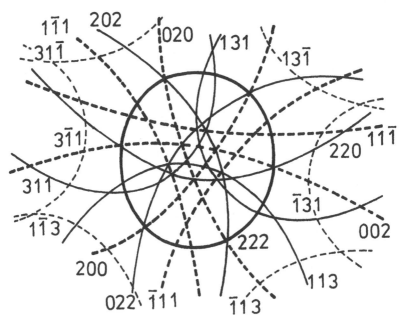

Figure 3.7   A backward Kossel diffraction image by HeNe beams for the suspension of latex SSH6 with 1.0 vol% concentration and Miller indices. The threefold symmetry of the pattern shows that the colloidal crystal has the normal FCC structure. All indices are either even or odd integers.

Table 3.2 Structure analysis of the Kossel pattern in Fig. 3.7 by the three-point method

x (mm)	y (mm)	θ (deg)	ϕ (deg)	d_{hkl} (nm)	l_a (nm)	Index
13.94	−1.41	20.60	−6.01			
−6.29	13.02	20.11	115.61	253.0	876.5	222
−8.33	−12.16	20.69	235.96			
−13.64	0.60	20.10	177.38			
8.32	11.51	20.06	54.52	252.9	876.0	222
6.47	−13.77	20.98	−65.02			
0.46	0.75	7.07	84.47			
35.84	6.90	45.51	14.24	531.9	921.2	$11\bar{1}$
−34.37	11.13	44.84	157.37			
0.34	1.35	2.08	75.86			
33.95	−8.53	44.33	−17.81	424.4	848.7	002
−26.06	−21.66	40.78	223.77			
−0.28	−4.10	6.11	266.09			
33.55	10.21	44.19	21.15	311.3	880.6	220
−32.46	14.72	43.94	150.44			
−1.15	−0.50	1.87	203.50			
−0.33	26.68	30.70	90.00	265.7	881.2	311
−29.24	−4.46	39.91	190.35			

Table 3.3 Structure analysis of the Kossel pattern in Fig. 3.7 by the two-ring method

Indices $(hkl - h'k'l')$	β (deg)	Θ (deg)	l_a (nm)
$002 - 11\bar{1}$	7.07 − 2.08	125.26	889.2
$002 - 220$	2.08 + 6.11	90.00	883.6
$220 - 113$	6.11 − 2.85	64.76	884.8
$202 - 131$	4.97 − 1.77	64.76	884.5
$022 - 311$	4.40 − 1.88	64.76	881.1
$\bar{1}13 - 222$	31.16 − 20.98	58.52	882.0

3.4 Crystal Growth

In comparatively concentrated suspensions ($2 < \phi < 10$ vol%) of latex paticles, the ordering process develops rapidly in its early stage and slows down markedly in its last stage. On the other hand, in the dilute suspensions ($\phi < 2$ vol%), the process advances very slowly in its early stage and then makes sudden and rapid progress in its last stage. As stated in the empirical law (Law 3) in the introduction, the crystallization process proceeds to the FCC structure for the concentrated suspensions and to the BCC structure for the dilute suspensions. The Kossel line analysis on colloidal crystals at the final stage of crystallization shows that the ratio of the nearest-neighbor distance to the particle radius R/a increases continuously as the volume fraction ϕ decreases (see Fig. 3.8).

Before reaching the thermodynamically stable cubic structure, the ordering process is confirmed to develop along the path of a rich sequence of phase transitions. It is relevant to separate the ordering process into early and later stages. The former is the stage where the so-called wall effects play an important role and is named the era of layer structures. The latter is the stage, called the era of cubic structures, where thermal agitation and interaction among colloid particles rectify gradually the anisotropy caused by the wall effects and complete the formation of crystals of cubic symmetry. As noted in the introduction, the index of the Kossel line in the early stage is given in terms of the FCC structure.

3.4.1 Era of Layer Structures

Three-dimensional isotropic symmetry degrades to two-dimensional symmetry in the region close to the cuvette surface where the global crystallization starts to take place. Direct observation by an optical microscope [15, 16] confirmed that the particles executing Brownian motion came close together and began

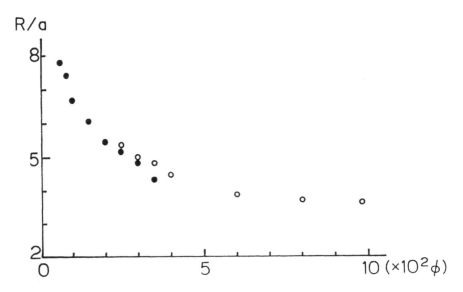

Figure 3.8   Concentration dependence of the ratio of nearest-neighbor distance R to particle radius a for suspensions of latex SS32. Filled circles and open circles are, respectively, for BCC and FCC structures. The nearest-neighbor distance decreases as the volume fraction ϕ increases.

to form a two-dimensional hexagonal arrangement in a plane parallel to the cuvette surface. Such hexagonal close-packed (hcp) planes increased sheet by sheet until the suspension began to exhibit faint iridescence, at which point the laser diffraction experiment became applicable. To get clear diffraction images in this era, we used specimens of concentrated suspensions that were filled up with rather small crystals.

Figure 3.9 shows the backward Kossel images taken by Ar laser beams for the first three stages of the three-dimensional ordering process. The single broad ring in Fig. 3.9 (a) is the Kossel diffraction ring from the hcp layers stacked parallel to the cuvette wall. The FCC index 111 is given to this Kossel ring, since those hcp layers were confirmed to develop to the (111) plane of the FCC structure in the later stage of crystallization. The fact that no diffraction signal appears other than the 111 ring implies that planes of adjacent layers have no lateral coherence with each other at all. Accordingly this earliest three-dimensional ordering in the colloidal suspensions can be identified with the *random layer structure*, which was first discovered for amorphous graphite [17, 18].

In Fig. 3.9 (b) two Kossel rings appear and cut across the 111 ring. The new rings have the same lattice constant as the 111 ring only if they acquire the indices 002 and $\underline{002}$. Here the underlined index shows the twin ring with respect to the (111) lattice plane. It is possible to explain this kind of diffraction pattern by a model [19, 10] in which the (002) and ($\underline{002}$) lattice planes are formed by freely sliding the two-dimensional hcp layer along one direction ([$1\bar{1}0$] or [$11\bar{2}$]), keeping the interplanar distance d_{111} constant. We call such a crystal structure the *layer structure with one sliding degree of freedom.*

Figure 3.9 (c) shows six Kossel rings crossing symmetrically over the 111 ring. By giving the indices 200, $\underline{020}$, 002, $\underline{200}$, 020, and $\underline{002}$ to these rings, we got a lattice constant $l_a = 432$ nm, which is approximately equal to the value $l_a = 419$ nm estimated from the 111 ring. This diffraction pattern is explained by the *stacking disorder structure* [20] in which two-dimensional hcp layers conventionally named A, B, and C stack up in a statistically random manner.

At the end of the era of layer structures there appears a crystallographically interesting diffraction image as shown in Fig. 3.10. In addition to the central 111 ring and {200} and {$\underline{200}$} families of Kossel lines, weak but definite signals of new {$\bar{1}11$} and {$\underline{\bar{1}11}$} families are recognized that accompany sheaves of fine and faint diffraction lines. The appearance of new {$\bar{1}11$} and {$\underline{\bar{1}11}$} families indicates that the correlation among layers increased beyond the stacking disorder structure. Namely, the 111 hcp layers A, B, and C began to arrange in a somewhat regular order. The bunches of faint diffraction lines imply that the layers arranged in many ordered units with different periods. We named this intermediate stage that of *stacking structure with multivariant periodicity.*

3.4.2 Era of Cubic Structures

The further advance of crystallization provides new richer structures of Kossel images as in Figs.3.11 – 3.13. Apparently the Kossel pattern in Fig. 3.11 (a) has

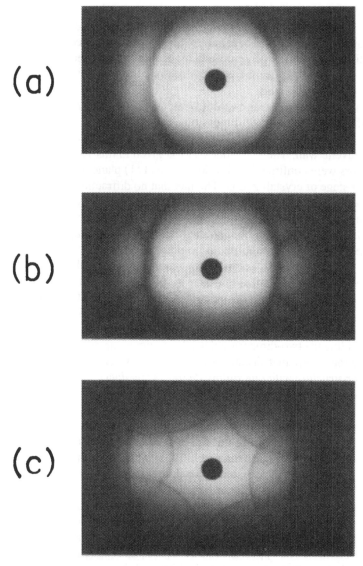

Figure 3.9   Kossel images taken by Ar laser beams showing layer structures of colloidal crystals at the early stages of order formation in suspensions of latex SS32. The lattice constant is estimated by assuming the modified FCC structure. (a) A single broad Kossel ring observed at $t = 20$ h after preparation of the suspension with particle concentration $\phi = 9.8$ vol% shows that the crystal has the random layer structure. The interplanar distance is estimated to be $d = 214$ nm. (b) A Kossel diffraction image observed at $t = 24$ h for a suspension of concentration $\phi = 9.8$ vol%. The crystal has the layer structure with one sliding degree of freedom, and the lattice constant is estimated to be $l_a = 376$ nm. (c) A Kossel image photographed at $t = 720$ h for suspensions of concentration $\phi = 6.0$ vol%. The crystal has the stacking disorder structure, and the lattice constant is $l_a = 419$ nm.

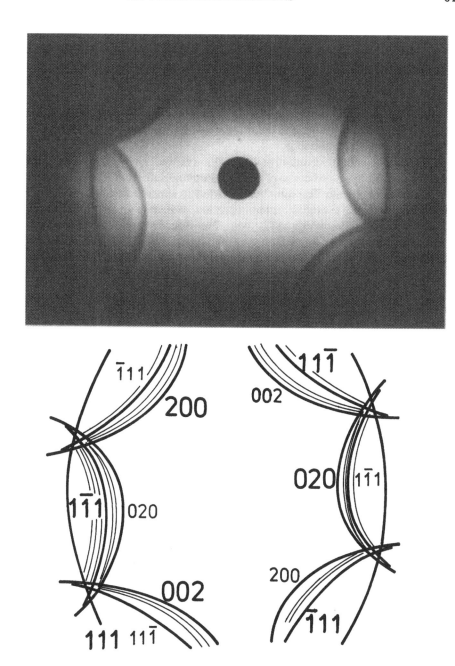

Figure 3.10   A Kossel pattern of the crystal with multivariant periodicity observed at intermediate stage from the era of layer structure and the era of cubic structure in the suspension of latex N100 with concentration 2.1 vol%. The photograph was taken by HeNe beams at $t = 3120\,h$. New Kossel lines $\{\bar{1}11\}$ and $\{\underline{\bar{1}11}\}$ appeared with sheaves of fine diffraction lines. The lattice constant is estimated to be approximately $l_a = 516\,nm$.

sixfold symmetry. The quick criterion in Subsection 3.3.3 shows that the crystal at this stage has the FCC structure with a (111) twin. In spite of crystallographic equality, the only twin plane observed so far was the (111) plane parallel to the widest cuvette surface to which the incident beams entered normally. The threefold symmetry of the diffraction pattern in Fig. 3.11 (b) proves that the crystal has the normal FCC structure. Our observation over several years has confirmed that the process of crystallization in the relatively concentrated suspensions ($\phi >$ 3 vol%) terminated at this stage.

The Kossel pattern recorded in Fig. 3.12 exhibits a very interesting phenomenon, that is, the coexistence of the FCC structure and the BCC twin structure in the same crystal grain. The solid lines indexed by uppercase letters and the broken lines indexed by lowercase letters represent, respectively, the FCC structure and the BCC structure with a $(1\bar{1}2)$ twin. Two structures coexist satisfying the Nishiyama conditions [21]: $(111)_{FCC} \parallel (110)_{BCC}$ and $[1\bar{2}1]_{FCC} \parallel [1\bar{1}0]_{BCC}$. This fact suggests that the phase change proceeds through a martensitic transformation [22]. The martensitic transformation is not accompanied by diffusion, so that it is rapid and easy for crystals with abnormally small elastic moduli. It is possible to interpret this diffraction image as recording the very moment of the phase transition from FCC to BCC twin.

At the last stage of crystallization, the BCC structure appears in dilute suspensions ($\phi < 2$ vol%) of small particles. Figure 3.13 (a) shows a Kossel pattern with doubly twofold symmetry indicating the BCC structure with a $(1\bar{1}2)$ twin. It is worthwhile to notice that, in contrast to the FCC twin structure, which has the (111) twin plane parallel to the cuvette surface, the twin plane of all the BCC twin structures observed so far is limited without exception either to $(1\bar{1}2)$ or $(\bar{1}12)$, both of which are normal to the cuvette surface. The Kossel pattern in Fig. 3.13 (b) has twofold symmetry, exhibiting the normal BCC structure. The normal BCC structure is thermodynamically favored in the dilute suspensions since it appeared at the end of the ordering formation. In general, the larger the size of colloidal particles, the longer it takes for crystals to develop into the BCC structure. The largest particle that was to date confirmed to form the BCC crystal is latex N200, with a diameter of 220 nm.

3.4.3 Ordering Processes

The results of our Kossel line analysis have established that the ordering formation proceeds in salt-free suspensions of monodisperse latex particles by way of the following intermediate processes:

embryos of three-dimensional ordering $\equiv$ two-dimensional hcp layers
$\rightarrow$ random layer structure
$\rightarrow$ layer structure with one sliding degree of freedom
$\rightarrow$ stacking disorder structure
$\rightarrow$ stacking structure with multivariant periodicity
$\rightarrow$ FCC structure with (111) twin

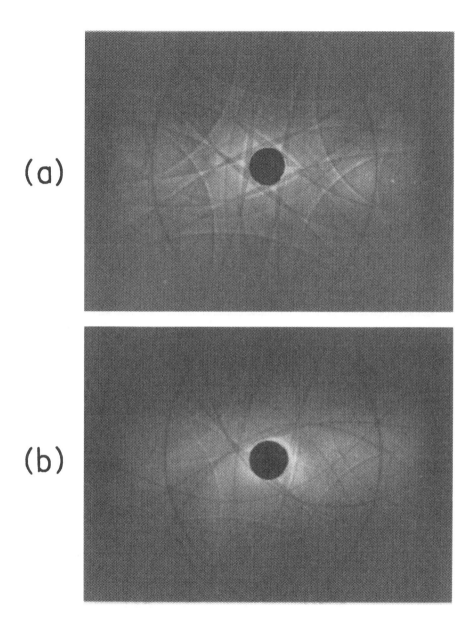

Figure 3.11 Kossel images observed by Ar beams at the first two stages of the cubic structure era. (a) The Kossel pattern with sixfold symmetry observed at $t \approx 900\,$h for a suspension of latex SS32 of concentration 0.33 vol% shows the appearance of the FCC twin structure with (111) twin plane and a lattice constant of $l_a = 774\,$nm. (b) The Kossel pattern with typical threefold symmetry photographed at $t = 1056\,$h for a suspension of latex N150 of concentration 0.4 vol% shows that the crystal has the normal FCC structure and a lattice constant $l_a = 768\,$nm.

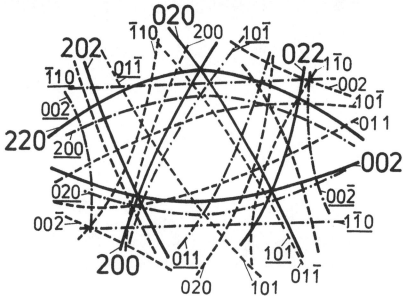

Figure 3.12 A Kossel image showing the coexistence of FCC and BCC structures, which was photographed just at the course of structure transformation, in a suspension of latex N150 of concentration 0.3 vol%. The photograph was taken with Ar beams at $t = 888$ h. In the lower diagram solid and dotted lines show the FCC and BCC structure, respectively. The lattice constants were calculated to be l_a(FCC) = 967 nm for the FCC structure and l_a(BCC) = 770 nm for the BCC structure, which led to nearly the same value for the interparticle distance as l_a(FCC)$/2^{1/2}$ = 680 nm and l_a(BCC)$3^{1/2}/2$ = 670 nm.

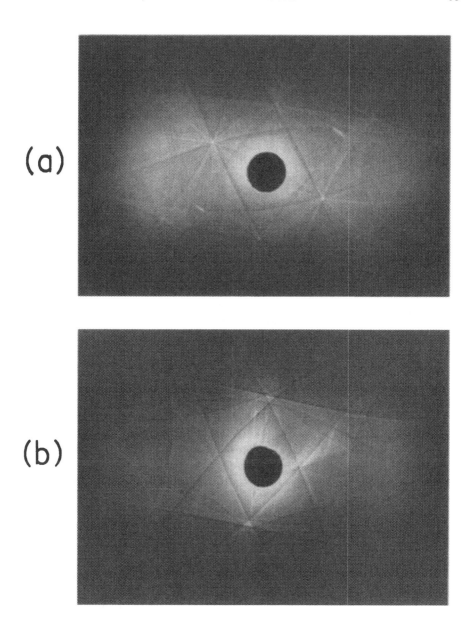

Figure 3.13 Kossel images taken with Ar beams for a suspension of latex SS32 of concentration 0.1 vol%. (a) The Kossel pattern photographed at $t = 4416$ h. A doubly twofold symmetry of the pattern shows that the crystal has the BCC twin structure with $(1\bar{1}2)$ twin plane and lattice constant $l_a = 807$ nm. (b) The Kossel pattern taken at $t = 9840$ h has a typical twofold symmetry showing the normal BCC structure. The lattice constant is estimated to be $l_a = 791$ nm.

→ normal FCC structure
→ BCC structure with ($1\bar{1}2$) or ($\bar{1}12$) twin
→ normal BCC structure.

The last two stages of BCC structures were observed only in dilute suspensions of colloidal particles with small diameter ($\leq 220\,\mathrm{nm}$). It should be noticed that, until now, the three-dimensional hcp structure was never observed in the process of crystal growth in colloidal suspensions.

3.5 Discussion

The changes of crystal structures were faithfully observed through the changes of the Kossel diffraction images. All the processes shown by the arrow marks (→) in the ordering sequence in Section 3.4.3 are nothing other than the phase transitions. Namely, the crystallization in colloidal suspensions was proved to develop through a many-faceted sequence of phase transitions. Five transition stages occur from the random layer structure to the normal FCC structure in relatively concentrated suspensions of colloid particles. Two further transition stages proceed from the normal FCC structure to the BCC twin structure through the martensitic transformation and then to the normal BCC structure in dilute suspensions of colloid particles with small diameter ($\leq 220\,\mathrm{nm}$).

It is natural to interpret that the entropy effect favors the BCC structure rather than the FCC structure in dilute suspensions of small particles. In fact a symmetry consideration has shown that crystals of spherical particles should exhibit the BCC structure near the melting point so long as the BCC crystal energy is not too large compared with that of the FCC [23]. Here it is worthwhile to note that there was no signal indicating the appearence of the three-dimensional hexagonal close-packed structure in the process of the ordering formation in colloidal suspensions. Along with the existence of multistages of phase transitions, this observation of the nonappearence of the hcp structure seems to provide important information on the interaction between highly charged colloidal particles. It seems very difficult to explain these characteristic features of the crystallization in colloidal suspensions from the naive viewpoint of the Alder-Wainwright mechanism [24, 25] which has been proposed in order to improve deficiencies in the classical DLVO theory [26, 27]. The results of our analysis seems to require reinvestigation of the interaction of highly charged particles in macroionic suspensions [28, 29]. In this connection, we should notice that, through a careful analysis of microscopic data, Ise et al. [16, 30] already discovered the existence of a long-range weak attraction in addition to the medium-range strong repulsion between highly charged macroions.

The wall effects that initiated the formation of the embryos of two-dimensional hcp arrangements cease to influence the crystal symmetry in the later era of cubic structure. However, the marks of the initial wall effects persist through all stages of crystal growth, one mark on the twin plane and the other mark on the orientation of the crystal grains. No twin plane other than the (111) plane parallel to the cuvette surface $[((1\bar{1}2)$ or $(\bar{1}12)$ orthogonal to the cuvette surface $)]$ is confirmed to exist in the FCC (BCC) structure. The grain of the FCC (BCC) crystal grows keeping its (111) plane $[((110)$ plane $)]$ parallel to the cuvette surface.

In this way, the charge-stabilized colloidal suspensions were confirmed to be unexcelled systems for the study of crystal growth and the Kossel line analysis turned out to be a very powerful method for their study. Note that it is the geometric features of the Kossel diffraction patterns that provide sufficient information on crystal structures. Beyond the geometry of the conical sections, however, the Kossel images carry far richer structures such as linewidths and bright-dark line structures, and gaps and anomalies at crossing points of lines [31]. These dynamical diffraction effects are so pronounced owing to the strong interaction between highly charged colloidal particles and laser beams. A detailed analysis of these effects by the dynamical theory of diffraction [32, 33] unveils a profile of the electric field in colloidal suspensions.

References

1. Kossel, W., Loeck, V., and Voges, H., *Zeit. für Physik* **94**, 139 (1935).
2. Crandall, R. S., and Williams, R., *Science* **198**, 293 (1977).
3. Lindsay, H. M., and Chaikin, P. M., *J. Chem. Phys.* **76**, 3774 (1982).
4. Kesavamoorthy, R., and Arora, A. K., *J. Phys.* **A18**, 3389 (1985).
5. Clark, N. A., Hurd, A., and Ackerson, B. J., *Nature* **281**, 57 (1979).
6. Pieranski, P., *Contemp. Phys.* **24**, 25 (1983).
7. Yoshiyama, T., Sogami, I. S., and Ise, N., *Phys. Rev. Lett.* **53**, 2153 (1984).
8. Yoshiyama, T., *Polymer* **27**, 827 (1986).
9. Monovoukas, Y., and Gast, A. P., *J. Colloid and Interf. Sci.* **128**, 533 (1989).
10. Sogami, I. S., and Yoshiyama, T., *Phase Transition* **21**, 171 (1990).
11. Hiltner, P. A., and Krieger, I. M., *J. Phys. Chem.* **73**, 2386 (1969).
12. Chang, S. L., *Multiple Diffraction of X-Rays in Crystal* (Springer-Verlag, New York, 1984), p. 9.
13. Kittel, C., *Introduction to Solid State Physics*, 5th Edition (John Wiley & Sons, 1976), p. 59.
14. Carlson, R. J., and Asher, S. A., *Applied Spectroscopy* **38**, 297 (1984).
15. Kose, A., Ozaki, K., Kobayashi, Y., and Hachisu, S., *J. Colloid and Interf. Sci.* **44**, 330 (1973).
16. Ise, N., Okubo, T., Sugiura, M., Ito, K. and Nolte, H. J., *J. Chem. Phys.* **78**, 536 (1983).
17. Warren, B. E., *Phys. Rev.* **59**, 693 (1941).
18. Franklin, R. E., *Acta Crystall.* **3**, 107 (1950).
19. Ackerson, B. J., and Clark, N. A., *Phys. Rev. Lett.* **46**, 123 (1981).
20. Wilson, A. J. C., *Proc. Roy. Soc. London* **A180**, 277 (1942).
21. Nishiyama, Z., *Martensitic Transformations* (Maruzen, in Japanese).
22. Wayman, C. M., *Introduction to the Crystallography of Martensitic Transformations* (Macmillan, 1964).
23. Alexander, S., and McTague, J., *Phys. Rev. Lett.* **41**, 702 (1978).
24. Alder, B. J., and Wainright, T. W., *J. Chem. Phys.* **27**, 1208 (1957); **31**, 459 (1959).
25. Wadati, M., and Toda. M., *J. Phys. Soc. Japan* **32**, 1147 (1972).
26. Derjaguin, B. V., and Landau, L., *Acta Physiochem.* **14**, 633 (1941).
27. Verwey, E. J. W., and Overbeek, J. Th. G., *Theory of the Stability of Lyophobic Colloids* (Elsevier, Amsterdam, 1948).
28. Sogami, I., *Phys. Lett.* **A96**, 199 (1983).

29. Sogami, I., and Ise, N., *J. Chem. Phys.* **81**, 6320 (1984).
30. Dosho, S. et al., *Langmuir* **9**, 394 (1993).
31. Yoshiyama, T., and Sogami, I. S., *Phys. Rev. Lett.* **56**, 1609 (1986).
32. Laue, M., *Röntgenstrahl–Interferenzen*, Dritte Auflage (Akademische Verlagsgesellschaft, Frankfurt, 1960).
33. James, R. W., *The Optical Principles of the Diffraction of X-Rays* (Bell, London, 1967).

4

Direct Imaging of the Local Dynamics of Colloidal Phase Transitions

David G. Grier and Cherry A. Murray

Over the past decade, charge-stabilized colloidal suspensions have come to be recognized as tremendously useful model condensed matter systems. Ensembles of uniformly sized colloidal spheres undergo phase transitions from fluids to solids, from FCC to BCC crystals, and from crystals to glasses. Single layers of colloidal crystal melt through two second-order phase transitions, while systems of a few layers display cascades of fascinating phase transitions. Unlike the atoms in conventional materials, the individual spheres in these colloidal structures can be imaged with a conventional light microscope on time scales compatible with standard video equipment. This combination of properties makes possible studies of the local dynamics of phase transitions with "atomic" resolution. The present study focuses on a comparison between the equilibrium fluid-FCC crystal interface and the nonequilibrium freezing front between a deeply supercooled colloidal fluid and an FCC crystal. The equilibrium interface is a single lattice spacing thick and fluctuates on length and time scales comparable to the in-plane correlation length and time of the fluid phase. The nonequilibrium interface, on the other hand, is kinetically roughened with a width of almost 10 lattice spacings. In both cases, we find substantial quantitative agreement with the Hansen-Verlet and Löwen-Palberg-Simon empirical criteria for freezing, and the Lindemann criterion for melting. The structure and dynamics of the interfaces are influenced by the presence of the smooth repulsive glass window through which we observe the suspension. Thus these investigations also provide insights into the phase-influencing properties of geometrical confinement at the local structural level.

4.1 Introduction: The Influence of a Smooth Hard Wall on Structure

Unlike virtually any other condensed matter system, the microscopic structure and dynamics of colloidal suspensions can be studied with "atomic" resolution. This experimental accessibility coupled with their rich and varied phase behavior set colloidal suspensions apart as an unusually powerful class of model systems for studying the microscopic processes underlying structural phase transitions [1, 2]. The commercially manufactured polystyrene sulfonate spheres we will discuss in this chapter have diameters around $2a = 0.3$ μm, with each sample monodisperse to within 1 percent, and thus are large enough to image with a conventional light microscope. If we characterize their motions by the time required to diffuse one radius through water, $\tau = \pi \eta a^3 / k_B T \sim 0.03$ s, then they move slowly enough to track with standard video equipment. Here η is the viscosity of water at the experimental temperature $T = 29^0$C.

The results discussed in this chapter were obtained for samples of charge-stabilized colloidal microspheres that have of the order of $Z = 10^5$ ionic groups chemically bonded to their surfaces (approximately one group for each 4 nm^2 of surface area). These groups dissociate in water, leaving charges on the spheres' surfaces that are screened by counterions in solution. Charge renormalization arguments [3] and comparison with simulations of Yukawa systems [4] suggest that more than a few Ångstroms from the surface of a sphere of this size, its effective charge is roughly $Z^* = 1000$ electron equivalents. In a well-deionized suspension, the Debye-Hückel screening length, κ^{-1}, can extend to 100 nm or more. Under these conditions, the intersphere interaction is dominated by the screened-Coulomb repulsion, which was first formulated for the macroion-counter ion system some 50 years ago by Derjauguin, Landau, Verwey, and Overbeek (DLVO) [5, 6]

$$U(r) = \frac{(Z^*e)^2}{4\pi\epsilon} \left(\frac{\exp(2\kappa a)}{(1+\kappa a)^2} \right) \frac{\exp(-\kappa r)}{r} \qquad (4.1)$$

Here r represents the center-to-center separation between spheres. The factor in large parentheses accounts for the volume of the spheres from which the counterions are excluded, and ϵ is the dielectric coefficient of the fluid. The pairwise interaction potential described by the DLVO theory has been the subject of some controversy and has only recently been tested directly [7-9]. While the DLVO interaction serves as a useful point of reference, none of the results discussed in this chapter is believed to depend strongly on even the qualitative form of the interaction potential.

The ensemble of spheres in such a strongly interacting monodisperse system can adopt a rich variety of structural phases including fluids, crystals with either face-centered cubic (FCC) or body-centered cubic (BCC) symmetry, and even glasses [10-14]. Reentrant fluid phases and phase separation in colloidal fluids also have been reported [15]. Typically, transitions among these phases are induced by varying the volume fraction, ϕ, of the spheres in solution rather than the temperature. This practical expediency avoids complications arising from the temperature dependence of the effective charge and Debye-Hückel screening length in the interaction potential. The qualitative aspects of the fluid-crystal region of this phase diagram also appear in molecular dynamics simulations of Yukawa systems [16–19], as well as in analytical treatments of the one-component plasma model [20, 21]. The observed glass transition, reentrant behavior, and other phenomena have not been accounted for within the DLVO theory.

Colloidal crystals have bulk and shear moduli many orders of magnitude too small to support their aqueous media against gravity [22–24]. They exist, therefore, only within the confines of containers. Our sample container, one variant of which appears schematically in Fig. 4.1, is constructed with extremely smooth, stringently cleaned glass surfaces. The transparent walls facilitate observations, and the glass develops a strong surface charge in contact with water, which repels the spheres. As is the common practice, we maintain a very low ionic strength in the solution with large reservoirs of mixed-bed ion-exchange resin, which act as getter pumps for contaminant ions.

Once a suspension is introduced into the sample cell, chemical gradients come to equilibrium with the sphere distribution in several hours. Gradients in the sphere concentration, on the other hand, require months to relax. We take advantage of these long-lived density gradients to study structures over a wide range of volume fractions in a single sample cell. While the structures are not precisely in equilibrium, the flux of spheres is small enough to be negligible over the field of view and duration of a typical experiment, as we will discuss in the following.

We follow the motions of individual colloidal spheres by focusing an inverted metallurgical microscope through the sample container's bottom wall and capturing the resulting images on video tape with a precision monochrome CCD camera. Our mercury arc illuminator is bandpass filtered against both infrared and ultraviolet light to avoid heating or otherwise damaging the sample. A typical image of one layer of a face-centered cubic colloidal crystal appears as part of Fig. 4.1. The field of view provided by the microscope's oil-immersion objective (140X, numerical aperture 1.3) depends on the total system magnification, but generally falls near $\Delta x \times \Delta y = 60 \times 45 \ \mu m^2$. For crystals with lattice spacings of the order of 1 μm, a typical field of view contains as many as 2000 spheres. The depth of focus, $\Delta z = 0.4 \ \mu$m, is comparable to a sphere diameter, making it possible to image a single layer of spheres. The ability to image single colloidal layers suggests that the areal density, $n = N/A$, of N particles in the field of view of area $A = \Delta x \Delta y$ provides a more convenient measure of local density than the local volume fraction, ϕ, which can only be inferred from images. Where appropriate in this chapter, we parameterize our results in terms of n rather than ϕ. The interested reader can convert to volume fractions using $\phi = 2\pi 3^{-1/4} a^3 n^{3/2}$ for single layers of triangular crystal.

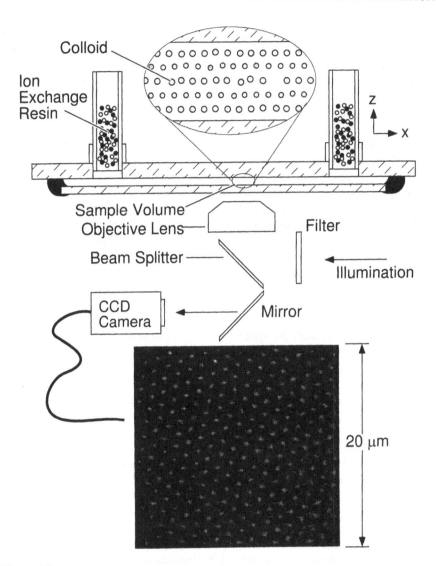

Figure 4.1 Schematic representation of the sample cell, not drawn to scale. The colloidal suspension is held in a thin hermetically sealed sample volume that is continuously cleaned by contact with ion-exchange resin. A high-power microscope objective images the colloidal particles onto a CCD camera whose output is recorded and digitized. The inset image shows a typical field of view.

The presence of an atomically smooth and flat repulsive wall influences the structure of both fluid and crystalline colloidal phases. Colloidal crystals, for example, always present their most closely packed faces to a glass wall. The FCC crystal in Fig. 4.1, therefore, is oriented with its (111) face to the wall. The effect in dense fluids near freezing is still more dramatic. By breaking the isotropy of space, the wall induces a density wave, or layering, in the fluid. The layering of fluids near hard walls has been observed in simulations [25] and is predicted by density functional theories [26]. We quantify the degree of wall-induced layering by scanning the focal plane of the microscope up through the suspension. The reflected light image of a sphere whose diameter is comparable to the wavelength of light both dims and broadens as the sphere moves out of the focal plane of the microscope. Sharply focused spheres produce brighter images on a darker background than do spheres out of focus. The variance of an image's brightness normalized by its mean brightness measures this contrast and therefore reflects the degree to which spheres lie in the focal plane. In a crystal, the relative variance reaches maxima when the focal plane coincides with crystal planes and minima when it falls between them. On the other hand, the uniform distribution of spheres in an isotropic homogeneous fluid should result in a relative variance independent of the focal plane's position. The overall contrast diminishes exponentially with distance from the surface, however, due to light diffusely scattered from other spheres.

Van Winkle and Murray [27] first exploited the depth dependence of the image contrast to study wall-induced structure in low-density colloidal fluids. In this case, turbidity does not suppress the contrast appreciably, and the resulting out-of-plane densitometry curves resemble very closely the measured in-plane correlation functions (see Figs. 4 and 5 of [27]).

While the dilute fluid's depth profile is fairly featureless, dense fluids near crystallization show clear evidence of long-range layering, as can be seen in Fig. 4.2. Each solid line in Fig. 4.2 is a fit to a sum of evenly spaced Gaussians multiplied by an overall exponentially decaying envelope. The length scale over which the contrast decays is found to be roughly 7 nearest-neighbor spacings at the volume fraction of this suspension. At an areal density considerably smaller than the first-layer melting value, $n = 0.88\ n_m$, a density wave already is clearly evident. The layer nearest to the smooth wall is sharpest. Subsequent layers become increasingly diffuse until, presumably, the bulk fluid isotropy is achieved far from the wall. As the fluid's density increases, the first layer moves closer to the confining wall, although the separation between fluid layers does not change. Surprisingly, the measured layer separation of 1.00 μm in the fluid corresponds almost exactly to the crystal's layer separation. The number of layers and their sharpness both increase as the volume fraction of the fluid increases toward the freezing point at n_m. At $n = 0.94\ n_m$, the first fluid layer's out-of-plane density profile is virtually indistinguishable from that of a crystal layer, although the in-plane dynamics remain fluidlike. Although the correlation length perpendicular to the wall grows as the suspension approaches the freezing transition, in-plane correlations remain as short as two nearest-neighbor spacings even in the first layer just before freezing. At n_m, the first fluid layer crystallizes while layers farther from the wall remain fluid. As the volume fraction increases, additional layers crystallize until, at the bulk melting point, the entire suspension crystallizes.

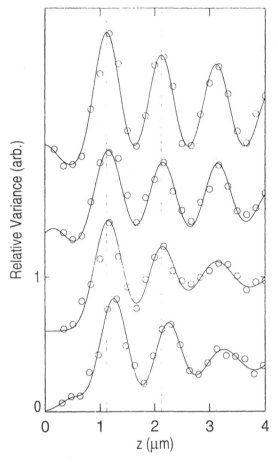

Figure 4.2 Relative variance of the image contrast as a function of depth into the suspension. The interface between the glass wall and the suspension is at $z = 0$ μm. The top trace is taken for a crystalline suspension at an areal density of $n = n_m$. The other traces, from the bottom up, show layered fluids at $n = 0.88$, 0.90, and 0.94 n_m. Traces have been displaced upward by 0.6 for clarity. Solid lines are fits to a sum of evenly spaced Gaussians with an exponential envelope function which accounts for the turbidity of the suspensions. These fits had a contrast decay length of roughly 7 μm. The dashed lines show the positions of the equilibrium crystal layers, spaced by 1.00 μm.

The remainder of this chapter focuses on phase transitions in the first layer of dense colloidal suspensions. In Section 4.2 we discuss the structure and dynamics of the first layer in equilibrium, with an emphasis on tests of various empirical criteria for melting and freezing. These considerations reappear in Section 4.3 in the context of fluctuations at the crystal-fluid interface in equilibrium. In

Section 4.4 we extend the insights drawn from the equilibrium system to the nonequilibrium case by studying the freezing of a supercooled colloidal fluid.

4.2 The Structure and Dynamics of the First Layer

4.2.1 Analytical Methods

We access the microscopic structure and dynamics of colloidal suspensions directly through quantitative analysis of digitized video images. As can be seen in Fig. 4.1, individual spheres appear as local maxima in the brightness field of the image. Simple algorithms suffice to locate these maxima quite rapidly, provided that geometric distortion in the imaging system, additive camera noise, and gradients in illumination have been corrected [28–32]. The image of an individual sphere subtends an area of more than 25 pixels so that fitting to a Gaussian surface of revolution further enables us to refine the centroid location to subpixel accuracy. Given a total system magnification around 100 nm/pixel, our resolution for centroid locations in the plane is as good as 25 nm. Following the argument of Section 4.1, the sphere's apparent width and brightness derived from the fit also provide information about its location along the z axis, out of the image plane. We calibrate this z dependence of a sphere's apparent width and brightness using spheres affixed to a glass wall, and thereby locate spheres to within 100 nm in that direction as well. A more detailed discussion of our methods will be presented elsewhere.

Having measured the distribution, $n(\mathbf{r}, t) = \sum_i^N \delta(\mathbf{r} - \mathbf{r}_i(t))$, of spheres located at $\mathbf{r}_i$ at time t, we have all the information we need to extract static measures of local order. We measure dynamical properties by tracking each colloidal particle in the field of view through consecutive video frames. Linking a sphere's image in one frame to its image in the next is performed with a maximum likelihood estimator algorithm. Although we have information on the particles' locations in three dimensions, we consider only the projections of the trajectories into the plane to avoid complications from the influence of the wall potential. Tracking particles through an isotropic fluid is impractical because of their strong out-of-plane motions. In the dense fluid just before crystallization, however, as many as 80 percent of the particles remain in the plane during a $\frac{1}{3}$ s measurement period despite their extended lateral motion.

We can gain a qualitative understanding of local ordering in a snapshot by plotting the lattice positions, $\mathbf{r}_i$, connected by their network of nearest-neighbor bonds. Of the many methods for determining such networks, the Delaunay triangulation [33] combines the benefits of being uniquely defined and of having an extremely efficient numerical implementation [34]. This network is the dual of the set of Wigner-Seitz cells for the particle distribution, and is known also as the Voronoi tesselation. The resulting triangulations, typical examples of which appear in Fig. 4.3, also are the natural choice for a system with intrinsic triangular

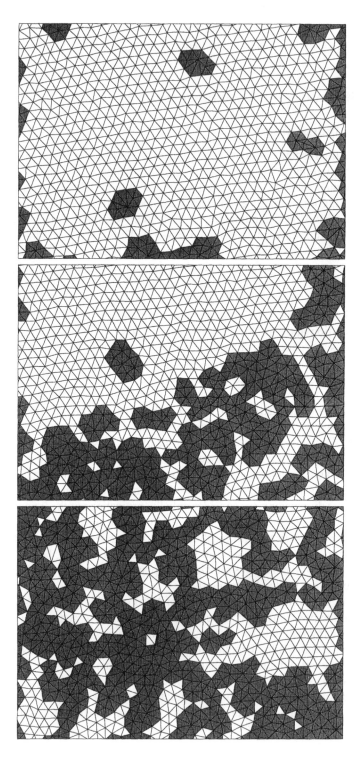

Figure 4.3 Delaunay triangulations of three snapshots at different positions along the density gradient ($n = 1.400 \times 10^{-2} a^{-2}$, $1.445 \times 10^{-2} a^{-2}$, and $= 1.500 \times 10^{-2} a^{-2}$) showing the transition from fluid to crystal structure. Triangles with at least one non-sixfold-coordinated vertex are shaded. Shaded triangles thus reflect the distribution of disclinations, or topological defects in the ensemble. Notice in particular the abrupt interface between disordered fluid region on the left and the relatively ordered crystal on the right.

symmetry. Counting nearest-neighbor bonds at each site provides an estimate of the density of topological defects. For example, those triangles in Fig. 4.3 with non-sixfold-coordinated vertices are shaded. Further quantitative insights derived from such triangulations will be discussed in Section 4.4.1.

The most generally useful measure of static order is provided by the pair correlation function

$$g(r) = n^{-2} \left\langle \int n(\mathbf{r} - \mathbf{x}) n(\mathbf{x}) d\mathbf{x} \right\rangle \tag{4.2}$$

where the angular brackets indicate an average over angles. Because we will consider only a face-centered cubic crystal whose (111) plane has triangular symmetry, we fit experimentally determined pair correlation functions to the correlation function for an ideal triangular lattice with nearest-neighbor spacing R, convolved with a Gaussian of half-width σ to account for measurement errors and elastic deformations, and suppressed by an exponential envelope that qualitatively accounts for topological disorder in the lattice:

$$g_{\text{fit}}(r) = \left[\int g_{\text{ideal}}(r - x, R) \exp\left(-\frac{x^2}{4\sigma^2}\right) dx - 1 \right] \exp\left(-\frac{r}{\xi}\right) + 1 \tag{4.3}$$

The correlation length ξ extracted from this three-parameter fit provides an estimate of the range of instantaneous positional ordering in a snapshot of the lattice. Typical pair correlation functions measured in the layered fluid and crystal phases appear in Fig. 4.4 together with fits to Eq. (4.3). The extracted lattice

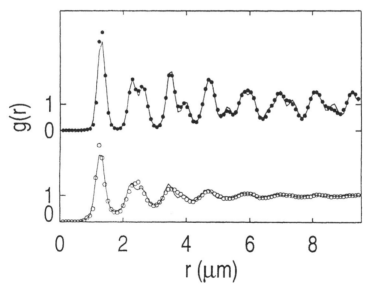

Figure 4.4 In-plane pair correlation functions for crystal (filled circles) at $n = 1.403 \times 10^{-2} a^{-2}$ and layered fluid (open circles) at $n = 1.500 \times 10^{-2} a^{-2}$. Solid lines are three-parameter fits to Eq. (4.3).

constant for the crystal, $R = 1.34$ μm, agrees well with the figure derived from the average separation between nearest neighbors of 1.36 μm. As might be expected, the Debye-Waller-like factor, $\sigma/R = 0.02$, is quite small, while the correlation length, $\xi/R > 40$, is as large as can be estimated from a sample 40 lattice spacings on a side.

Figure 4.5 shows the correlation length in the first fluid and crystal layers as a function of areal density. We estimate the areal density at the freezing transition, n_f, from the divergence of the correlation length. Above the freezing point, crystalline and fluid regions coexist. At densities higher than the melting point, n_m, the system is entirely crystalline. The identification of n_m is supported by the appearance of a line of topological defects in the Delaunay triangulation of the spheres at this areal density (central panel of Fig. 4.3). This line of defects marks the crystal-fluid interface at the first layer. The appearance of such an interface is consistent with the first-order nature of melting in three dimensions and does not appear in single-layer (two-dimensional) suspensions [35]. Measurements of dynamical quantities on either side of the interface also support this conclusion and will be discussed in Section 4.3.

Calculating the freezing points of simple liquids from first principles is one of the outstanding problems in condensed matter physics. In the absence of a complete microscopic theory for melting and freezing, several empirical criteria have been proposed that appear to hold quite generally. The next sections briefly outline three of these criteria and their applicability to colloidal suspensions in

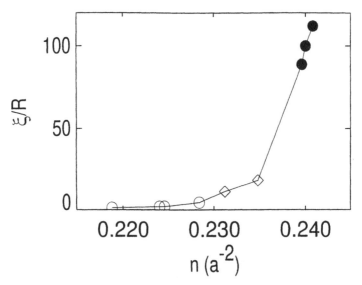

Figure 4.5 Correlation lengths estimated from fits such as those in Fig. 4.4 expressed in units of nearest-neighbor spacings as a function of areal density. Open circles represent fluid samples, filled circles crystal, and diamonds coexisting fluid and crystal. The coexistence point and melting point are determined from this plot.

equilibrium. We will revisit them in more detail with respect to nonequilibrium phase transitions in Section 4.4.

4.2.2 Hansen-Verlet Freezing Criterion

Hansen and Verlet recognized in 1969 that some features of freezing in Monte Carlo simulations of Lennard-Jones systems appear reproducibly enough from run to run to be considered generic to the freezing process [36]. To monitor the development of order, they studied the evolution of the azimuthally averaged structure factor for the system

$$S(Q) = n^{-2} \langle | \int n(\mathbf{r}) \exp^{-i\mathbf{Q}\cdot\mathbf{r}} d\mathbf{r} |^2 \rangle \qquad (4.4)$$

during freezing to a FCC crystal. The first peak of the structure factor at wave vector $Q = Q_0$ encodes information on long-range ordering and consistently achieves the value 2.85 when a three-dimensional system reaches freezing. The corresponding value in two dimensions is 4.0. Mean-field calculations by Ramakrishnan [37, 38] support the contention that these numerical criteria for freezing depend only weakly on the form of the interaction. Their applicability to Yukawa systems in particular has been demonstrated by lattice dynamics calculations [18]. Experimental tests on materials such as Ar [39-41], Na [42], Rb [43], and Pb [44] reveal that, to within experimental errors, the first peak in the structure factor of a fluid appears to reach universal values near 2.8 and 3.0 for fluids freezing into FCC and BCC crystals, respectively.

The first published test of the Hansen-Verlet criterion in colloidal suspensions of which we are aware [45] was performed using light scattering to measure $S(Q)$ directly. The onset of crystallinity in this case was determined by the appearance of Bragg peaks in the scattered light and so would tend to overestimate the volume fraction at crystallization and thus overestimate $S(Q_0)$. The observed value of $S(Q_0) = 2.1$ at freezing suggests that the colloidal lattice is considerably softer than anticipated by the Hansen-Verlet criterion. However, the quoted 5% polydispersity in radius implies a corresponding polydispersity in charge, which in turn may account for the observed suppression of long-range ordering in the crystal phase [46]. Indeed, this result suggests the need to extend the Hansen-Verlet result to the multicomponent case.

Figure 4.6 shows structure factors calculated from measured particle distributions in our monodisperse suspension ranging in density from fluid at $n = 1.403 \times 10^{-2} a^{-2}$ to crystal at $n = 1.505 \times 10^{-2} a^{-2}$. The different curves are parameterized by volume fraction. As the density increases and the suspension approaches freezing, the height of the first peak grows and its width diminishes. By interpolating between the peak heights at the measured densities, we estimate that $S(Q_0)$ attains the value 2.8 ± 0.1 at n_m, consistent with the Hansen-Verlet assertion.

The two-dimensional melting transition is found to consist of two second-order phase transitions with a hexatic fluid intervening between the isotropic

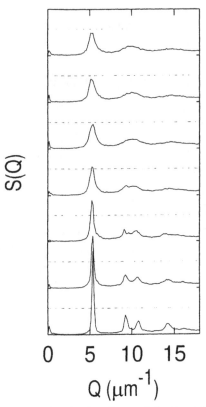

Figure 4.6   Structure factors for the first layer of the three-dimensional suspension. Running from the top to the bottom, the areal densities are 1.368×10^{-2}, 1.400×10^{-2}, 1.404×10^{-2}, 1.428×10^{-2}, 1.445×10^{-2}, 1.468×10^{-2}, and 1.498×10^{-2}, all measured in units of inverse radii squared. Consecutive plots are offset by 5 for clarity. The horizontal dashed lines at $S(Q) = 2.85$ represent the Hansen-Verlet freezing criterion for FCC crystals. The first peak at $S(Q_0)$ reaches 2.85 at a density of $1.438 \times 10^{-2} a^{-2}$, which coincides with the appearance of crystal in the field of view.

fluid and quasi-long-range-ordered crystal [47]. The first peak in the structure function reaches 4 at the hexatic fluid-crystal phase transition. Again, this result is consistent with the Hansen-Verlet freezing criterion.

4.2.3 Lindemann Melting Criterion

As early as 1910, Lindemann [48] postulated that thermally excited lattice vibrations can destabilize a crystal if they achieve amplitudes comparable to the equilibrium spacing between atoms. In the ensuing 80 years, scattering experiments in conventional materials have supported Lindemann's melting hypothesis and have found that crystals melt when the relative rms fluctuations of the atoms in a lattice about their equilibrium positions

$$\frac{\delta r}{R} = \frac{\langle |\mathbf{r}_i(t) - \mathbf{r}_i|^2 \rangle^{1/2}}{R} \tag{4.5}$$

fall in the range 0.10 - 0.20.

Given the microscopic particle locations and their trajectories, we can calculate the rms lattice fluctuations in two ways. From a single snapshot, we can estimate the ideal lattice that most closely resembles the experimental lattice, assign a lattice site to each particle in the field of view, and calculate the rms average displacement of particles from this lattice. From an ensemble of trajectories, on the other hand, we can calculate each particle's mean location and fluctuations about this mean, and then average these fluctuations over the field of view. In this case, the degree of perfection of the lattice of mean particle locations provides a measure of the degree of crystallinity. Furthermore, this dynamical approach works in a system with topological defects since it does not rely on there being a single underlying ideal lattice. The equilibrium lattice spacing, in either case, can be estimated from the average nearest-neighbor spacing.

Figure 4.7 shows the rms lattice fluctuations as a function of in-plane density for the first layer of the three-dimensional suspension near melting. These data suggest that indeed the first layer of the three-dimensional lattice melts when

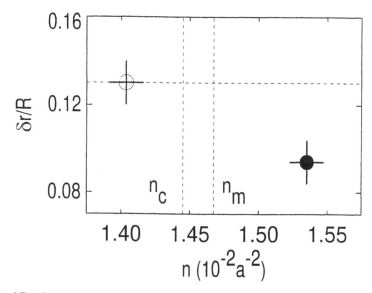

Figure 4.7 In-plane fluctuations about mean particle positions in units of the mean nearest-neighbor spacing as a function of areal density. Vertical dashed lines connote the coexistence point at n_c and the melting point at n_m. The horizontal dashed line at $\delta r/R$ = 0.13 represents the Lindemann criterion for hard-sphere FCC crystals. Note that the fluid sample (open circle) at density $n = 1.400 \times 10^{-2} a^{-2}$ has fluctuations larger than the Lindemann value, while the crystal sample (filled circle) has smaller fluctuations.

$\delta r/R = 0.13$, consistent with Lindemann criterion. Perhaps coincidentally, the Lindemann ratio for the melting line of hard-sphere FCC crystals is $\delta r/R = 0.133(2)$ [49].

4.2.4 Löwen-Palberg-Simon Dynamical Freezing Criterion

From the trajectories $\mathbf{r}_i(t)$ of each of the 1400 spheres in the central field of view, we calculate the in-plane displacement distribution, $N(r, t)$. These distributions, typical examples of which appear in Fig. 4.8, are found to be well represented by the Brownian form

$$N(r, t) = \frac{r}{2D(t)t} \exp\left(-\frac{r^2}{4D(t)t}\right) \tag{4.6}$$

where $D(t)$ is the time-dependent self-diffusion coefficient. The fit values for the spheres' mean square in-plane displacements satisfy the familiar Einstein-Smoluchowski form

$$\langle |\mathbf{r}_i^2(t) - \mathbf{r}_i^2(0)|^2 \rangle = 4D(t)t \tag{4.7}$$

and appear as functions of time in Fig. 4.9. The mean square displacements are linear in time to within experimental error after approximately one collision time, $\tau \sim 0.03$ s. Part of the offset can be accounted for by the 40 nm random measurement errors in the initial sphere locations. The rest arises from the rapid

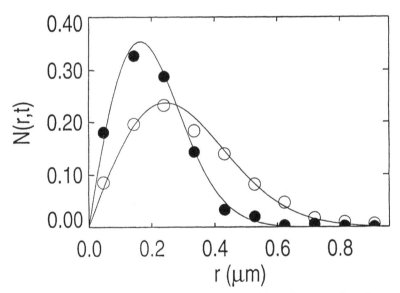

Figure 4.8   Probability of a particle diffusing in the plane a distance r from its starting point in time t. Data points are shown for $t = 1/6$ s. Filled circles came from trajectories in a crystal sample at $n = 1.500 \times 10^{-2} a^{-2}$, while the open represent fluid data at $n = 1.403 \times 10^{-2} a^{-2}$. Solid lines are one-parameter fits to Eq. (4.6).

effectively free diffusion of the spheres at short times. At the longer times shown in Fig. 4.9, the spheres' motions are already impeded by the "cage" formed by their neighbors. The so-called "cage effect" [50, 51] manifests itself in these dense fluids in roughly one collision time. From the subsequent slopes of the mean squared displacements, we estimate the late-time self-diffusion coefficients.

Löwen, Palberg, and Simon [52] recently have postulated that colloidal fluids crystallize when their average self-diffusion coefficient falls to 0.095 of the free particle value, D_0. Whereas their results were obtained by light scattering for bulk suspensions, our measurements probe the local dynamics and thus provide a complementary test of this new freezing criterion. When comparing the self-diffusion coefficients measured in the first fluid layer to the free value, we must take into account the influence of the nearby smooth wall [53]. For a sphere displaced from such a wall by 4 diameters, $D_0 = 0.89 D_S$, where $D_S = k_B T / 6\pi\eta a$ = 1.3 μm^2 s^{-1} is the Stokes diffusion coefficient. The Löwen-Palberg-Simon criterion appears as a dashed line in Fig. 4.10, and the agreement with our data is quite satisfactory. More compelling support for this dynamical freezing criterion is provided in Section 4.4.4 for the case of freezing in a supercooled colloidal crystal.

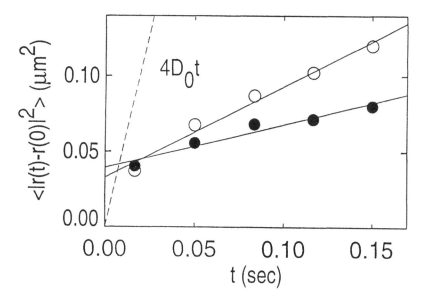

Figure 4.9 Mean square in-plane displacements taken from fits such as those in Fig. 4.8 as a function of time. Open circles come from a fluid sample, while filled circles come from a crystal. The slopes of the superimposed best-fit lines provide estimates for the self-diffusion coefficient of spheres in these phases. The diagonal dashed line shows the diffusion of a single particle in the presence of a nearby smooth wall.

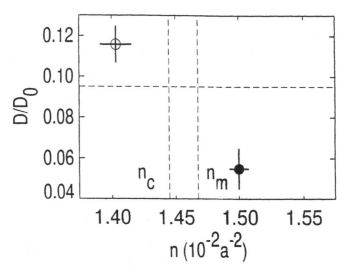

Figure 4.10   Self-diffusion coefficients for spheres in fluid (open circle) and crystal (filled circle) estimated from fits in Fig. 4.9 and plotted in units of the free-particle self-diffusion coefficient. Vertical dashed lines denote the coexistence and melting points at n_c and n_m, respectively. The horizontal dashed line at $D/D_0 = 0.095$ represents the Löwen-Palberg-Simon dynamical freezing criterion. In accord with this criterion, particles in the fluid diffuse more freely than $D/D_0 = 0.095$ and particles in the crystal less freely.

4.3 The Crystal-Fluid Interface in Equilibrium

The observation of interfaces such as that in the central panel of Fig. 4.3 provides an opportunity to examine in microscopic detail the dynamics of phase boundaries very near equilibrium. Such studies in conventional materials are extraordinarily difficult, and consequently few results have been forthcoming either from simulations or theory. As a starting point, therefore, we compare the fluctuations in the solid-fluid interface with corresponding processes in the pure phases on either side.

To create a crystal-fluid interface reproducibly, we start with an FCC crystal and inject excess pure water at one side. Although the mean volume fraction of the suspension is still above the freezing point, a gradient in the sphere density is set up that takes almost a year to relax by diffusion alone. The colloid at the low-density side melts and the melting front propagates into the suspension until it reaches a region whose density supports the crystalline phase. At this point, the interface stops. The results we report here were obtained with a $2a = 0.305$ μm diameter colloid at a volume fraction of $\phi = 0.0088$ approximately 1 month (10^9 collisions) after the initial gradient was set up. In the high-density fluid, the gradient of density across the imaging region is at most 1%. The total thickness of the colloid for these runs was approximately 200 μm along z, perpendicular to the wall. The results we obtain, therefore, should represent properties of the bulk suspension and not simply the walls' influence. The 3D fluid just before freezing has an in-plane density $n = 1.405(1) \times 10^{-2}$ in units of inverse radii squared. This is about 6% lower than that of the (111) face of the FCC crystal, which nucleates at the glass surface and reflects the pronounced layering in the fluid. We image a region of size $\Delta x \times \Delta y \times \Delta z = 59 \times 46 \times 0.4 \ \mu m^3$ in the center of a sample that is 10^2 to 10^3 times larger in linear extent.

4.3.1 Fluid Structure and Dynamics near the Interface

By counting nearest-neighbor bonds in the Delaunay triangulations of spheres on the fluid side of the crystal-fluid interface, we find that only 67% are sixfold coordinated, on average. In a two-dimensional fluid at the same areal density, as many as 84% of the spheres have perfect sixfold coordination. The difference in instantaneous topology is reflected by the degree of order measured by the correlation lengths: $\xi_{3D} = 2R$ while $\xi_{2D} = 5R$. Locally ordered clusters within the fluid with radii of two or three nearest-neighbor spacings are clearly evident in the triangulation of Fig. 4.3. Disorder in the two-dimensional fluid is distributed more evenly. At this density, the single-layer fluid is in the intermediate state between isotropic fluidity and crystallinity known as a hexatic fluid [54]. Hexatic ordering has also been seen in liquid crystal films [55] and magnetic fluid lattices in high-temperature superconductors [56, 57] and is characteristic of second-order Kosterlitz-Thouless melting in two dimensions. Despite its strong anisotropy, the first layer of a three-dimensional colloidal fluid displays a distinct phase boundary and lacks the orientational ordering of a hexatic fluid. Three-dimensional effects such as out-of-plane motion and coupling between planes dominate the fluid layer's behavior.

A useful probe of the time scale of fluctuations in the fluid is provided by the in-plane intermediate scattering function [58]

$$S(Q, t) = \frac{1}{n(t)n(0)} \left\langle \int n(\mathbf{r}, t) e^{i\mathbf{Q}\cdot\mathbf{r}} d\mathbf{r} \int n(\mathbf{r}', 0) e^{-i\mathbf{Q}\cdot\mathbf{r}} d\mathbf{r}' \right\rangle \qquad (4.8)$$

a typical example of which appears in Fig. 4.11. The angle brackets in Eq. (4.8) indicate an average over angles in reciprocal space. The function at zero delay, $S(Q, 0)$, is simply the structure factor. The decay of the first peak at wave vector $Q = Q_0$ gives an estimate of the lifetime of density fluctuations at the crystal row spacing $d_r = \sqrt{3}R/2 = 2\pi Q_0$. These fluctuations are roughly the size of the correlation length ξ determined from $g(r)$. We define a "correlation time" τ_{Q_0} as the lifetime of the ordered patches determined from the slope of $\ln S(Q_0, t)$ versus t for $0 \leq t \leq 0.2$ s. We find $\tau_{Q_0} \sim 0.28$ s in the first layer in three dimensions and $\tau_{Q_0} \sim 1$ s for a two-dimensional layer, or 8 and 30 collision times, respectively.

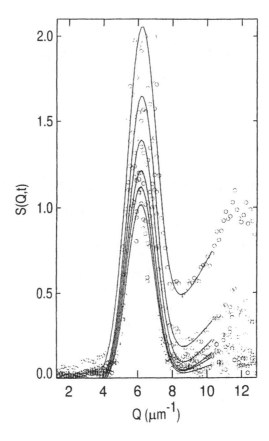

Figure 4.11   In-plane intermediate scattering function as a function of wave vector plotted at $\frac{1}{30}$ s intervals over a total period of $\frac{1}{6}$ s. Data points are shown for a fluid sample at $n = 1.400 \times 10^{-2} a^{-2}$. Solid lines are background-corrected fits to the first peak of $S(Q, t)$ at wave vector $Q_0 = 6.35 \ \mu m^{-2}$.

4.3.2 The Equilibrium Interface

Figure 4.12 shows the Delaunay triangulation of a snapshot at the crystal-fluid interface. The overall density gradient in the cell is directed upward along the [100] lattice direction. The crystal was probably shear-aligned when the gradient was set up. In this light, the presence of a microfacet along the comparatively loosely packed [110] direction, indicated by the dashed line in Fig. 4.12, is surprising. Equally noteworthy are the atomic-scale roughness of the interface and the presence of large ordered regions in the nearby fluid. Over any 1 s period, one of these ordered domains is likely to orient with respect to the crystal and thereby join the crystal-fluid interface. In this way, small-scale structural rearrangements frequently lead to large-scale fluctuations of the position of the

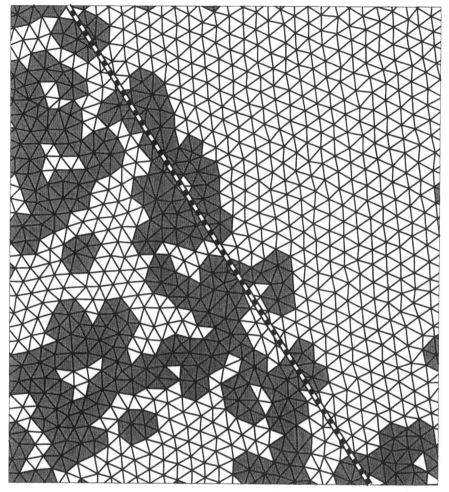

Figure 4.12 Delaunay triangulation of a snapshot of the crystal-fluid interface. Shaded triangles represent topological defects in the (111) plane of the FCC crystal. The density gradient in the sample runs vertically in this picture, along the [100] direction. The dashed line is a guide to the eye and represents the orientation of a (110) plane.

crystal-fluid interface. This mechanism naturally leads to interfacial fluctuations on length and time scales set by fluctuations in the fluid structure.

4.4 The Nonequilibrium Freezing Front

Because the shear moduli of colloidal crystals are so much smaller than those of conventional materials, the bulk fluid motion generated by tumbling the sample container suffices to melt a colloidal lattice. Moreover, because the heat capacity of the surrounding water carries off the latent heat of the phase transition, the resulting fluid is generated at the temperature and volume fraction of the original crystal. In other words, the fluid is supercooled against the formation of the original crystal [59]. The crystalline phase nucleates both homogeneously in the bulk of the suspension and heterogeneously at the walls of the cell and propagates through the entire sample volume. The probability of a nucleation event happening in the field of view is very small. Nevertheless, the propagation of the freezing front through the field of view provides an opportunity to follow the chain of events leading up to crystallization and to examine the microstructure of the nonequilibrium freezing front.

The degree of supercooling in a shear-melted colloidal fluid can be measured by the dimensionless quantity

$$\Delta_s = (\phi^{1/3} - \phi_m^{1/3})/\phi_m^{1/3} \tag{4.9}$$

In the long-screening-length limit of the interaction potential, this parameter is approximately equal to the supercooling parameter for convential materials, $(T_m - T)/T_m$, where T_m is the melting temperature. This connection is made by noting that the dimensionless interaction strength $\Gamma = U(R)/k_B T$ for a system with nearest-neighbor separation $R \propto \phi^{-1/3}$ is fixed by experimental conditions immediately after shear melting so that $U(\phi)/k_B T_m = U(\phi_m)/k_B T$.

In this experiment, the parallel glass walls of the sample cell were separated by 18 μm, which is sufficient for 16 crystal layers to form at the experimental volume fraction of $\phi = 9.8 \times 10^{-3}$. This geometry resembles very closely that chosen for molecular dynamics simulations of freezing in confined Lennard-Jones fluids by Ma et al. [60]. The volume fraction at melting for this colloid is about $\phi_m = 8.6 \times 10^{-3}$, so that the undercooling coefficient is $\Delta_s = 0.05$. This is comparable to a supercooling of about 13°C in water.

4.4.1 Three-Stage Solidification

The snapshots in Figs. 4.13(a–d) provide an overview of a typical run's progress, with frame (a) taken shortly after shear melting, (b) just before recrystallization, (c) just after recrystallization, and (d) after the crystal has fully reconstituted. The number of particles in the full field of view ranges from about 100 in the fluid to more than 500 in the equilibrium crystal. The Delaunay triangulations in Figs. 4.13(e–h) depict the uniquely defined nearest-neighbor network for the spheres located in Figs. 4.13(a–d), respectively. Only those particles determined to be in the volume of the first crystal plane (see Section 4.1) are included in these maps and in the analysis that follows.

After a long latency period characterized by large fluctuations both in observed density and local ordering, the shear-melted fluid suspension abruptly regains long-range triangular order in the plane. To quantify the time scales involved, we refer again to the characteristic time τ required for a free sphere to diffuse its own radius. Even a rough estimate of the time for freezing from Fig. 4.13

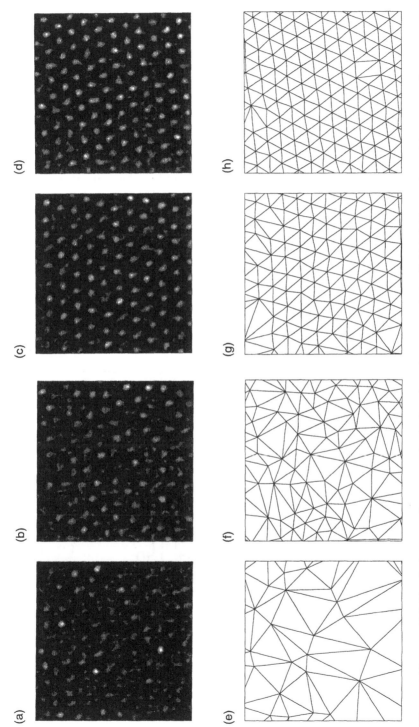

Figure 4.13 Small fields of view at various stages in the freezing of a supercooled colloidal fluid, together with corresponding Delaunay triangulations. (a) and (e): Isotropic fluid at $t_0 = 10490.3$ s after cessation of shearing. (b) and (f): Dense layered fluid 550 seconds after t_0. (c) and (g): Recrystallized first layer 610 s after t_0. (d) and (h): Crystal 1000 s after t_0.

suggests that more than 10^5 such time steps are required before the suspension begins to form FCC-like layers, but only 1000 more before the restoration of intralayer order runs to completion. These times are consistent with onset times observed in molecular dynamics simulations of crystallization in Yukawa [18] and classical one-component plasma [20, 21] systems. In the following paragraphs, we examine in detail the local structure and dynamics to refine this picture of the nonequilibrium freezing transition.

Perhaps the most obvious change that occurs during crystallization is the increased number of particles in the field of view of the microscope. Because the microscope is focused at the midplane of the first triangular crystal layer with an effective depth of focus $\Delta z = 0.35$ μm, the observed areal density of spheres in an isotropic fluid phase, n_f, should be smaller than that of a crystal, n_x, at the same volume fraction by a factor accounting for the inter-lattice-plane volume

$$n_f = n_x \frac{\Delta z}{R} \sqrt{3/2} = 0.33 n_x \tag{4.10}$$

where R is the nearest-neighbor spacing for the triangular crystalline layer. This large effect was not seen in the dense equilibrium fluids discussed in Section 4.3 because of their high degree of layering.

For the supercooled fluid, the areal density of spheres in the plane provides a straightforward probe of the phase transition's progress. Furthermore, Schätzel

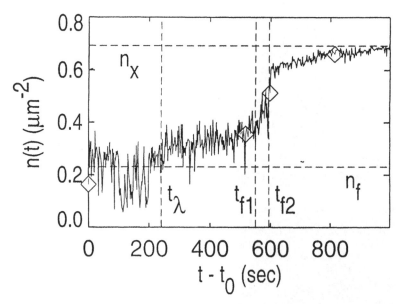

Figure 4.14   Areal density in the volume of the first crystal layer as a function of time after cessation of shearing. Diamonds correspond to the snapshots in Fig. 4.13. Horizontal dashed lines at 0.694 and 0.231 μm^{-2} represent the areal densities of the equilibrium crystal plane and of a slice of an isotropic homogeneous fluid at the same density, respectively.

and Ackerson [61] found in their light scattering studies that fluctuations in crystalline ordering are strongly coupled to fluctuations in the local density during homogeneously nucleated freezing of hard-sphere suspensions. Figure 4.14 shows the areal density of spheres in the volume of the crystal plane nearest to the wall as a function of time. In the remaining figures and in all references to the laboratory time that follow, a base time of $t_0 = 10490.3$ s has been subtracted from the reported times. This is the period between the cessation of shearing and the beginning of the reported data. The horizontal dashed lines in Fig. 4.14 indicate the measured density of the equilibrium crystal lattice, n_x, and the corresponding density for an isotropic fluid, n_f, calculated according to Eq. (4.10). Data points represent the number of spheres counted in individual video frames digitized at 2 s intervals. We identify the large steplike increase in areal density that occurs at time $t_{f2} = 595$ s with the completion of the freezing transition. This identification is supported by the simultaneous decrease in density fluctuations. In the fluid state, these fluctuations arise from particles entering and leaving the experimental volume in all directions. After crystallization, particles no longer hop out of the plane. The remaining fluctuations are dominated by counting errors at the edges of the field of view. The steady increase in density after freezing can be ascribed to particles appearing at the edges of the field of view and so is driven by hydrostatic compression of the sample volume by crystallization in the surrounding volume. In this period, the mean nearest-neighbor bond length decreases from 1.39 to 1.30 μm. The identification of $t_{f1} = 550$ s with the onset of crystallization in the field of view follows from the interpretation of other measures of order, which we will discuss in the following.

The extremely long latency period of approximately $10^5 \tau$ observed for crystallization in confined colloid differs markedly from the immediate onset of wall nucleated crystallization observed by Aastuen et al. [62, 63] in suspensions contacting only one wall. Although these bulk suspensions formed BCC crystals in equilibrium while ours form FCC crystals, the difference in latency time may arise from the additional confining wall in our system. The presence of the second wall is observed to impose more layering in the fluid than might be expected from observations of layering in bulk fluids near a single wall, which are observed to have approximately four layers just before freezing in equilibrium [27, 35, 64]. Thus the observation of a supercooled colloidal fluid that is metastable against crystallization in a confined geometry is reminiscent of the enhanced undercoolings possible for conventional fluids contained in pores.

Crystallization is preceded by another more subtle increase in areal density at time $t_\lambda = 230$ s. The areal density at times between and t_λ and t_{f1} is consistently 40 percent higher than we would expect for an isotropic fluid, and the fluctuations correspondingly lower. The image in Fig. 4.13(b) was taken during this stage. While there are some signs of local ordering, the structure is clearly disordered at long length scales. The increased density in the field of view necessitates a corresponding decrease in density elsewhere. We postulate, therefore, that this precursor to crystallization is the layered fluid described in Section 4.1.

Identifying t_{f2} with the freezing transition is supported by the development of structural order in the system. Despite the dramatic transformations that are taking place, the instantaneous radial distribution function $g(r)$ is still well modeled by Eq. (4.3), as can be seen in Fig. 4.15. To minimize the effect of noise, we average five consecutive correlation functions, separated in time by 2 s, before fitting. Figure 4.16 shows the evolution of the inverse correlation length measured from the radial pair distribution function $g(r)$. An inverse correlation length larger than one inverse lattice constant indicates disorder, while the transition to long-range order is signified by an asymptotic decline to zero. The transition to long-range order as measured in this fashion occurs between t_{f1} and t_{f2}. Before this rather abrupt transition, the correlation length gradually extends to about two equilibrium lattice spacings in the period we have associated with layered fluid formation. The growth of positional ordering starting at t_{f1} provides one indication that crystallization begins at that time.

Another measure of crystalline ordering is provided by the sixfold bond-orientational order parameter [54],

$$\Psi_6(\mathbf{r}) = \sum_j \delta(\mathbf{r} - \mathbf{r}_j) N_j^{-1} \sum_{k \in \{j\}} \exp(i 6 \theta_{jk}) \tag{4.11}$$

Here $\{j\}$ denotes the set of N_j nearest-neighbors of lattice site j, and θ_{jk} is the angle made by the bond to its kth nearest-neighbor with respect to a fixed axis. Each site in a perfect triangular lattice has the same exponential factor, so that the average over all lattice sites, $< \Psi_6(\mathbf{r}) >$, has a magnitude of 1. The

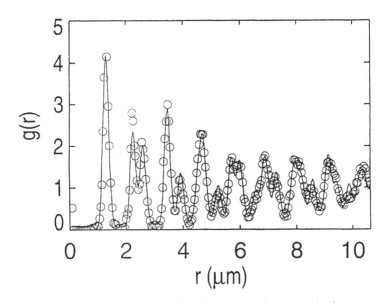

Figure 4.15   In-plane pair correlation function, averaged over angles, for one snapshot shortly after freezing. The solid line is a three-parameter fit to Eq. (4.3).

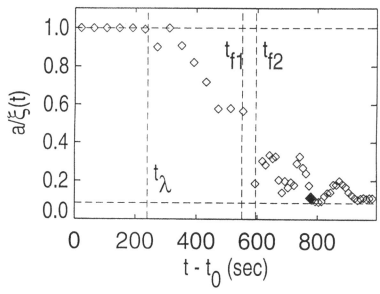

Figure 4.16 Inverse of the correlation length in units of the sphere radius as a function of time. Correlation lengths are estimated from fits such as that in Fig. 4.15. The fit in Fig. 4.15 appears here as a solid diamond. The horizontal dashed line at $a/\xi = 1$ is the limit of complete disorder, while the other near $a/\xi = 0$ represents long-range order, limited in resolution by the size of our field of view. Correlation functions from five consecutive video frames were averaged together before fitting to improve statistics.

corresponding average over a disordered ensemble has a magnitude near zero. We find from an examination of $\Psi_6 = |<\Psi_6(\mathbf{r})>|$ in Fig. 4.17 that the magnitude of the average bond-orientational order parameter begins to rise at t_{f1}. This completes our identification of t_{f1} with the commencement of freezing. The rate at which orientational order develops actually declines at t_{f2}, after freezing is complete. Once the lattice has achieved rigidity, misplaced spheres become topological defects such as vacancies, interstitials, and edge dislocations. Rather than simply flowing away through the suspension, they either must jump out of the first layer, or else must be transported by large collective rearrangements of the lattice. The relative slowness of these processes explains the observed decrease in the rate of orientational ordering after the recrystallization transition. In the equilibrium sample discussed in Section 4.2, Ψ_6 reaches 0.85 at n_m, which is somewhat greater than the value $\Psi_6 = 0.7$ achieved in the nonequilibrium case. Isolated vacancies and bound edge dislocations account for the disorder in both cases.

In this context, the marked drop in orientational ordering about 200 s after t_{f2} is somewhat surprising. While the abruptness of the freezing transition and the subsequent compression of the crystal attests to the rapid domain growth, the sudden decline in orientational ordering corresponds to the appearance of a

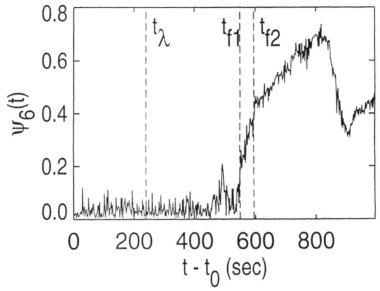

Figure 4.17   Magnitude of the sixfold bond-orientational order parameter averaged over the field of view as a function of time.

single grain boundary that propagates across the field of view. The subsequent value $\Psi_6 = 0.5$ is appropriate for two nearly equal-sized crystal domains in the field of view.

In their light scattering studies of homogeneously nucleated freezing of hard-sphere suspensions, Schätzel and Ackerson [61] observed that isolated crystallites grow rapidly until they come into contact with their neighbors. The resulting shear forces fracture the crystals into smaller crystallites, which subsequently grow slowly, probably through coarsening. This scenario is consistent with the observed real-space behavior of our system. While our field of view encompasses less that one full crystallite, the observed cracking of the domain attests to external shear forces acting on the field of view.

Thus the supercooled colloidal fluid regains equilibrium in three stages. First the isotropic homogeneous fluid develops a density wave perpendicular to the confining wall. The crystal nucleates in the first fluid layer and propagates rapidly through the system. When domains meet, the long-term growth of order proceeds through slow coarsening processes. The present study does not address the evolution of order in the direction normal to the first layer of spheres.

4.4.2 Hansen-Verlet Freezing Criterion

Complementary insights are offered by the structure factor $S(Q)$. Figure 4.18 shows the height of the first peak, $S(Q_0)$, as a function of time calculated for our nonequilibrium freezing transition. The small values of $S(Q_0)$ before t_{f1}

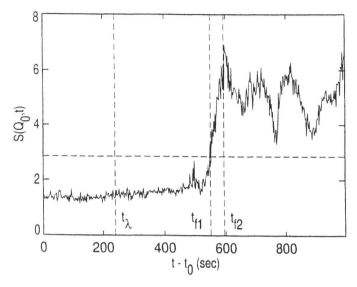

Figure 4.18 Height of the first peak of the structure factor as a function of time. The horizontal dashed line at $S(Q_0) = 2.85$ represents the Hansen-Verlet freezing criterion for the fluid-FCC crystal phase boundary.

are consistent with the instantaneous disorder of the fluid state. The observed value of 2.80 ± 0.10 at t_{f1} corresponds closely with the Hansen-Verlet criterion and supports our identification of this time with the initiation of crystallization in our field of view. This close agreement in the present case suggests that our suspensions may be considered to be in steady state on the time scales over which we measure structure factors, in other words, the camera's shutter period of $1/100$ s. The large fluctuations after t_{f2} arise from elastic and plastic deformations and, eventually, fracturing as the crystal lattice relaxes.

4.4.3 Lindemann Melting Criterion

Sphere trajectories taken over $\frac{1}{3}$ s intervals 2 s apart reveal the evolution of lattice fluctuations shown in Fig. 4.19. The horizontal line demarks the Lindemann criterion for hard spheres at 13 percent of the equilibrium lattice spacing. Once again, the quantitative correspondence between our experimental results on a nonequilibrium system and an empirical criterion deduced for systems in equilibrium is quite good.

4.4.4 Löwen-Palberg-Simon Dynamical Freezing Criterion

As in Section 2.4, we estimate the average single-particle self-diffusion coefficient $D(t)$ from the trajectories of particles in the field of view through the two-dimensional Einstein-Smoluchowski relation [Eq. (4.7)]. We assume that the

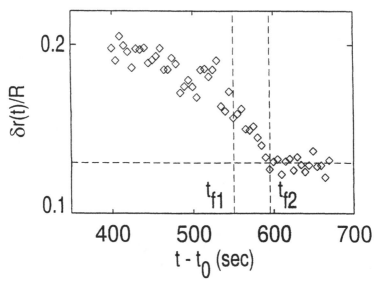

Figure 4.19 Average fluctuations about mean trajectory locations in units of the mean nearest-neighbor separation as a function of time. The horizontal dashed line at $\delta r/R =$ 0.13 corresponds to the Lindemann melting criterion for crystals of hard spheres. The plateau beyond t_{f2} reflects the dynamics of the crystal and not the resolution of our measurement system.

diffusion coefficient varies slowly enough before and after freezing that it may be treated as constant during the $\frac{1}{3}$ s over which each set of trajectories is calculated. Representative plots of the mean squared displacement as a function of time appear in Fig. 4.20. The slopes of the best-fit lines reflect the long-time self-diffusion coefficient. As in the equilibrium case, these fits do not pass through the origin both because of measurement error and also because of the cage effect.

The diffusion coefficient's evolution during nonequilibrium freezing appears in Fig. 4.21. Although small, the distinct jump discontinuity in the time evolution of the self-diffusion coefficient $D(t)$ at t_{f2} signals the onset of rigidity in the lattice and is consistent with the first-order nature of this phase transition. The observed crossover near the onset of crystallization at t_{f1} is noteworthy particularly since the criterion was postulated for systems in equilibrium. Furthermore, these results suggest that this new criterion may hold locally and not simply in the aggregate.

4.4.5 The Nonequilibrium Freezing Front

Because the freezing front propagates through the field of view in these recrystallization experiments, we have an opportunity to study the structure of the phase boundary far from equilibrium. Whereas the equilibrium crystal-fluid interface is reasonably well defined, the nonequilibrium freezing front, depicted at various stages in Fig. 4.22, is comparatively nebulous and is clearly above its roughening

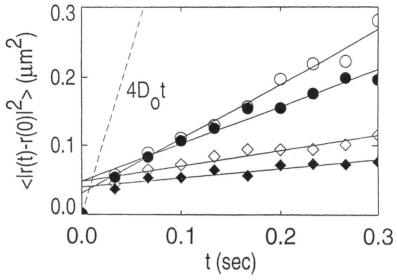

Figure 4.20 Einstein-Smoluchowski plots for four different times during the freezing process. The slopes of the best-fit straight lines to these plots provide estimates of the time-dependent self-diffusion coefficient. The diagonal line shows the result for free diffusion of particles 0.325 μm in diameter.

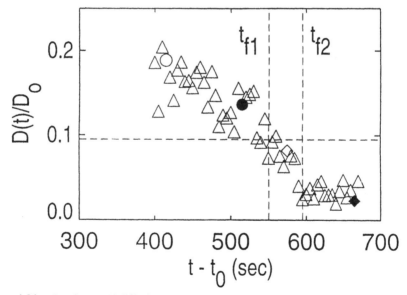

Figure 4.21 In-plane self-diffusion coefficient as a function of time measured in units of the free particle self-diffusion coefficient. Data from the fits in Fig. 4.20 appear with the same plot symbols here. The small yet distinct jump discontinuity at t_{f2} is consistent with the first-order nature of the freezing transition in three dimensions. The horizontal dashed line at $D/D_0 = 0.095$ shows the Löwen-Palberg-Simon dynamical freezing criterion. The data cross this line near t_{f1}, which we independently identify with the onset of freezing in the field of view.

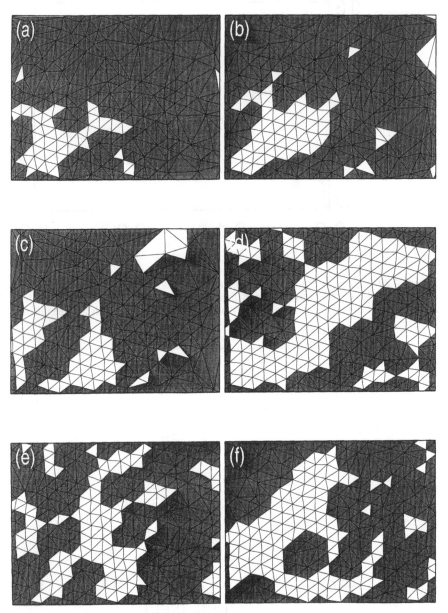

Figure 4.22   The nonequilibrium freezing front. The crystalline region advances from the lower left corner toward the upper right, more or less along the [100] direction. The times for image (a) through (f) are 570, 580, 590, 600, 610, and 620 s, respectively.

transition. Unlike the equilibrium interface, there are few indications of preferential facets from the ensemble's instantaneous topology at least on the scale of the present study. The crystalline region advances from the lower left corner in the triangulations of Fig. 4.22 toward the upper right. This direction corresponds, more or less, with the [100] lattice direction and suggests that as in the equilibrium case the (110) face is preferred for the fluid-crystal interface. Comparison with images taken before and after sequence suggests that the dense fluid may achieve a state very close to a perfect crystal before the actual freezing front passes through. Defects such as vacancies and interstitials, which maintain the layer in the fluid state then disappear through communication with the next layer. Clearly, though, considerably more work remains before we can hope to achieve a complete understanding of the local processes by which such interfaces advance.

Acknowledgements

We would like to acknowledge thought-provoking conversations with T. Witten and P. Pincus. John Crocker developed the particle tracking code and has contributed valuable insights to the data analysis. The work at The University of Chicago was supported by the NSF Materials Research Laboratory through grant number DMR-88-19860.

References

1. Pieranski, P., Strzlecki, L., and Pansu, B., *Phys. Rev. Lett.* **50**, 900 (1983).
2. Russel, W. B., Saville, D. A., and Schowalter, W. R., *Colloidal Dispersions* (Cambridge University Press, Cambridge, 1989).
3. Alexander, S., Chaikin, P. M., Grant, P., Morales, G. J., Pincus, P., and Hone, D., *J. Chem. Phys.* **80**, 5776 (1984).
4. Robbins, M.O., private communication (1994).
5. Verwey, E. W. J., and Overbeek, J. Th. G., *Theory of the Stability of Lyophobic Colloids* (Elsevier, Amsterdam, 1948).
6. Derjaguin, V., and Landau, L., *Acta Physicochimica (USSR)* **14**, 633 (1941).
7. Crocker, J. C., and Grier, G. G., *Phys. Rev. Lett.* **73**, 252 (1994).
8. Calderon, F. L., Stora, T., Monval, O. M., Poulin, P., and Bibette, J., *Phys. Rev. Lett.* **72**, 2959 (1994).
9. Kepler, G. M., and Fraden, S., *Phys. Rev. Lett.* **73**, 356 (1994).
10. Sirota, E. B., Ou-Yang, H. D., Sinha, S. K., Chaikin, P. M., Axe, J. D., and Fujii, Y., *Phys. Rev. Lett.* **62**, 1524 (1989).
11. Kose, A., Ozaki, M., Takano, K., Kobayashi Y., and Hachisu, S., *J. Colloid Int. Sci.* **44**, 330 (1973).
12. Monovoukas, Y., and Gast, A. P., *J. Colloid Int. Sci.* **128**, 533 (1989).
13. Furusawa, K., and Yamashita, S., *J. Colloid Int. Sci.* **89**, 574 (1982).
14. Kesavamoorthy, R., Rajalakshmi, M., and Baburao, C., *J. Phys.: Condens. Matter* **1**, 7149 (1989).
15. Arora, A. K., Tata, B. V. R., Sood, A. K., and Kesavamoorthy, R., *Phys. Rev. Lett.* **60**, 2438 (1988).
16. Hone, D., Alexander, S., Chaikin, P. M., and Pincus, P., *J. Chem. Phys.* **79**, 1474 (1983).
17. Kremer, K., Robbins, M. O., and Grest, G. S., *Phys. Rev. Lett.* **57**, 2694 (1986).
18. Robbins, M. O., Kremer, K., and Grest, G. S., *J. Chem. Phys.* **88**, 3286 (1988).
19. Meijer, E. J., and Frenkel, D., *J. Chem. Phys.* **94**, 2269 (1991).
20. Ogata, S., and Ichimaru, S., *J. Phys. Soc. Japan* **58**, 356 (1989).
21. Ogata, S., and Ichimaru, S., *J. Phys. Soc. Japan* **58**, 3049 (1989).

22. Williams, R., and Crandall, R. S., *Phys. Lett.* **48A**, 225 (1974).
23. Dubois-Violette, E., Pieranski, P., Rothen, F., and Strzelecki, L., *J. de Physique* **41**, 369 (1980).
24. Lindsay, H. M., and Chaikin, P. M., *J. Chem. Phys.* **76**, 3774 (1982).
25. Magda, J. J., Tirrell, M., and Davis, H. T., *J. Chem. Phys.* **83**, 1888 (1985).
26. Evans, R., in *Microscopic Theories of Simple Fluids and their Interfaces*, Eds. J. Carolin, J. F. Joanny, and J. Zinn-Justins. (Les Houches Session XLVIII, 1988, Liquids at Interfaces) (Elsevier, Amsterdam, 1989)
27. Van Winkle, D. H., and Murray, C. A., *J. Chem. Phys.* **89**, 3885 (1988).
28. Gonzalez, R. C., and Wintz, P., *Digital Image Processing* (Addison-Wesley, Reading, 1987).
29. Inoué, S., *Video Microscopy* (Plenum, New York, 1986).
30. Jain, A. K., *Fundamentals of Digital Image Processing* (Prentice-Hall, Englewood Cliffs, 1989).
31. Jähne, B., *Digital Image Processing: Concepts, Algorithms, and Scientific Applications* (Springer-Verlag, Berlin, 1991).
32. Pratt, W. K., *Digital Image Processing* (John Wiley & Sons, New York, 1991).
33. Preparata, F. P., and Shamos, M. I., *Computational Geometry* (Springer-Verlag, New York, 1985).
34. Fortune, S., *Algorithmica* **2**, 153 (1987).
35. Murray, C. A., Sprenger, W. O., and Wenk, R. A., *Phys. Rev.* **B 42**, 688 (1990).
36. Hansen, J. P., and Verlet, L., *Phys. Rev.* **184**, 151 (1969).
37. Ramakrishnan, T. V., and Yussouff, M., *Phys. Rev.* **B 19**, 2775 (1979).
38. Ramakrishnan, T. V., *Phys. Rev. Lett.* **48**, 541 (1982).
39. Gingrich, N. S., and Tompson, C. W., *J. Chem. Phys.* **36**, 2398 (1962).
40. Page, D. I., Egelstaff, P. A., Enderby, J. E., and Wingfield, B. R., *Phys. Lett.* **29**, 296 (1969).
41. Yarnell, J. L., Katz, M. J., Wenzel, R. G., and Koenig, S. H., *Phys. Rev.* **A 7**, 2130 (1973).
42. Greenfield, A. J., Wellendorf, J., and Wiser, N., *Phys. Rev.* **A 4**, 1607 (1971).
43. Copley, J. R. D., and Rowe, J. M., *Phys. Rev.* **A 9**, 1656 (1974).
44. North, D. M., Enderby J. E., and Egelstaff, P. A., *J. Phys. C: Solid State Phys.* **1**, 784 (1968).
45. Kesavamoorthy, R., Tata, B. V. R., Arora, A. K., and Sood, A. K., *Phys. Lett.* **138A**, 208 (1989).
46. Barrat, J. L., and Hansen, J. P., *J. de Physique* **47**, 1547 (1986).
47. Murray, C. A., and van Winkle, D. H., *Phys. Rev. Lett.* **58**, 1200 (1987).
48. Lindemann, F. A., *Z. Phys.* **11**, 609 (1910).
49. Young, D. A., and Alder, B. J., *J. Chem. Phys.* **60**, 1254 (1974).
50. Pusey, P. N., and Tough, R. J. A., *J. Phys. A: Math Gen.* **15**, 1291 (1982).
51. Nägele, G., Medina-Noyola, M., Klein, R., and Arauz-Lara, J. L., *Physica A* **149**, 123 (1988).
52. Löwen, H., Palberg, T., and Simon, R., *Phys. Rev. Lett.* **70**, 1557 (1993).
53. Happel, J., and Brenner, H., *Hydrodynamics at Low Reynolds Numbers* (Martinus Nijhoff Publishers, Boston, 1986).
54. Zippelius, A., Halperin, B. I., and Nelson, D. R., *Phys. Rev.* **B 22**, 2514 (1990).
55. Geer, R., et al., *Phys. Rev. Lett.* **66**, 1322 (1991).
56. Murray, C. A., Gammel, P. L., and Bishop, D. J., *Phys. Rev. Lett.* **64**, 2312 (1990).
57. Grier, D. G., Murray, C. A., Bolle, C. A., Gammel, P. L., Bishop, D. J., Mitzi, D. B., and Kapitulnik, A., *Phys. Rev. Lett.* **66**, 2270 (1991).
58. Hansen, J. P., and McDonald, I. R., *Theory of Simple Liquids* (Academic Press, London, 1986).
59. Grier, D. G., and Murray, C. A., *J. Chem. Phys.* **100**, 9088 (1994).
60. Ma, W. J., Banavar, J. R., and Koplik, J., *J. Chem. Phys.* **97**, 485 (1992).
61. Schätzel, K., and Ackerson, B. J., *Phys. Rev. Lett.* **68**, 337 (1992).
62. Aastuen, D. J. W., Clark, N. A., Cottler, L. K., and Ackerson, B. J., *Phys. Rev. Lett.* **57**, 1733 (1986).
63. Aastuen, D. J. W., Clark, N. A., Swindal, J. C., and Muzny, C. D., *Phase Transitions* **21**, 139 (1990).
64. Murray, C. A., Sprenger, W. O., and Wenk, R. A., *J. Phys.: Condens. Matter* **2**, SA385 (1990).

5

Colloidal Dispersions Studied by Microscopy and X-Ray Scattering

Norio Ise, Kensaku Ito, Hideki Matsuoka, and Hiroshi Yoshida

Optical microscopy, in combination with video-enhanced image analysis, allows one to study the behavior of latex particles in dispersions. Brownian motion, ordering of the particles, lattice vibration, lattice defects, concerted motion of lattice planes in colloidal crystals can all be visualized. Such studies show that, at low latex concentrations, non-space-filling ordered structures are found in co-existence with disordered particles (the two-state structure). In addition, large stable voids are also observed. Pair correlation functions for latices in dispersion can be obtained from the micrographs and can be Fourier transformed into scattering profiles. One can follow crystal growth using this procedure, and the results show that the Ostwald ripening mechanism is valid for crystal growth. Parallel and supporting information on the structure of dispersions can also be obtained by other means. For example, ultra-small-angle X-ray scattering (US-AXS) experiments give several orders of Bragg diffraction from a colloidal silica dispersion, thus confirming the existence of non-space-filling structures with a sixfold or a fourfold symmetry. The sharp upturn at low angles observed in these experiments is consistent with structural inhomogeneities such as localized ordered structures or voids. Studies using a laser scanning microscope show that negatively charged latex particles in salt-free dispersions are found to be concentrated in the distance range between 5 and 50 μm from negatively charged glass surfaces. This positive adsorption disappears when NaCl is added to 10^{-4} M. The two-state structure, void structure, the Ostwald ripening mechanism, positive adsorption of charged entities near similarly charged surfaces, reentrant phase separation, and vapor-liquid condensation in colloidal systeems testify to the existence of an electrostatic attractive interaction between colloidal particles (though similarly charged), which has been ignored in the traditional view in present-day colloidal science.

5.1 Introduction

5.1.1 Microscopy and Colloidal Systems

One of the most famous experiments using colloidal particles and a microscope was by Perrin, who studied (1) sedimentation equilibrium, (2) Brownian displacement, and (3) rotational diffusion in dispersions of colloidal particles such as gamboge particles [1]. He counted the number of particles at various heights in the dispersion by microscopy in experiment (1). The average displacement of Brownian particles on the focal planes was directly measured by microscopy in experiment (2). He took the advantage of the presence of minute structural inhomogeneities in the particles, which were used as markers to determine the velocity of particle rotation in experiment (3). These three microscopic investigations led Perrin to obtain the values for the Avogadro number N_A as 6.82, 6.88, and 6.50×10^{23}. These values are surprisingly close to the presently accepted value (6.02×10^{23}), if one takes into account the precision of the instruments used and the monodispersity of the particles. The basic equation for sedimentation equilibrium is given by

$$N_H = N_0 \exp[-4\pi N_A g a^3 (\rho - \rho_0) H / 3 R^g T] \tag{5.1}$$

where a is the particle radius, N_H and N_0 are the particle numbers at a height H from a reference height and at the reference height, respectively. g, R^g, and T are the acceleration due to gravity, gas constant, and temperature, respectively, and ρ and ρ_0 are the densities of the particle and the dispersion medium.

As mentioned by Perrin in the conclusion of his book, the fact that almost constant values for N_A were derived from diversified, independent measurements proves the existence of molecules and atoms. This conclusion has been accepted, and no one questions it or the magnitude of N_A. However, it appears that the fact that Perrin used a bare sphere radius, directly determined from microscopic observation, for the N_A determination has been ignored in the prevailing view in present-day colloid science. In other words, if other values (such as an enlarged effective radius) were used for a, Perrin could not have obtained a reasonable value for N_A.

Another important contribution by microscope was by Zsigmondy, who established the method of ultramicroscopy [2,3]. By this method, one can detect the presence of particles smaller than the wavelength of visible light. With some special device, particles of diameter 0.004 μm could be seen. The importance of the microscopic method in colloidal systems was again demonstrated by Hachisu et al., who successfully photographed ordered structures of polymer latex particles by using a metallurgical microscope [4].

Microscopy provides direct information on particle distributions in real space and time, which is highly useful in elucidating various behaviors of particles in dispersions. When use is made of polystyrene-based latex particles, whose monodispersity and chemical structures can be relatively easily controlled in comparison with other polymer systems, in combination with water, heavy or light, a technical problem arises: The dispersions are so turbid that microscopic observation of particles inside the dispersion is difficult. Because of this, observation is limited to particles near the cover glass-dispersion interface or to very dilute dispersions. The wall effect is often considered to be highly serious; so one often presumed it to be the source of various phenomena that were unanticipated according to an established concept. Naturally, an effort must be made to eliminate or minimize this effect. However, it should be recalled that, in his sedimentation equilibrium experiments, Perrin measured particle numbers at four focal planes, 5 and 35 μm away from the top and bottom interfaces of a glass container, and obtained the reasonable N_A value from these. This success implies that the wall effect is already negligible at a distance of 5 μm from the glass, as far as free particles at low concentrations are concerned.

The wall effect may be circumvented by using confocal laser scanning microscopy (LSM) to a large extent. It will be shown below that void structures in latex dispersions, which were once considered to be due to a wall effect caused by an unclean glass surface, can be observed more easily inside the dispersion by the LSM than near the interface. This shows that the wall effect is not so important; the void structures are physically real, bulk effects.

5.1.2 Microscopy and Scattering Methods

Another way to avoid the wall effect is to use scattering techniques. Because of the large interparticle separation to be studied in latex dispersions, small-angle X-ray scattering (SAXS) could not generally be employed: Small-angle neutron scattering (SANS) and light scattering (LS) were frequently used to analyze the structure of latex dispersions. Now with the Bonse-Hart-type diffractometer one can deal with large distances up to 8 μm, and hence it can be used for the study of colloidal structure.

One characteristic of the scattering techniques is that the information, obtained in Fourier space, is statistically averaged over a large number of configurations. This has both advantages and disadvantages with respect to microscopic methods. To obtain a correct understanding of colloidal systems, it is necessary carefully to utilize information from both microscopic and the scattering methods while paying due attention to the respective limits of the two methods.

It is worth mentioning here that the density function $\rho(\mathbf{r})$ in real space and the amplitude of the scattering profile $A(\mathbf{Q})$ in Fourier space are related by the following equation

$$A(\mathbf{Q}) = \int \rho(\mathbf{r}) \exp[-2\pi i (\mathbf{Q} \cdot \mathbf{r})] \, d\mathbf{r} \qquad (5.2)$$

It is of particular significance for polymer latex particles that the $\rho(\mathbf{r})$ is obtained in the form of a micrograph so that the scattering profile can be computed by Eq. (5.2). The computer-aided 2D Fourier analysis of micrographs of particle distributions in dispersions was undertaken in 1987 [5]. The study clearly showed the basic features shown in Table 5.1.

Figure 5.1 shows the Fourier analysis of particle distributions at three different salt concentrations. The ordered structures of latices, which form at very low salt concentrations, become less stable with increasing ionic strength and gets destroyed at high salt concentrations (above 10^{-3}M NaCl, in this particular case). From Fig. 5.1(i), it can be seen that a large number of small ordered regions are maintained in the deionized dispersion. At 10^{-4}M, the ordered structures are largely destroyed, and only small clusters are found, which is consistent with the very vague Fourier ring (ii-b). At [NaCl] = 1.2×10^{-3}M, the Fourier pattern is more faint.

It should be recalled that the above Fourier analysis was carried out on the literally "real" distribution of particles, which have 3D degrees of freedom. This is in contrast with studies of 2D dispersion of particles, which are in contact with "wall(s)." The behavior of the particles is then possibly somehow distorted by the wall effect. Furthermore, the present study [5] should be distinguished from the famous optical experiments performed on artificially constructed patterns by Hosemann [6]. The optical transform method is certainly very helpful in interpreting diffraction patterns of real atomic and molecular systems, as discussed by Lipson [7]. The patterns are results of intellectual imagination and may agree only qualitatively or semi-quantitatively with the real distribution. On the other hand, the binary images of latices [5] are real in the sense that these are pictured in real space. Thus it should be pointed out that the Fourier analysis of latex systems enables one to scrutinize the interpretation of scattering data in general on the basis of more fundamental principles with the least number of assumptions and parameters.

Table 5.1 Computer-aided 2D Fourier analysis of latex distribution in dispersions

Real space	Fourier space
Ordered	Spots
Ordered and disordered	Broad ring(s)
Disordered	No spot(s) or ring(s)

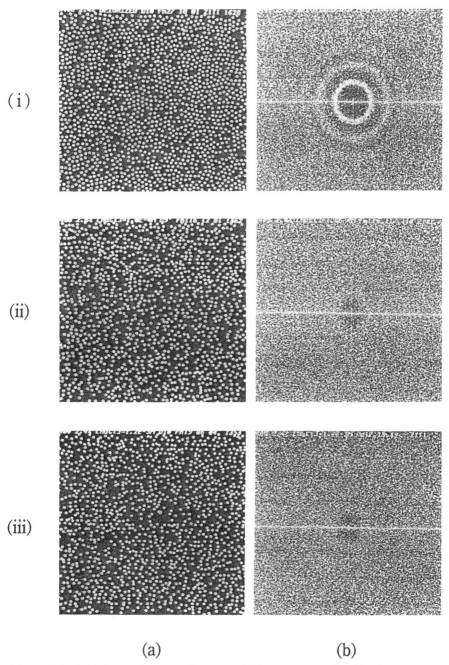

(a) (b)

Figure 5.1 (a) Binary microscope images and (b) the corresponding Fourier patterns at various concentrations Latex: N400 (diameter d: 4000 Å, charge density ρ_e: 6.9 $\mu C/cm^2$), [latex]: 2.4%, [NaCl]: (i) deionized; (ii) 1.0×10^{-4}M; (iii) 1.2×10^{-3}M. The micrographs were taken by a Carl-Zeiss reversed-type microscope; the pictures show particle distributions in a horizontal focal plane [5].

Following these introductory remarks, this chapter discusses recent experimental results on colloidal dispersions of polystyrene-polystyrene sulfonate copolymer particles, unless otherwise stated. Furthermore, the solvent used was light water, unless otherwise stated. In some cases, where the influence of gravity had to be strictly eliminated, a mixture of D_2O and H_2O was used and a density-matching condition was realized. In Section 5.2, microscopic data on Brownian particles are discussed. The historical results obtained by Perrin and Zsigmondy are reconfirmed in the recent studies, except that an image data analyzer and optical instruments with higher precision were used. It is surprising that Perrin's experiments contradict a view that has been widely accepted in traditional analysis used in present-day colloid science. In Sections 5.3 and 5.4, the microscopic method, in combination with the image data analyzer, of studying the structure of colloidal crystals is described. Various phenomena of atomic solids that were believed to exist are displayed in reality. The two-state structure, the coexistence of localized ordered structures, and Brownian motion of particles are photographed. The process of colloidal crystal growth is followed by using the Fourier analysis mentioned above. Stable and large voids in apparently homogeneous latex dispersions are described. Scattering at very low angles displays a sharp upturn, in conformity with the structural inhomogeneities such as the two-state structure. Section 5.5 describes the ultra-small-angle X-ray scattering (USAXS) apparatus is used to determine the particle diameter, its distribution, and structural symmetry of colloidal crystals. At least five orders of Bragg diffraction are found for the first time in colloidal silica dispersions. Finally, the controversy on the interparticle interaction is addressed in Section 5.6.

5.2 Brownian Particles

5.2.1 Average Displacement, Sedimentation Equilibrium, and the Effective Sphere Concept

The displacement of free particles was studied using a microscope by Ottewill et al. [8] and Ise et al. [9], who used a video digitizer for 20 particles. Although the number of particles was very small so that the experimental uncertainty was large, the studies clearly show that the Einstein theory of Brownian motion is satisfactory except at high latex concentrations and low salt concentrations. The displacement was also found to increase with increasing temperature in water. Figure 5.2 shows the viscosity dependence of the root-mean-square displacement of latex particles in aqueous sucrose solutions.

Recently it has been claimed that colloidal phenomena can be satisfactorily explained by the repulsion-only assumption [10]. This assumption was embodied, though in a highly arbitrary manner, by the enlarged effective hard-sphere concept, for example by assuming the spheres to have a bare radius a, plus the so-called Debye length $(1/\kappa)$ [11]. This was questioned by Hirtzel and Rajagopalan [12], who asserted that this mental exercise is purely ad hoc. Ito et al. [13] also

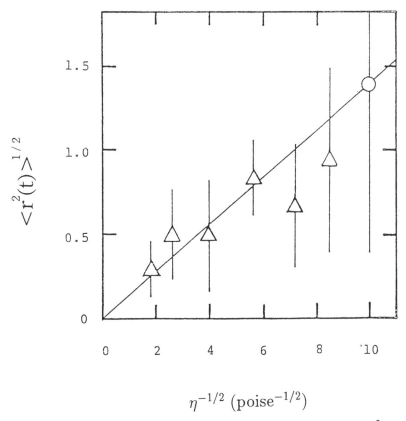

Figure 5.2 Viscosity dependence of the root-mean-square displacement $< r^2(t) >^{1/2}$ of latex particles in aqueous sucrose solutions: latex: SS-37 (d: 0.45 μm, ρ_e: 4.4 μC/cm^2), [latex]: 0.01%, [NaCl]: 10^{-5}M, room temperature. The circle denotes the $< r^2(t) >^{1/2}$ value in water without sucrose [9].

pointed out that it is the bare sphere radius that Perrin used to obtain the correct N_A. The role of the Debye length could not be examined from the Perrin's experimental data, because the number of charges on the gamboge particles is not known, and hence it is not possible to estimate $1/\kappa$ for this case. The sedimentation equilibrium experiment has been repeated with much better characterized polymer latex particles using an LSM and particle numbers at various heights in a latex dispersion were determined [13]. The experimental data obeyed Eq. (5.1), as is shown in Fig. 5.3.

The particle radius can be calculated from the slope of Fig. 5.3 by using the accepted value of N_A (6.02×10^{23}), and is tabulated in Table 5.2, together with the radius value (a^*) determined by transmission electron microscopy (TEM). The agreement between a and a^* is satisfactory and indicates that the sedimenting particles can be regarded as "bare" spheres. This is in line with the analysis by

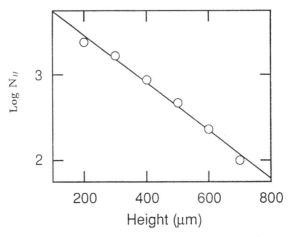

Figure 5.3 Particle distribution as a function of height at sedimentation equilibrium. Latex: PS-S5A (d: 5000 Å, ρ_e: 2.54 $\mu C/cm^2$), [latex]: 0.01%, microscope: confocal laser scanning microscope, Carl Zeiss, Oberkochen, Germany [13].

Table 5.2 Particle radius determined from the sedimentation equilibrium, the Debye length and the Avogadro number obtained from the enlarged effective sphere concept.

[NaCl] (M)	a (μm)	$1/\kappa$ (μm)	$a^* + 1/\kappa$ (μm)	κa^*	N_A $\times 10^{-23}$
0	0.43	0.17	0.65	2.8	1.9
10^{-6}	0.43	0.13	0.61	3.7	-
10^{-5}	0.42	0.08	0.56	5.9	2.9
10^{-4}	0.43	0.03	0.51	16	-
10^{-3}	0.44	0.01	0.49	48	4.4

Latex: N1000 (a^*: 0.48μm, surface charge density, ρ_e: 6.4 $\mu C/cm^2$)
[latex]: 0.01 vol%.

Perrin. On the other hand, if one follows the effective sphere concept by adopting ($a^* + 1/\kappa$) as the particle radius, the value for N_A obtained are as shown in the sixth column of Table 5.2. Obviously, in the absence of added salt, when the enlarged sphere concept should be most effective, N_A turns out to be 2×10^{23}, which is incorrect. This consequence questions the interpretation in terms of the Debye length. The use of the effective sphere concept also deprives the Onsager limiting law [14] for the conductance of simple electrolyte solution of its physical significance.

5.3 Ordered Particles

5.3.1 Oscillation of Lattice Points at Various Latex Concentrations and Temperatures

By using a video device coupled with an image data analyzer, the centers of the latex particles in the ordered regions were followed as a function of time [15]. The lattice oscillation was quantified by the largest length of the (noncircular) area (D_{max}) the centers occupied in 8.3 s and the standard deviation of the inter-particle spacing (ΔR_{exp}). The results are shown in Table 5.3. Clearly, the D_{max} value is larger at lower latex concentrations and decreases with concentration by a factor of six when the concentration is increased from 1.0% to 8.0%. The ratio of ΔR_{exp} to R_{exp} decreases by a factor of two in the corresponding concentration range. Although a statistically meaningful number of particle pairs have to be treated before one can make a significant statement, this difference, together with the noncircular nature of the occupied area, suggests a directional oscillation of the lattice plane. This aspect will be discussed in the following.

The temperature dependence of the thermal motion is also given in Table 5.3. It appears that not only D_{max} but also $\Delta R_{exp}/R_{exp}$ and R_{exp} decreased with increasing temperature and, after passing through a minimum, increased above 50°C. These trends are understandable if one accepts that the Coulombic attraction between the particles is intensified with temperature, since the dielectric constant of water becomes smaller. Above 50°C, the kinetic energy contribution

Table 5.3 Thermal motion of latex particles in the ordered structure.

(a) Latex concentration dependence

[latex](%)	$D_{max}(\mu m)$	$\Delta R_{exp}/R_{exp}$	$R_{exp}(\mu m)$
1.0	0.50±0.08	0.048	1.08±0.08
2.0	0.35±0.07	0.029	1.07±0.07
4.0	0.21±0.07	0.021	0.87±0.07
8.0	0.08±0.01	0.022	0.73±0.03

Latex: N400. The numerical data were obtained by averaging 240 measurements for each particle.

(b) Temperature dependence

$T(°C)$	ϵ	$D_{max}(\mu m)$	$\Delta R_{exp}/R_{exp}$	$R_{exp}(\mu m)$
10	83.33	0.26±0.06	0.024	0.85± 0.08
20	80.20	0.14±0.04	0.023	0.83± 0.06
30	76.55	0.11±0.02	0.011	0.75± 0.04
50	69.51	0.12±0.03	0.007	0.73± 0.04
60	66.81	0.24±0.06	0.048	0.75± 0.08

Latex: SS-43 (d: 4100 Å, ρ_e: 9.0 $\mu C/cm^2$)

would be so influential that D_{max} and $\Delta R_{exp}/R_{exp}$ become larger and the ordered structure starts to dilate. It is noted that the temperature dependence mentioned here is consistent with our earlier observation [9].

5.3.2 Oscillation of Lattice Planes

The trajectory of particles in neighboring lattice planes for 1 s is demonstrated in Fig. 5.4 [16]. Note that the interparticle spacing was about 1 μm. As seen from Fig. 5.4(a), the amplitudes of vibration of the particles 1 to 6 were about $0.1 \sim 0.05$ μm. Although not at all clear from (a), the motion of the particles was seen to be coupled with that of neighboring particles. The displacement versus time curves (b) have similar shapes and phases, particularly for particles 1, 2, and 3. It is to be remarked that the interparticle interaction is so powerful that it affects another particle 2 μm away, as far as the present observation is concerned. According to the Fourier analysis, the coupled motion is an oscillation of period 1 s. Note the correspondence of this oscillation to that in atomic crystals, although the oscillation in colloidal crystals is not quantized [17].

5.3.3 Lattice Defects

Like crystalline atomic solids, colloidal crystals are also not perfect. It is not difficult, particularly at low latex concentrations, to observe lattice defects such as edge dislocations and Frenkel and Schottky-type defects [18, 19]. It may be mentioned that the lattice vibration around point defects is highly anisotropic and demonstrates that the restoring force on a particle in the direction toward a defect is weaker than that in other directions, as was discussed previously [16]. Qualitatively speaking, this is exactly what is assumed in the Einstein model of lattice vibration [20].

5.3.4 Localized Ordered Structures (the Two-State Structure), Radial Distribution Function, and Structure Factor

The coexistence of ordered structures with free Brownian particles, namely the two-state structure, in ionic polymer solutions and in silica dispersions, has been suspected to exist on the basis of the fact that the Bragg spacing (R_{exp}) is smaller than the average spacing (R_0) calculated from the polymer concentration [21, 22]. In the case of polymer latices, the localized ordered structure, or the two-state structure, has been photographed [15]. These clearly show the coexistence of ordered regions of a higher number density (with an interparticle spacing of around 1 μm), and disordered regions having a lower density (see Fig. 1 of Ref. 15). The clear difference between the ordered particles and Brownian particles can be demonstrated by trajectory analysis. Figure 5.5 is an example in which the particle trajectory in 11/15 s is determined by video imagery combined with

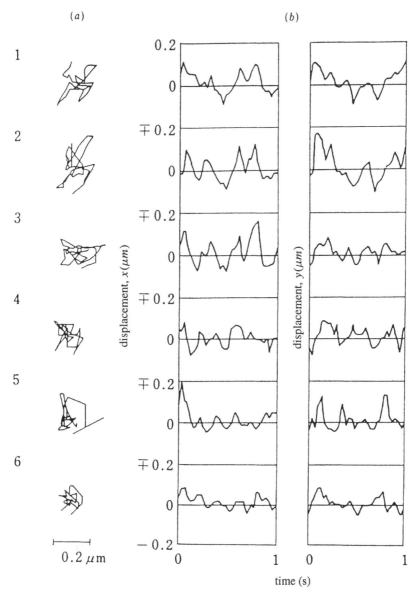

Figure 5.4 Coupled vibration of particles on neighboring planes. Latex: N400, [latex]: 0.5%, room temperature. (a) Trajectory in 1 s. (b) Displacement resolved in the x and y directions versus time. The distance between the two neighboring particles was about 1 μm. [16]

an image data analyzer. The ordered particles show a small displacement around the lattice points, whereas the Brownian particles display substantial movement. Note from Fig. 5.5 the difference in the number densities in the two regions.

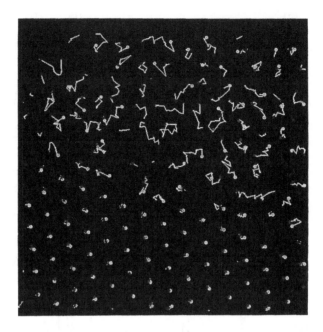

Figure 5.5   Trajectories of latex particles in the ordered and disordered regions. Latex: N300 (d: 3000 Å, ρ_e: 1.3 μC/cm^2) [latex]: 2%. The dispersion was observed from the side of the observation cell by fixing an Olympus reversed-type microscope in a horizontal direction. Thus the trajectory shows particle motion on a vertical (focal) plane. The information on positions of particle centers was stored in an image data analyzer, reproduced in one new frame and connected by lines with the analyzer. To avoid too complicated a picture, not of all the information obtained in 11/15 s was used [15]

From microscopic observation, it is possible to conclude that a spectrum of diffusion coefficients must be found for latex particles in dispersions, even when the dispersions appear to be homogeneous. From Fig. 5.5, one can notice at least two diffusion coefficients, a large one corresponding to the Brownian particles and a small one to the ordered particles. While the video imagery provides the most direct information of the multiple diffusion constants for latex particles, it is relevant to mention that at least two diffusion coefficients have been found for solutions of "invisible" ionic polymers. In this respect, Schurr et al. were the first to observe that the diffusion coefficient of poly-L-lysine hydrobromide sharply dropped with decreasing NaBr concentration [23]. Schmitz et al. observed two diffusion coefficients ($D_S = 8 \times 10^{-7}$ cm^2/s and $D_L = 4 \times 10^{-8}$ cm^2/s) for a poly-L-lysine sample at a KCl concentration of 10^{-3}M [24]. Subsequently other ionic polymers were studied by dynamic light scattering, and various interpretations were presented. We like to refer the the reader to a recent monograph by Schmitz for detail [25].

The localized ordered structure is thus proven to exist in apparently homogeneous dispersions in a most direct way for latex dispersions. An important fact is that there is no mechanical boundary (wall) between the ordered and disordered regions by which these two regions can be distinguished. In other words, particles are easily observed to "evaporate" or "condense" between the ordered and disordered regions. Thus the size of the localized ordered structure is time dependent, as far as the initial stage of crystallization is concerned.

The localized, non-space-filling ordered structure without a mechanical "wall" suggests that *similarly charged particles attract each other at a distance of at least 1 μm, whereas they naturally repel each other at much shorter distances.* The present understanding is that the particles in the ordered structure sit in the potential well formed by these two components. As will be mentioned later, the depth of the potential minimum is rather small so that particles in the boundary regions easily "evaporate" from the ordered regions due to the thermal energy. For the same reason, colloidal crystals melt easily when they are irradiated by strong laser light for a short period.

Rather small, fluorescent polymethylmethacrylate latex particles (MC-6, d: 1400 Å, ρ_e: 0.75 μC/cm^2) were studied by fluorescence microscopy to determine the particle distribution [26]. Dynamic light scattering was performed concurrently to obtain the structure factor $S(Q)$ directly. In Fig. 5.6(a), the computer-aided micrograph showing the distribution of particle centers is given. It clearly shows an ordered structure (pointed by the arrow) in a "sea" of disordered particles. The coordinate information was Fourier transformed by Eq. (5.2) to obtain the scattering profile, which is shown in Fig. 5.6(b). The radial distribution function $g(r)$ was directly determined from the coordinate information for 35,000 particles by an image data analyzer and is shown in Fig. 5.6(c). The $g(r)$ was further transformed into the structure factor $S(Q)$ by the relationship

$$S(Q) - 1 = n_p \int [g(r) - 1] \left[\frac{\sin(Qr)}{Qr} \right] 4\pi r^2 \, dr \qquad (5.3)$$

where Q is the scattering vector, r the distance, and n_p the number of particles per unit volume. The $S(Q)$ thus obtained was compared with the structure factor determined by the DLS measurement in Fig. 5.6(d). Clearly $S(Q)$ has two broad peaks. It should be recalled that this $S(Q)$ corresponds to a real-space structure containing randomly distributed particles and a localized ordered structure as large as about 5 μm in diameter (5/0.14 = 35 particle diameters). This implies that the general belief [27] that there is no correlation between molecules in liquids (characterized by a small number of peaks of the structure factor) when they are separated from each other by a distance of 5 molecular diameters is unwarranted. It is clear that solute distribution cannot be judged on the basis of the shape and the number of peaks of the structure factor. In other words, the so-called "liquidlike" structure factor does not imply a lack of a long-range order.

In this respect, it should be recalled that the structure factor is determined by (1) paracrystalline distortion, (2) the size of the ordered structure, and (3) the Debye-Waller effect, according to a paracrystal theory for three-dimensional

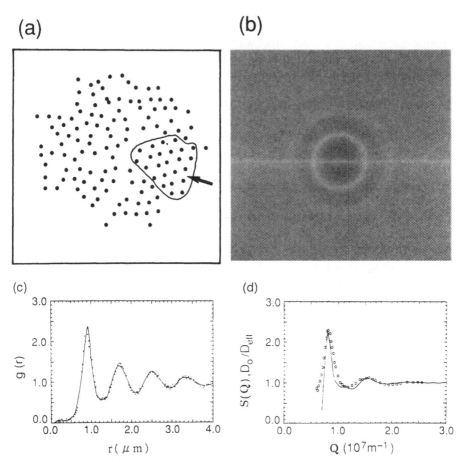

Figure 5.6 Structure in a latex dispersion containing localized ordered structure with free Brownian particles: Latex: MC-6, a polymethylmethacrylate latex containing Coumarin 6 (maximum absorption at 4580 Å), [latex]: 0.15%, T: 25°C. (a) Computer-treated micrograph of particle distribution. The localized ordered structure was conveniently surrounded by a closed curve and pointed by an arrow. Remember that the interparticle distance in the ordered region is about 1 μm. (b) The Fourier pattern obtained from (a). (c) $g(r)$ determined by measuring interparticle distances for 35,000 particles. (d) Interference function D_0/D_{eff} determined by the DLS measurement at 6328 Å (open circles) and the structure factor $S(Q)$ (curve) derived from the $g(r)$ shown in (c) [26].

cubic systems developed by Matsuoka et al. [28, 29]. This means that the change of the scattering curve cannot necessarily be ascribed to a variation either in the size of crystalline regions or in the distortion. It might be possible that a large, highly distorted crystal structure gives rise to the same structure factor as that of a small, less distorted structure. Due attention must be paid to a discussion of the crystal growth based on the time evolution of the scattering profiles.

5.3.5 Crystal Growth

As mentioned, the size of the localized ordered structure varies with time. This dynamic aspect was studied by Yoshida et al. using video imagery and 2D Fourier transformation, as discussed in Section 5.1.2 [30]. The procedure was as follows: First, latex dispersions were purified, and the ordering of the latices was confirmed by microscopy. Then NaCl was added to the purified dispersions (about 10^{-4} M in the present case) to destroy the ordered structure completely. A measured amount of well-washed, mixed-bed ion-exchange resin beads was then added to the dispersion, and the microscopic observation was started. The concentration of the added NaCl decreased rapidly after the onset of deionization. After 10 min, the conductance became almost equal to that before the salt addition. The coordinates of the centers of gravity of the particles were determined by a digital image processor. This information was 2D Fourier transformed to determine the particle spacing (R_{exp}) in the ordered structures. Furthermore, the coordinate information was used by a work station to perform elementary unit analysis and radial distribution function (RDF) analysis. In view of the fact that, under the present conditions, FCC symmetry was often found microscopically, a regular triangle with a side of R_{exp} (1 ± 0.15) was defined as an elementary unit of the ordered structure, and all particle pairs having a distance of $R_{exp}(1 \pm 0.15)$ were searched out and picked up. A localized structure consisting of the elementary units with common sides was defined as a cluster, and its number and size were also analyzed by computer. By measuring the distances between 4,000 and 10,000 particles, the 2D radial distribution function $g(r)$ was determined directly.

Figure 5.7 shows the time evolution of the number fraction of the particles forming various cluster sizes. Clusters consisting of three particles (elementary units) were formed first, and then larger clusters (five-particle clusters) followed. Finally clusters of 21 and more particles were organized. It is clear that, after 2 h, the fraction of three-particle clusters decreased and that of the larger clusters increased. The larger clusters grew at the expense of the smaller ones; that is, Ostwald ripening occurred [31]. *This tendency is understood without introducing ad hoc assumptions if one considers the presence of an attractive interparticle interaction.* Because of this attraction, particles inside the ordered structure are more strongly stabilized than particles in the boundary regions, since the latter are attracted only by interior particles, not from all directions. From this difference the situation arises that the total free energy of the system tends to decrease by losing a large number of small clusters (thereby decreasing an unfavorable contribution from a less stable surface) and producing a small number of large clusters.

The particle distribution and the elementary unit distribution are shown together with the directly determined RDF in Fig. 5.8. The $g(r)$ at 20 min displays the trend typical for completely noninteracting systems. The corresponding particle distribution looks homogeneous though it still contains a small number of elementary units. On the other hand, the $g(r)$ at 180 min is typical of highly

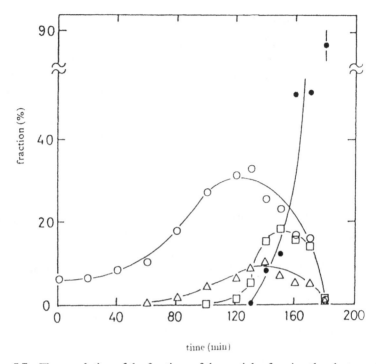

Figure 5.7   Time evolution of the fractions of the particles forming the clusters of three particles (○), five particles (△), 10-20 particles (□), and more than 21 particles (●). Latex: N300, [latex]: 1.0%, temp.: 25°C, solvent: H_2O [30].

"crystalline" states. At 130 min, a liquidlike RDF is seen, whereas clusters as long as 4 μm (equivalent to 12.5 particle diameters) are found. Here again the liquidlike $g(r)$ does not imply a lack of long-range order. It may also be mentioned that the use of D_2O-H_2O did not appreciably affect the RDF, as far as the time span covered is concerned.

5.4 Study of Internal Structure by Confocal Laser Scanning Microscopy: Voids, Ordered Structures, and Positive Adsorption of Particles

When the observed interparticle spacing (R_{exp}) is smaller than the average spacing (R_0), a simple stoichiometric consideration shows that there must exist regions of a high particle density and those having a low density in the dispersions. As an example of the latter case, it is not difficult to find void structures in which practically no particles can be found. Hachisu et al. were the first to report such a structure by microscopy [4]. Thereafter, Ise et al. [32] and Kesavamoorthy et al. [33] confirmed the presence of void structures for polystyrene-based latex

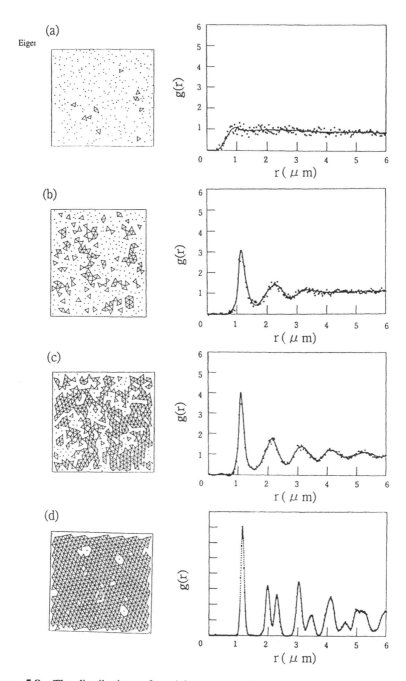

Figure 5.8 The distributions of particle centers and elementary units with the corresponding 2D radial distribution functions directly determined at (a) 20, (b) 130, (c) 160, and (d) 180 min after deionization was started. Latex: N300, [latex]: 1.0%, temp. 25°C, solvent: H$_2$O [30].

particles in water by using a metallurgical microscope. Because of the refractive index difference between the solute particles and solvent, only particles in the region near the dispersion-cover glass interface (at most 10 μm) could be observed by this technique. It was therefore possible that the void structure is an artifact due to an unclean glass surface. Ito et al. employed a confocal laser scanning microscope (LSM), which made it possible to study particles inside the dispersion [19, 34-35]. Figure 5.9 shows the time evolution of the void structure in apparently homogeneous dispersions at three different horizontal planes.

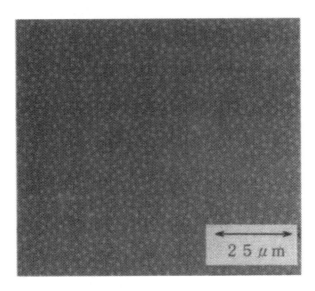

(a)

Figure 5.9 Confocal laser scanning micrographic image of void structures in a horizontal focus plane in a latex dispersion. (a) Random distribution of latex particles before the void structures developed. The latex particles are seen as white dots. (b) Void structures at various times and distances from the dispersion-cover glass interface. Direct photographs were image processed to obtain stronger contrast between particles and background. This processing caused distortion (for example, in the particle size) in the final images. The scale given applies to all nine pictures. Latex: N1000 (d: 9600 Å, ρ_e: 12.4 μC/cm^2), [latex]: 2%, dispersant: D$_2$O-H$_2$O. The measurements were made as follows: First the latex dispersion was extensively washed by ultrafiltration, and purified ion-exchange resin particles were added. Then the dispersion was brought into a Pyrex cell together with Bio-rad ion exchange resin particles. The homogeneity of the particle distribution at this stage was confirmed by LSM. After the dispersion has been left standing for 24 h, void formation was observed. The void-containing dispersion was again shaken, randomizing the particle distribution, as is seen from (a). The dispersion was then allowed to stand on the micrograph platform (this was defined as time $t = 0$). Thereafter the LSM micrographs were taken of particles that were sufficiently distant from the resin particles (> 0.5 cm). The density was matched by selecting a D$_2$O–H$_2$O mixture. The dispersion temperature was kept constant with a thermostatted air bath [35].

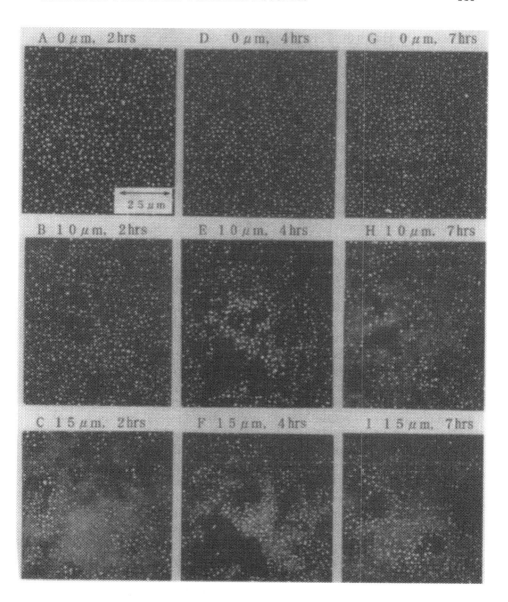

(b)

It can be concluded from Fig. 5.9 that void structures can be more easily found inside the dispersion rather than in the region near the dispersion-cover glass interface. This supports the view that the void formation has no bearing on the wall effect. Second, the voids may be fairly large; for example, at 15 μm inside and at 4 h, a void of about 20×30 μm^2 was seen. Another void with a cross section of 150×50 μm^2 was observed, whose vertical length was about 50 μm (See Fig. 6 in Ref. 19). This size is too small for naked eye observation and too large to study by scattering techniques. This could be one reason why void structures in latex dispersions have not been considered previously. One of the reasons why the void structure could be more easily found inside the dispersion would be the positive adsorption of anionic particles near a negatively charged glass surface, as will be discussed in the following.

Another observation to be noted is that the void structures observed for latex N1000 discussed in Refs. 19, 34, and 35 were found to coexist with Brownian particles, whereas those reported previously [16, 32, 33] were observed together with ordered structures. Although other factors cannot yet be ruled out, it seems plausible that the large diameter of N1000 slows down the particle motion so that its ordered structure could not be formed on the time scale of the void observation. As a matter of fact, Ito et al. observed that macroscopically discernible void structures are formed in coexistence with the ordered structure after the dispersion (shown in Fig. 5.9) had been standing for 2 months [36]. The existence of such a structural inhomogeneity clearly indicates the presence of an attractive interaction between the latex particles. If only a repulsive interaction were prevailing, the void structures could not be maintained, since the particles at the boundary of the void must then be pushed into the voids. Thus the void structure is strong evidence for the interparticle attraction. If the attraction is strong enough, void structures may be observed for other systems as well. Actually, Matsumoto et al. [37] and Ringsdorf et al. [38] reported similar structural inhomogeneities in Langmuir-Blodgett films by using an electron microscope and an atomic force microscope. It seems that these inhomogeneities are one of the characteristics of ionic systems, though the time scale varies from solute to solute.

Another interesting observation by the LSM is the concentration gradient of latex particles near the dispersion-cover glass interface. Although this feature has been reported earlier [19], it is worth discussing again with reference to more recent observations. Figure 5.10 shows the principle of the LSM observation of crystal growth. As mentioned, the coordinate information was Fourier transformed to determine the interparticle spacing (R_{exp}) in the ordered structure, with which the elementary units were drawn. The R_{exp} did not change with depth in the range studied (1 to 40 μm). As is clear from Fig. 5.10, the cluster size became smaller the deeper the focal plane was shifted inside the dispersion because the crystallization proceeded, layer by layer, from the region close to the interface into the internal region. This anisotropy is in qualitative agreement with the conclusion from Kossel line analysis [39] that colloidal crystal growth can be classified into two stages, namely the layer structure stage and the cubic structure stage. In other words, Fig. 5.10 indicates that the crystallization at 6 h is still in

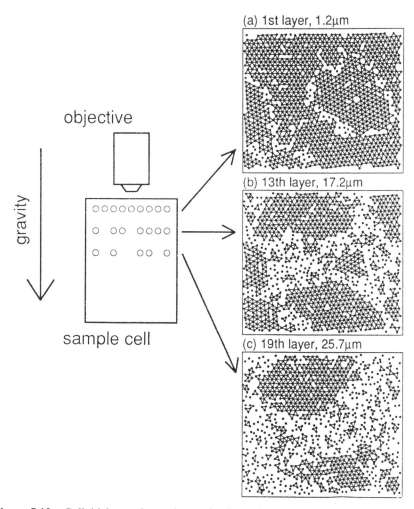

Figure 5.10   Colloidal crystal growth near the dispersion-cover glass interface as studied by a confocal laser scanning microscope. Computer-assisted micrographs showing particle distribution and the elementary units (▲) of ordered structures on three focus planes at various depths from the top of a latex dispersion under a density-matching condition. Latex: N300, [latex]: 0.7% in a D_2O–H_2O mixture. Photographs were taken 6 h after crystallization started. The circles in the sample cell denote latex particles [19].

the layer structure stage, in which the correlation between the layers is not strong. It seems that the anisotropy is partly due to the positive adsorption of negatively charged latex particles in the region close to the negatively charged glass surface. Though very strange at first sight, this adsorption is real if one compares the particle numbers at the three focal planes shown in Fig. 5.10. Since the LSM observation was carried out under the density-matching condition, the number concentration on a focal plane is expected to be independent of the depth. This is not the case. A similar adsorption was confirmed by the neutron reflection study of Thomas et al. for an ionic surfactant solution [40] and by the Ito et al. microscopic study for latex particles [41]. According to Thomas et al., the structure of the adsorbed layer of tetradecyltrimethylammonium bromide at a concentration 50 times the critical miceller concentration consists of the usual monolayer of surfactant, an aqueous layer, and a further layer of surfactant with about a 5% excess over the bulk solution. The amount of surfactant in the subsurface layer and its dimensions strongly suggest that this layer consists of micelles at about double the bulk concentration. The adsorption indicates that there is an effective electrostatic attractive interaction between the ionic particles or micelles and the similarly charged wall. The positive adsorption discussed previously is essentially due to the same mechanism as the attraction between like charged latex particles.

5.5 Ultra-Small-Angle X-Ray Scattering of Colloidal Systems

5.5.1 Structural Inhomogeneities in Colloidal Dispersions

Recent technical advances are revealing structural inhomogeneities in various systems such as Langmuir-Blodgett films and superconductors [37, 38, 42]. Colloidal dispersions are not an exception. In addition to the direct microscopic observation discussed, Yoshiyama and Sogami reported photographic recording of the coexistence of crystalline regions and random regions after allowing a latex dispersion to stand for several months by using the Lang method devised in X-ray topography [43]. They further identified the lattice system of a crystalline region by taking backward Kossel images using a focused laser beam. The structural inhomogeneities in colloidal systems are unambiguous.

The specific problem of colloidal systems is the slow time scale, which is caused by the large dimensions of the colloidal particles. Because the particle motion is slow in comparison with simple ions, for example, one can see the void structures or the localized ordered structures with a microscope. For the same reason, they persist over a period long enough for thermodynamic measurements. This feature has not been considered before in theoretical approaches, so that more detailed investigation is required if a thorough understanding of the colloidal phenomena is to be achieved. The microscopic method is reliable, but it furnishes only local information. To get statistically significant information, we have to resort to scattering methods. However, the dimensional scale is fairly large so that conventional scattering techniques are not useful. The extension of the measuring range of scattering methods to a larger length scale is necessary, and hence an ultra-small-angle X-ray scattering (USAXS) setup has been fabricated. In the following first the dimensional problems are considered along with a description of the USAXS apparatus, and then a few preliminary results on colloidal systems are discussed.

5.5.2 Preliminary Determination of Cluster Size in Sodium Polystyrenesulfonate Solutions by Neutron Scattering

Small-angle X-ray scattering produced a single broad peak for solutions and dispersions of ionic polymers including biopolymers and colloidal particles, when solutes are electrically charged and the peak was attributed to an ordered structure of the ionic solutes [22]. This interpretation was substantiated by a paracrystal theory of cubic lattice systems [28, 29]. According to the theory, when the paracrystalline distortion becomes large, higher-order peaks become indiscernible, while the peak position itself is not largely affected. Thus the single broad peak implies highly distorted paracrystalline structure, provided the Debye-Waller effect and the size of the structure remain unchanged. Therefore it is legitimate to calculate the spacing between macroions (strictly the distance between lattice planes) in the ordered structure from the position of the single peak by using the Bragg equation. The Bragg spacing (R_{exp}) thus obtained was found to be smaller than the average spacing (R_0) obtained from the concentration [22]. The difference is especially marked when the macroions are highly charged, their concentration is low, and the molecular weight of the solutes is not low. For example, a fractionated sodium polyacrylate of a degree of polymerization of 1470 had a R_{exp} of 88 Å at a polymer concentration of 0.02 g/ml and 22°C, whereas R_0 was 222 Å. Since the experimental uncertainty in the spacing was about 10%, the difference (between 88 and 222 Å) is physically significant. This inequality relation implies that the ordered structure under consideration is not space filling, but localized. If the above data of the spacing are correct, the total volume of the ordered structures must be $(88/222)^3 (= 0.06)$ of the total solution volume, although the volume of the individual ordered structure is not known. In other words, a large portion of the solution volume is occupied by voids containing no macroions. Thus it appeared to be important to estimate the volume of the localized ordered structure and also the size of the voids.

In this respect, the following preliminary information obtained for sodium polystyrenesulfonate in D_2O is useful [44]. The neutron scattering showed a steep upturn at low angles, as shown in Fig. 5.11. This upturn is not new, but it was identified by Boue et al.[45], who attributed it to impurities in the solutions. This interpretation appears to have been recently retracted [46]. The upturn could also arise due to the presence of a large density fluctuation in the solution. On the basis of the micrographic images of apparently homogeneous dispersions (See Fig. 5.5), one finds that the localized ordered structures of polystyrenesulfonate ions do exist. The density difference between the localized ordered structures and the rest of the solution or dispersion gives rise to the upturn. Following this interpretation, the Guinier method was applied to the upturn, although its application should be limited to homogeneous entities. Thus, the following data should be judged as giving qualitative, not highly precise, information. From the slope of the Guinier plots in the very small angle region, the radius of gyration, R_G, of the scatterer was obtained and is listed in Table 5.4. It looks as if the cluster size (R_G) and the number of macroions (N_c) in one cluster become larger

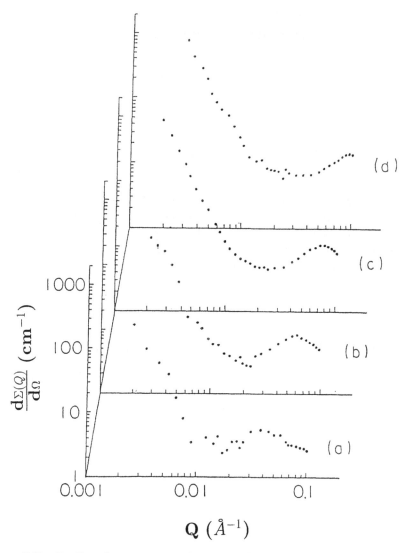

Figure 5.11 Small-angle neutron scattering curves of sodium polystyrenesulfonate D$_2$O solutions at different polymer concentrations without added salt. [Polymer]: (a) 0.01, (b) 0.02, (c) 0.04, (d) 0.08 g/ml [44].

with increasing polymer concentration, while the R_G stays almost constant and the N_c value decreases with increasing salt concentration. Although the details should be referred to elsewhere [47], similar trends were also concluded from the Debye-Bueche analysis. The polymer concentration dependence is easy to understand while the salt independent, constant cluster size is unexpected. The

Table 5.4 SANS data on NaPSS-D$_2$O solutions

[NaPSS] (g/ml)	[NaCl] (M)	R_G (Å)	R_{RG} (Å)	R_{exp} (Å)	R_0 (Å)	N_c
0.01	0	407	525	166	255	130
0.02	0	516	666	103	202	1100
0.04	0	686	886	85	160	4700
0.08	0	630[a]	813	64	128	8600
0.04	0.05	664	857	94	-	3200
0.04	0.1	683	882	132[b]		1250[b]
0.04	0.3	671	866			
0.04	0.5	631	814			

R_{RG}: radius of the spherical cluster obtained from R_G,

N_c: the number of macroions in one cluster calculated from R_{exp} and R_{RG},

[a] The linearity of the Guinier plot was not so satisfactory.

[b] The Bragg peak was not sharp.

tendency of N_c to decrease with increasing salt concentration is acceptable, since the intermacroion attraction is progressively weakened.

As is well known from the Babinet principle, it is impossible to determine by simple scattering experiments whether the density difference is positive or negative. In other words, the upturn cannot be ascribed solely to clusters of a high polymer density. There remains a possibility that void structures have caused the upturn. As mentioned for latex dispersions, void formation is real. Thus it is possible to analyze the SANS upturn for NaPSS-D$_2$O solutions in terms of void structures. Without going into detail, we mention here that, by assuming a model containing ordered macroions and spherical voids, the number of voids turned out to be in the range between 840×10^{12} and 52×10^{12} ml^{-1}, decreasing with increasing polymer concentration (from 0.01 to 0.08 g ml^{-1}) [47]. The R_{exp} estimated from the observed peak position was consistent with earlier SAXS results for the same polymer and also smaller than the R_0, which substantiates the two-state structure.

In the case of latex particles, no upturn has been found so far by light scattering and SAXS; the lowest Q value studied seems to be 5×10^{-4} Å^{-1} by Brown et al. [48] This corresponds to a scatterer size of about 1 μm. If ten of their 0.05 μm particles form an ordered domain with an interparticle distance of 0.5 μm in one direction, the domain size is already too large to examine by the scattering technique. This is a very low estimate according to the present microscopic observations. In other words, one would not be able to use light scattering and SAXS techniques for determining the cluster size in latex dispersions.

5.5.3 Study by Ultra-Small-Angle X-Ray Scattering: Rocking Curve and Calibration by Standard Particles

From the preceding considerations, ultra-small-angle X-ray scattering (USAXS) is expected to be more promising for colloidal systems. The USAXS apparatus I and II were constructed in Kyoto [49] and in Fukui [50], respectively, according to the Bonse-Hart principle [51]. Detailed accounts of the apparatus I and preliminary investigations of various systems were given in Ref. [52]. Figure 5.12 shows the rocking curves of the apparatus II with Ge and Si crystals [50]. The rocking curve of apparatus I, reported in an earlier paper [52], is similar to that in Fig. 5.12 with an intensity five times weaker. Distinct maxima were observed in the region between 100 and 600 seconds of arc for polystyrene latex particles (as a powder and in an ethanol dispersion) of a diameter of 3000 Å [49]. Consider first, the size of the particles, which could not be studied by previous SAXS or light scattering methods. Their peak positions were in excellent agreement with

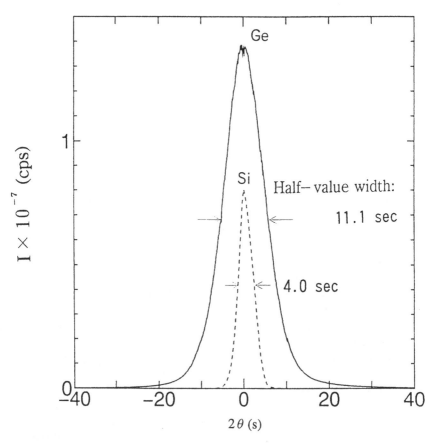

Figure 5.12 Rocking curves of the apparatus II with Ge and Si crystals. X-ray generator: Rigaku Rotaflex RU-300, source power: 60 kV, 300 mA, wavelength: 1.540562 Å.

those of the theoretical scattering intensity of an isolated sphere (radius a), which can be expressed by the following equations:

$$I(Q) = \text{const} \times a^6 \Phi^2(Qa) \tag{5.4}$$

$$\Phi(Qa) = 3[\sin(Qa) - Qa\cos(Qa)]/(Qa)^3 \tag{5.5}$$

When the distribution of the particle radii is characterized by the distribution function $P(a)$, the intensity $I(Q)$ is given by:

$$I(Q) = \text{const} \times \int P(a)a^6 \Phi^2(Qa)\,da \tag{5.6}$$

If the distribution is Gaussian, $P(a)$ is given as follows:

$$P(a) = \frac{1}{\sqrt{2\pi}\sigma} \exp\left(-\frac{(a - a_0)^2}{2\sigma^2}\right) \tag{5.7}$$

where a_0 is the number-average radius and σ is the standard deviation. The z-average radius a_z is given by:

$$a_z = \frac{\int P(a)a^7\,da}{\int P(a)a^6\,da} \tag{5.8}$$

Since a_z satisfies the following relation

$$a_z \times (Q \text{ values at the maximum position}) = 5.76, 9.10, 12.3, \ldots \tag{5.9}$$

the radius value estimated from the scattering maximum is a_z.

Figure 5.13 gives smeared and desmeared scattering curves for colloidal silica powder [50]. The desmeared curve gave satisfactory agreement with the theoretical curve for an isolated sphere, when 1480 Å and 7.5% were adopted for a_z and σ, respectively. The estimated size was in a good agreement with the value provided by the producer (d: 3000 Å).

Time Evolution of USAXS Curve of Colloidal Dispersions and Upturn at Low Angles

Figure 5.14 is the time evolution of the scattering curves of a latex dispersion in a water-ethanol mixture containing ion-exchange resin particles. As deionization proceeds, interparticle-interference peaks appear, indicating that the ordered structure is growing. It is interesting to note that the first peak splits into two peaks 2 weeks after sample preparation. At this time, the second peak is higher than the first peak. The structure at 4 weeks may be face-centered cubic (FCC) according to the relative position of the first and second peaks ($3^{1/2}$: $4^{1/2}$). The Bragg spacing (R_{exp}) was found to be 5600 Å, whereas the average interparticle spacing (R_0) was 5700 Å. The agreement seems reasonable for such low-charge-density particles (ρ_e: 0.31 μC/cm^2). (Remember that R_{exp} becomes smaller with

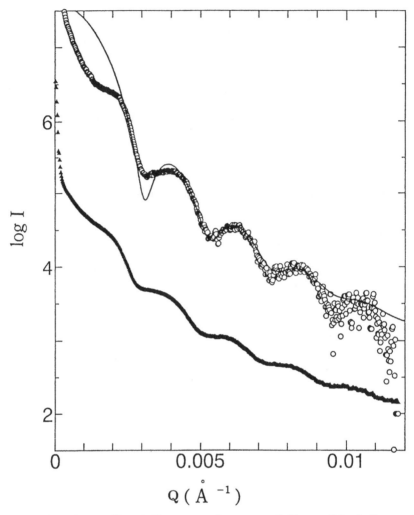

Figure 5.13 Ultra-small-angle X-ray scattering curves of silica particles in the powder state. Triangle: smeared; circles: desmeared. The data were obtained by using the USAXS apparatus I (50 kV- 200 mA) at 25°C. The silica particles were the product of Nippon-Shokubai Co. Ltd. (lot number S-03518C, d: 3000 Å).

increasing number of charges on the particles, though this is in contradiction to the widely accepted view that two similarly charged particles simply repel each other. Though the details will be discussed later, the role of counterions in between the particles is ignored in the repulsion-only assumption.) The change in the heights of the first two peaks may be construed as indicating that the three curves do not reflect the same structure. Though further study is in progress, one

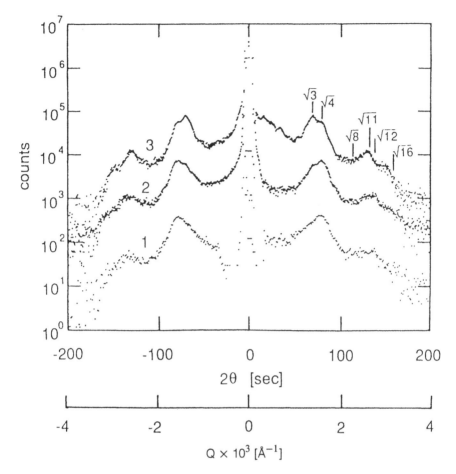

Figure 5.14 Time evolution of the USAXS curve for a 3.7% latex dispersion in ethanol/H_2O (2:3) mixture. Latex: SS-121 (d: 2100 Å, ρ_e: $0.31\mu C/cm^2$). In the bottom of the USAXS capillary cell, ion-exchange resin particles were introduced to remove ionic impurities continuously from the dispersion. As deionization took place, the particles formed an ordered arrangement. Curves 1, 2, and 3 were obtained at 1, 2, and 4 weeks after onset of the deionization. The curves were shifted vertically by a factor of 10 [49].

of factors to be considered may be the evolution from a two-dimensional hexagonal layered structure to the normal FCC structure, as was made clear by the Kossel line analysis of Sogami and Yoshiyama [39]. A similarly complicated diffraction pattern was obtained for a silica particle dispersion and will be discussed in the following.

Due to a higher contrast of electron densities than that between polystyrene and water, polymethylmethacrylate-based latex particles can be studied in H_2O at much lower concentrations by the USAXS apparatus [53]. Since the overall rate of crystallization is smaller at lower particle concentrations according to Yoshida et al. [30], no scattering peaks could be observed in USAXS measurements performed a short time after onset of deionization at low latex concentrations [54]. Thus deionization of a 1.3% dispersion of polymethylmethacrylate latices was effected for 6 weeks and the USAXS curves were measured on days 0, 5, and 9 thereafter. On the ninth day, interference peaks were observed, consistent with an FCC symmetry. R_{exp} was estimated by assuming this symmetry and is compared with R_0 in Fig. 5.15. R_{exp} is found to be approximately equal to R_0 at first and decreases with time, in other words with decreasing ionic strength. This trend implies that there must be an attraction between similarly charged particles, which is enhanced more and more with decreasing salt concentration, since the shielding effect of coexisting salt is accordingly weakened. The spacing R_{exp} thus becomes smaller. This trend is in contradiction to the widely accepted repulsion-only assumption [55]. This theory cannot be valid: if it were correct, the R_{exp} must increase with decreasing salt concentration because the shielding effect becomes less and less influential.

It is worth examining why the crystallization proceeds slowly in Fig. 5.15 in comparison with cases studied by microscopy (Fig. 5.7). There are several factors to be considered: In the USAXS study, ion-exchange resin particles were

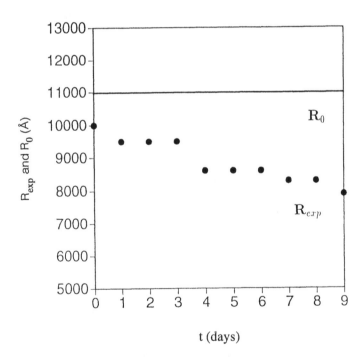

Figure 5.15 The interparticle spacing R_{exp} with time. Latex: MC-8 (polymethylmetha-crylate-based latex, d: 3000 Å, ρ_e: 1.9 $\mu C/cm^2$), [latex]: 1.3% [54].

placed in the lower part of a long glass capillary so that deionization might take place slowly, or the microscopic method provided information in regions near the glass-dispersion interface where crystallization is more rapid than in the interior, whereas in a USAXS study information from the bulk of dispersion is obtained. Certainly, these problems should be investigated in a systematic manner.

It is pertinent to point out here that an upturn of the scattering profile was observed by the apparatus I for a 1.3% water dispersion of polymethylmethacrylate-based latex (MC-8, d: 2900 Å, ρ_e: 1.9 $\mu C/cm^2$) on days 0, 6, and 12 after 6-week purification by ion-exchange resin [56]. Figure 5.16 shows the data on the sixth and twelfth days. The radius of gyration R_G of the cluster computed from the Guinier plot was 2.2 μm on day 0, and increased to 2.6 μm on day 12. This implies that the cluster grew. On the other hand, the relative position of the first and second peaks was found to be $3^{1/2}:4^{1/2}$, indicating that the lattice system was FCC. R_{exp} was then found to be about 1 μm when the USAXS measurement

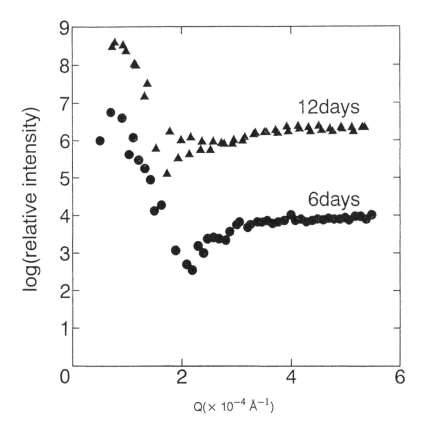

Figure 5.16 The upturn of the USAXS curves of a latex dispersion in water. Latex: MC8 (polymethylmethacrylate-based latex, d: 2900 Å, ρ_e: 1.9 $\mu C/cm^2$), [latex]: 1.3% [56].

was initiated, which was very close to R_0, and R_{exp} decreased to 0.79 μm on the ninth day. This implies that two-state structures grew even in the interior part of the dispersion. From R_{exp} and R_G, one spherical cluster is concluded to contain on the average about $(2.6/0.8)^3$ ($= 36$) particles at 1.3%. Further study is in progress under various experimental conditions.

Figure 5.17 is the scattering curve obtained recently by USAXS apparatus II for a 3.76% water dispersion of silica particles (KE-P10W of Nippon Shokubai Co., Ltd., Osaka; stated diameter: 0.10 μm). A silica dispersion was dialyzed against ultrapure water for 16 d, introduced into a glass capillary of diameter of 2 mm together with Bio-Rad ion-exchange resin particles, and kept standing

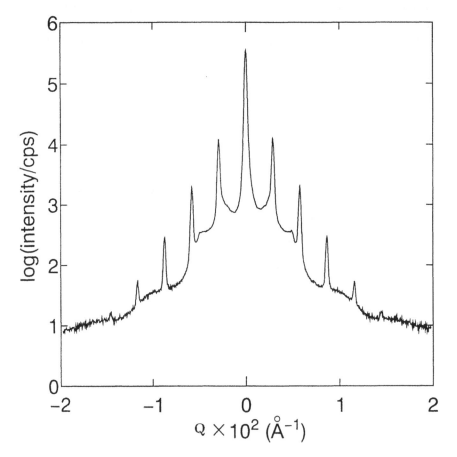

Figure 5.17   USAXS curve for a silica dispersion. Silica particle: KE-P10W (Nihon-Shokubai) (stated diameter: 1000 Å, ρ_e: 9.70 μC/cm^2), [silica]: 3.76%. USAXS apparatus II with a power source of 60 kV, 300 mA. The silica particle dispersion was dialyzed against Milli-Q water for 16 d and introduced into a capillary cell (inner diameter: 2 mm) with Bio-Rad ion exchange resin particles, which was stopped with silicon rubber. The dispersion was allowed to stand for 84 d and then studied by the USAXS apparatus.

for 84 d. The particle radius and its standard deviation were 570 Å and 8%, respectively, which were determined by fitting the observed USAXS profile under salt-containing conditions with the theoretical prediction for isolated spheres. It is to be noted that five orders of diffraction were observed at multiple angles of 60 degrees when the capillary tube was rotated around the capillary axis. To authors' knowledge, this is the first observation of several orders of diffraction from colloidal crystals. Although quantitative analysis is in progress, the sharp peaks appear to suggest the presence of fairly large and highly ordered structures. These are too large to determine by the Hosemann plot.

5.6 Repulsion-Only Assumption or Long- Range Attraction-Repulsion Assumption?

5.6.1 Experimental Facts Supporting the Presence of a Long-Range Attraction

Although the controversy on the interparticle interaction has been discussed earlier [16, 19], it is readdressed in this section with reference to recent developments.

First the main points must be clarified: (1) There exists an electrostatic attraction between similarly charged particles or macroions, in addition to the widely recognized repulsion; (2) the attraction manifests itself at very large distances (about 1 μm for the latex particles under consideration in the present chapter); and (3) this attraction is generated through the intermediary of counterions in between the particles or the macroions, according to present understanding.

The experimental facts, which have led to the recognition of the electrostatic attractive interaction in addition to the widely recognized repulsion, are as follows:

1. Two-state structures in apparently homogeneous dispersions or solutions;
2. Voids in colloidal dispersions;
3. Ostwald ripening in colloidal crystal growth.

These three phenomena have been discussed in detail in earlier papers [16, 19], and hence one can now restrict discussion to recent developments.

4. Positive adsorption of charged entities around similarly charged surfaces.

With the help of a laser scanning microscope, it has been observed that the concentration of negatively charged particles is higher in the region near a negatively charged glass surface and decreases with increasing distance from the surface [19], as shown in Fig. 5.10. While this phenomenon is being studied more quantitatively for latex dispersions [41], Thomas et al. found a similar effect for ionic surfactant solutions above the cmc in a neutron reflection study [40]: The structure of the adsorbed layer at a concentration 50 times the cmc was demonstrated to consist of the usual monolayer of surfactant, an aqueous layer without surfactant (55 Å thick), and a further layer of micellar surfactant (about 90 Å from the center of the monolayer) at about double the bulk concentration. *The net positive adsorption of micelles in the region below the monolayer indicates that there is an effective interaction between micelles and the "wall,"* according to Thomas et al. Although hydrophobic interactions cannot be ruled out since rather short distances are under consideration, another plausible interpretation is that the positive wall (monolayer) attracts the cationic spherical micelles through the intermediary of counterions in the aqueous layer, in a manner similar to the particle-particle attraction [19].

5. Reentrant phase separation.

Arora et al. discovered the separation of originally homogeneous dispersions into concentrated and dilute phases in a certain range of salt concentration [57]. This reentrant phase separation was studied by the Monte Carlo simulation technique [58] using the Sogami potential [59], which contains a short-range repulsion and a long-range attraction. The radial distribution function obtained showed peaks at distances much shorter than R_0, which suggested the possibility of either clustering or formation of a condensed phase. The size of the cluster and its distribution were derived. This computer simulation was clearly in agreement with observation, namely (1) phase separation upon deionization (at low salt concentrations), (2) an increase of the volume occupied by the dense phase as the salt concentration is further lowered, and (3) an eventual reentrant transition under much lower salt conditions. It is easily anticipated that these phenomena cannot be accounted for in terms of the widely accepted repulsion-only assumption, namely, a purely repulsive potential such as the Yukawa potential or DLVO potential. It is worth mentioning that, by this simulation, voidlike inhomogeneities could be reproduced (see Fig. 7 in Ref. 58).

6. Vapor-liquid condensation in colloidal dispersions.

Tata et al. [60] observed a novel phase transition in dilute colloidal dispersions, which is analogous to vapor-liquid condensation in atomic systems. When the salt concentration was lowered, the dispersion exhibited a concentrated phase macroscopically separated from the dilute phase. The structure factors $S(Q)$ were determined by light scattering, from which it was concluded that the dilute phase has no structural ordering whereas the concentrated one has a liquidlike order. The particle concentration, which was determined directly by the evaporation method, differed in the two phases by several times. Tata et al. concluded that the result can be considered to be direct evidence for an attractive component in the effective interaction potential. An alternate explanation for this phenomenon was presented by Palberg and Würth [61], which has been subsequently shown to be invalid [62].

5.6.2 Nature of the Attraction and the Sogami Potential

On the basis of the experimental data accumulated so far for latex dispersions, the dependence of the attraction on experimental conditions can be summarized as follows. For this purpose the position of the potential minimum (R_{exp}), which is created by the attraction and the short-range repulsion, is discussed.

An important point that should be mentioned here is the effective charge number on the colloidal particles or ionic polymers. It is a widely used practice to estimate the effective charge number (Ze) of latex particles by a renormalization procedure. In other words, Ze is determined so as to obtain the best fit of theoretical treatments to experimental results. Obviously this procedure would be acceptable if the theory under consideration has been proven to be correct. In reality, the best fit does not justify the theory nor the Ze value obtained. One can in principle measure the Ze directly by a transference measurement developed some time ago by Wall et al. [63]. One can then discuss the consequences of the theory using the Ze value thus measured, which is more reliable than the renormalized charge. According to the transference measurements of latex dispersions [64], the fraction of free counterions (f) (the ratio of dissociated counterions and to the number of analytical dissociable groups) was about 0.04 to 0.1 for latex particles and decreased with increasing surface charge density of the particles [65], and the f value for linear polyelectrolytes was about 0.4 [63, 64]. However, the transference measurements have not been frequently carried out, except in the case of latex dispersions without coexisting salt [64]. It is strongly desired to extend the measurements to various salt concentrations, dielectric constants, and temperatures, at which no transference measurements were done so that reliable Ze values have not yet been obtained. It would be useful to mention briefly relevant experimental results obtained for polyacrylate as follows: (1) The f value decreases with increasing analytical charge density (or degree of neutralization) [63, 66]; (2) it is substantially independent of temperature from 0 to 42°C [67]; (3) it is about the same for Na^+ and K^+ [67]; (4) the fractional number of associated counterions per ionizable group (j/s in the notation of Wall et al.) slightly increases with increasing concentration of NaCl [66]. It may be mentioned that these changes affect the effective charge number of macroions (latex paticles), which would in turn determine the intermacroion (interparticle) interaction through Ze and κ. Thus the experimental facts to be mentioned are correct, but their comparison with theory (including the Sogami theory) should be regarded as qualitative.

5.6.3 Salt Concentration Dependence

It is well recognized that latex particles form ordered structures in dispersion only at low salt conditions; the addition of a large amount of salt destroys the ordered structure. *The spacing between particles in the ordered structure, R_{exp}, has been found to decrease with increasing salt concentration as long as low salt conditions are maintained* [5, 68], but it cannot be measured when more

Table 5.5   Observed interparticle distance at various concentrations of coexisting salt

[NaCl] (M)	R_{exp} (μm)
0	1.27
1.71×10^{-6}	1.26
1.71×10^{-5}	1.13
6.84×10^{-5}	0.80
1.37×10^{-4}	No ordering

The spacing was measured by ultramicrograph. Its standard deviation was $\pm 0.01 \sim 0.03$ μm.

salt is added, since the structure is then destroyed. A typical example from an ultramicroscopic study [68] is shown in Table 5.5.

The trend of R_{exp} to decrease with salt concentration can be accounted for in terms of the DLVO potential. However, as the salt concentration further increases, its potential minimum becomes deeper and deeper as the position of the minimum becomes smaller and smaller; in other words, the ordered structure is more and more stabilized. This is not the case experimentally. The DLVO theory is thus not valid even qualitatively. On the other hand, the Sogami potential [Eq. (5.10)] [59] gives at least a qualitatively satisfactory explanation, which is expressed by

$$U_S(r) = \epsilon^{-1} \left(\frac{Ze \sinh(\kappa a)}{\kappa a} \right)^2 \left(\frac{1 + \kappa a \coth(\kappa a)}{r} - \frac{\kappa}{2} \right) \exp(-\kappa r) \quad (5.10)$$

where $U_S(r)$ is the Gibbs pair potential, ϵ the dielectric constant, Ze the effective charge of the particle, κ the inverse Debye screening length, and r the interparticle distance. As shown in Fig. 5.18, Eq. (5.10) predicts at a fixed Ze that, as the salt concentration (and hence κ) increases, the potential minimum becomes deeper and, after passing through a minimum, becomes shallower while its position shifts monotonically toward smaller distances. This tendency is in accord with the observation. [1]

5.6.4 Charge Number Dependence

When the charge number Ze is increased, R_{exp} becomes slightly smaller. This trend, which was surprising, was observed microscopically for a Dow latex, D1A92 (d: 4970 Å, ρ_e: 2.8 μC/cm^2) and SS-45 (d: 5000 Å, ρ_e: 13.3 μC/cm^2) [69]. Table 5.6 gives another set of data for smaller latices [68], which was obtained

---

[1] The measurements by Wall et al. demonstrated that the fractional number of associated counterions per ionizable group slightly increases with salt concentration [66]. This would imply a decrease in Ze. It is then suspected that the shift of the potential minimum position, or the interparticle spacing, with salt concentration is different from what was discussed in the text. Further systematic transference measurements are necessary.

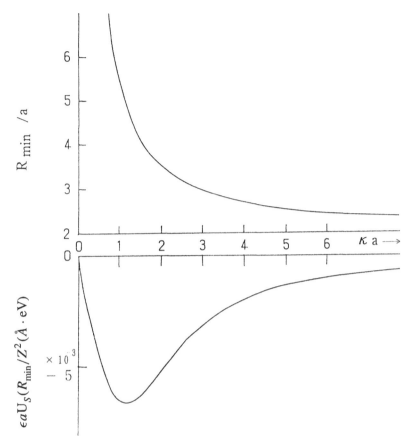

Figure 5.18 The position of the minimum of the Sogami pair potential ($R_{\min}$) versus κa (above) and the minimum value of the potential $\epsilon a U_S(R_{\min})Z^2$ versus κa (below) [59].

from the peak position of $S(Q)$ measured by dynamic light scattering. The same trend was observed for bovine serum albumin (BSA) by SAXS measurements at various pH values [70], for both positively and negatively charged BSA.

Table 5.6 Charge Number Dependence of the Interparticle Spacing ($R_{\exp}$) as Studied by Dynamic Light Scattering.

Latex	$2a(\mu m)$	ρ_e ($\mu C/cm^2$)	$R_{\exp}$ (μm)
L1	0.14	4.2	0.98
SS-31	0.14	6.4	0.94
L4	0.14	16.5	0.90

The tendency of R_{exp} to decrease with increasing charge number is difficult to accept in the framework of the DLVO theory, since the repulsion should then simply intensify with increasing charge number. On the other hand, the Sogami potential predicts the observed tendency. This can be understood from Fig. 5.18, since κa becomes larger as Ze increases. [2] It is reminded that the effective charge number determined by the transference measurements increases slowly with increasing analytical number of charges, while the f value was found to decrease sharply with increasing analytical charges [64]. Thus the charge number dependence discussed in the present section is more reliable than other cases.

5.6.5 Dielectric Constant Dependence

The microscopic measurements (Fig. 5.19) show that R_{exp} *increases with increasing ϵ* [9]. The same trend was found in a dynamic light scattering study of a rather small polymethylmethacrylate latex dispersion in water-ethanol mixtures [71]. Though the observed tendency is also predictable by the Sogami potential, a quantitative comparison is not possible since the Ze value has not been experimentally determined as a function of ϵ.

5.6.6 Temperature Dependence

The spacing has been measured for various latices by direct measurements on micrographs, by direct determination of radial distribution functions, by Fourier transforming the microscopic information, and by dynamic light scattering [9, 15, 26, 71]. It was found that R_{exp} *decreases slightly with increasing temperature.* It is important to mention that the DLVO potential predicts the decreasing tendency of R_{exp} as a result of the decrease in ϵ with increasing temperature. This led Asher et al. [72] to claim that the colloidal lattice contracts upon local heating as a result of repulsions arising from the surrounding unheated parts of the lattice. On the other hand, it has been shown recently that this effect is also predicted by the Sogami potential [73], as was demonstrated earlier [59]. It is recalled that the counterion association was found to be insensitive to temperature for ionic polymers by Wall et al. [66]. If this is the case with latex dispersions, the theoretical comparison mentioned is more meaningful. This problem will be discussed again in the following.

---

[2]Figure 5.18 shows that the potential well becomes shallower at high κa, implying that the ordered structure becomes less stable for high-charge-density samples. This trend is consistent with the observation that ordering phenomena can be less easily observed for high-charge-density particles (about $\rho_e \approx 10\ \mu C/cm^2$) than for low-charge-density ones.

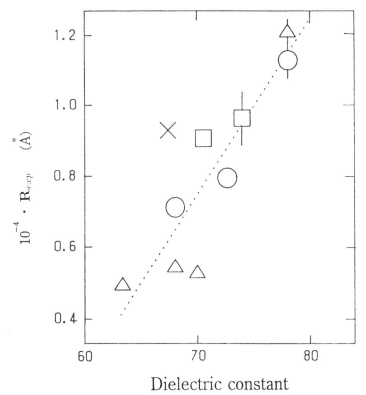

Figure 5.19   The interparticle spacing R_{exp} in the binary mixtures of ethylene glycol (o), methanol ($\times$), dimethylformamide ($\triangle$), and dimethyl sulfoxide ($\square$). Latex: G5301 (d: 3690 Å, ρ_e: 7.2 μC/cm^2), [latex]: 1.3%, room temperature. The dotted line represents the optimal relation between R_{exp} and ϵ [9].

5.6.7 Remarks on Earlier Arguments in Favor of the Repulsion-Only Assumption

It has been widely claimed that colloidal phenomena including crystallization can be described by a purely repulsive interparticle interaction. Until recently the repulsion-only assumption (or the DLVO theory) appeared to provide satisfactory agreement with observation. However, the recent experimental facts enumerated in Section 5.6.1 show that the repulsion-only assumption is not adequate. The mental exercise of replacing colloidal spheres by enlarged effective spheres, which is basically in line with the repulsion-only assumption, was demonstrated in Section 5.2.1 to lead to the unphysical conclusion that the Avogadro number may not be a universal constant. Hence one needs to understand why many investigators continue to believe the repulsion-only assumption. One of the main reasons was uncovered by Tata et al. [74], who carried out Monte Carlo and Brownian dynamics simulations to calculate the structure factor $S(Q)$ for charged sphere dispersions. It is remarkable to see that both the purely repulsive DLVO potential and the repulsive plus attractive Sogami potential give satisfactory agreement with the observed $S(Q)$. While the fact that these two potentials lead to practically the same $S(Q)$ is probably due to the shallowness of the potential well (only $k_B T$ per one degree of freedom in their simulation), this implies that the DLVO potential cannot be claimed to be the exclusively correct potential. Ito et al. also demonstrated that the elastic modulus of colloidal crystals, which was asserted to be explicable in terms of the Yukawa potential by Lindsay and Chaikin [75], can be almost equally satisfactorily accounted for by using these two potentials [76]. Most recently, it was further made clear that the experimental data reported by Asher et al. [72] on photothermal compression of colloidal crystals can be better explained by the Sogami potential than by the DLVO potential [73]. Actually the Sogami potential gave a more reasonable explanation for the non-space-filling nature of the crystals with an experimentally realistic value for the effective particle charges. [3] These considerations indicate that a dialectical error as discussed by Eigen [77], was involved in previous arguments in favor of the repulsion-only assumption. In other words, the repulsion-only assumption was presumed to be correct without disproving its alternate interpretation, namely, the long-range attraction-repulsion assumption. It should be re-emphasized that, in order to assert the validity of the repulsion-only assumption, it would have been logically necessary to assume the Sogami potential or a Sogami-like potential (containing an attractive tail) and to demonstrate its inadequacy in comparison with observed facts, not with the apparently established concept. Furthermore, the repulsion-only assumption is imperfect in that it fails to account for recent observations. The advantage of the long-range attraction-repulsion assumption can be appreciated only if one takes a global view of the properties of colloidal systems without ignoring recent developments.

The Sogami theory was criticized by Overbeek, who claimed that, if the role of the solvent is taken into account, the attractive tail in the theory exactly disappears [78]. This criticism turned out to violate the Gibbs-Duhem relation for multicomponent systems [19, 79] and to lead to the implausible conclusion that "there is no energy associated with the electrical double layers" [80]. Furthermore, if Overbeek's argument were correct, it would assert that the remaining purely repulsive term, which Overbeek appears to regard as correct, is inconsistent with the Gibbs-Duhem equation. Obviously it is incorrect. [4]

---

[3]In the case of the Asher et al. sample, the fit of the observed compression data to the Sogami theory gave $f = 0.13$, which is reasonable for latices in light of the transference data [64], and the renormalized charge number based on the DLVO potential corresponded to $f = 0.5$ [72]. The latter figure is not acceptable for latices, according to the transference experiments [64]. This makes the arbitrariness of the renormalization procedure very clear.

[4]It is worth remembering that the DLVO theory has not been tested in light of thermodynamic data (such as the activity or activity coefficient of solute and solvent, the heat of dilution, etc.). To the contrary, the Debye-Hückel theory has been thoroughly examined in terms of these quantities.

Finally, a few more points are necessary about the attraction. Recently, direct force measurements are claimed to be possible. Atomic force microscopy (AFM) was used to measure the force between an ionic colloidal particle and a charged plate by Larson et al. [81], who detected a repulsive interaction at distances up to 500 Å for titanium dioxide. The distance range covered was too narrow, although one anticipates tremendous difficulties at large distances. At such short distances the Sogami theory also predict a repulsive interaction. This means again that it is not possible to decide from the measurement whether the measured repulsion is the repulsive part of the DLVO potential (as the authors believe) or that of the Sogami potential. The AFM study does not contradict the long-range attraction under consideration, since this attraction manifests itself at distances not covered in the force measurements.

Another attempt to derive the interaction potential from microscopic observations was reported by Versmold et al. [82]. They concluded that no indication was obtained on the presence of a purely attractive interaction. However, their samples have an effective charge of only -190. As mentioned earlier, the attraction under consideration becomes stronger with increasing charge number, which can be accounted for in terms of the Sogami theory, but definitely not by the DLVO concept. Thus if the given value of the effective charge is correct, the particles can be concluded not to be charged sufficiently for the attraction to manifest itself. One thus agrees with Versmold et al., who pointed out that further experiments under various conditions such as variation of the particle size and charge, etc., are necessary before something conclusive can be obtained.

Although there are reasons to believe that the Sogami theory provides a satisfactory, basic explanation for the attraction, there have been several theoretical attempts to formulate it, as recently reviewed by Schmitz [25]. Furthermore, it is necessary to pay attention to the theoretical approach of Ninham [83]. It is worth scrutinizing these theories in light of experimental data that have been and are being accumulated using well-characterized particles. It is hoped that, in this way, "one of the three fallacies in the theories of colloidal structure" that Langmuir addressed [84] can be readdressed.[5]

Systematic research to inspect in detail the DLVO theory at such a basic level with well-characterized materials is necessary.

[5] According to Langmuir, one of the fallacies is to ignore or neglect the attraction between charged micelles and the ion atmosphere of opposite sign although it exceeds the repulsive force between the micelles.

5.7 Recent Advances

5.7.1 Positive Adsorption of Negatively Charged Latex Particles near a Negatively Charged Glass Surface

Very recently, Ito et al. [41] systematically studied by microscope the particle distribution in a latex dispersion as a function of the distance from a cover glass, that is, the bottom of a cylindrical observation cell. The cover glass was first treated with acid and then extensively washed with water before it was used as the bottom of the glass cell. The zeta potential of the glass surface was -90 mV in the absence of salt, which was obtained by measuring the electrophoretic mobility of latex particles in a rectangular glass cell according to the method of Mori et al. [85]. This value appears to be in good agreement with the one reported in the literature [86]. The particle number in the latex dispersion was counted by a confocal laser scanning microscope (reversed type) at various distances (5 to 300 μm) from the cell bottom. The results in the range of 5 to 80 μm are shown in Fig. 5.20 at three salt concentrations. It is clear that, at zero salt concentration, the number of negatively charged particles at distances of 5 to 40 μm is distinctly *higher* than the average number (the shadowed area), which can be obtained from the initial concentration of latex particles and from the focal volume. The particle number *decreased* with increasing distance and became equal to the average value above 50 μm. Obviously the *negatively* charged particles were *positively* absorbed onto the *negativlely* charged glass surface. It is quite interesting that the positive adsorption became less pronounced with increasing salt concentration and disappeared at 10^{-4}M. The simplest explanation is that the driving force for the positive adsorption is of electrostatic origin. The positive adsorption was also observed for density-matched dispersions, ruling out the possibility that the adsorption is due to gravitational sedimentation.

The observed fact is consistent with the observation of Thomas et al. for ionic surfactant solutions [40] and, in addition, with the scheme proposed earlier (mentioned above) for the electrostatic attraction between similarly charged spherical particles, namely, counterion-mediated interparticle attraction: In other words, the negativly charged glass surface attracts anionic latex particles through the intermediatery of counterions. These latex particles in turn attract neighboring particles by the same mechanism. On the other hand, the observation is in contradiction to the widely accepted double-layer interaction theory, which claims that the concentration of anionic particles should simply increase with increasing distance from the negative surface according to the exponential law (as schematically shown in the inset of Fig. 5.20). The observed fact cannot be accounted for even qualitatively in terms of the widely accepted repulsion-only assumption and implies that the double-layer interaction theory is not generally correct or that the theory is valid only at very small distances, where the number of counterions is not large enough to generate the interparticle attraction.

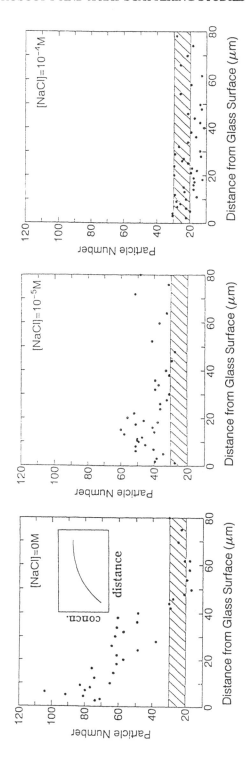

Figure 5.20 The particle distribution near a glass surface as studied by a confocal laser scanning microscope. Latex: MSS-3 (styrene-styrenesulfonate copolymer), d: 0.34 μm, ρ_e: 1.1 $\mu C/cm^2$, [latex]: 0.01%, solvent : water, microscope: LSM 410, Carl Zeiss, Oberkochen, Germany, with an oil immersion objective of 40X, light source: Kr-Ar (647, 564, 488 nm). The inset shows the qualitative feature of the distance from a negatively charged surface. The cover glass of the observation cell was treated with sulfonic acid and extensively washed with water. Its zeta potential was -90 mV by the method discussed in the text. The micrograph images were treated with BioScan OPTIMETRIC, BioScan, Inc., Edmond, Washington, to obtain high-contrast pictures and to count the number of particles in the images. Attempts were made to attain better particle images by changing contrast so that the depth of focus was not fixed but was in the range of 2-3 μm. By using the depth of focus, the focal volume, and hence the average number concentration (the shadowed area) was estimated. Under the present experimental conditions, it was rather difficult to carry out the observation in the vicinity of the glass surface (within 5 μm) and also in the region distant (> 100 μm) from the surface because of high noise levels. The sedimentation velocity is negligibly small under the experimental condition for the present latex particles.

5.7.2 Higher Orders of Diffraction from Colloidal Silica Crystals with Six- and Fourfold Symmetries

The silica particle dispersion, which gave several orders of Bragg diffraction the rotation angles of $(60 \times m)°$ (m: integer) in Fig. 5.17, was recently found to display similarly sharp peaks at the scattering angles of $(85 \times n)$ seconds (n: integer) when the capillarly was rotated by $(30 + 60 \times m)°$. From all this information, it was concluded that, under the present experimental conditions, a BCC lattice was maintained, with its [1$\bar{1}$1] direction being vertically upward and parallel to the capillary axis. Remembering that the detector was rotated on a horizontal plane, it was inferred that the profile at the rotation angles of $(60 \times n)°$ (Fig. 5.17) corresponds to diffraction from the (110) planes while that at $(30 + 60 \times m)°$ to (101) and/or (020) planes. The shift of the (101) reflection from 150″ to 85″ is due to the fact that the angle between the (101) plane and the vertical axis is 54.7°. The first peak of the (020) reflection is at 170″.

When the same colloidal silica dispersion was introduced in a different capillary tube, the USAXS study gave different profiles [87]. Distinct peaks were observed at angles of $(149 \times n'')$ that appeared when the capillary was rotated around its axis by $(90 \times m)°$. A different profile was found at $(45 + 90 \times m)°$ with the first peak at 109″. This observation implies that a body-centered cubic lattice with a fourfold symmetry was maintained with the [001] direction being vertically upward and parallel to the capillary axis. For both cases, the lattice constant obtained was 3000 Å, with $d_{110} = 2100$ Å. The closest interparticle distance (R_{exp}) was 2600 Å, whereas the average interparticle distance (R_0) from the particle concentration was 2900 Å. The value of the lattice constant and the crystal structure was confirmed using Kossel line analysis by Yoshiyama and Sogami [88]. It is obvious that the crystal is not space filling, but localized: The crystal occupies $(2600/2900)^3 = 0.72$ of the total dispersion volume, and the rest (0.28) contains voids and/or free particles. It may be described as the crystalline part being *contracted* during crystallization. The fact that a BCC symmetry was formed is consistent with the results obtained from Kossel line analysis that crystallization proceeds to BCC structure for dilute dispersions (above 2%) of small particles [39].

The fact that different symmetries were observed from the same dispersion in different capillaries indicates the important role of the capillary wall in colloidal crystal growth. It is most implausible that a small crystal that was created in the interior of the dispersion grows further to a single crystal in such a way that its [1$\bar{1}$1] direction or [001] direction happens to be vertically upward, since the small crystal is randomly directed. It is also implausible that many small crystals are born simultaneously with the same direction near the glass surface and grow to a single crystal with a definite direction. A plausible interpretation is as follows: The particles are concentrated near the glass surface by the positive adsorption discussed in the previous paragraph, so that rather large nascent crystals with specific orientations are likely to be formed near the surface. Then by an Ostwald ripening mechanism some of the large crystals continue to grow at the expense of smaller crystals near the surface and also toward the interior, and finally the largest crystal becomes the single crystal. There remains a question of how one direction (or symmetry) is selected from the two directions. That the stabilization due to the attractive interaction between the glass surface and the particles (and hence the planar surface of the crystal) is larger for the sixfold symmetry than that for a fourfold symmetry seems to play a role.

Acknowledgements

Our sincere thanks are due to co-workers whose names are cited in the references. We are very grateful to Martin V. Smalley, Polymer Phasing Project, JRDC, for interesting discussions and kind help in manuscript preparation. A large part of our work reviewed in the present article was supported by the Ministry of Education, Science and Culture (Grants-in-Aid for Specially Promoted Research).

References

1. Perrin, J., *Les Atomes* (Libraire Felix Alcan, Paris, 1913).
2. Siedentopf, H., and Zsigmondy, R., *Ann Phys.* **10**, 1 (1903).
3. Zsigmondy, R., *Nobel Lectures* (Elsevier, Amsterdam, 1966), p. 45.
4. Kose, A., Ozaki, M., Takano, K., Kobayashi, K., and Hachisu, S., *J. Colloid Interface Sci.* **44**, 330 (1973).
5. Ito, K., and Ise, N., *J. Chem. Phys.* **86**, 6502 (1987).
6. Hosemann, R., *Polymer* **3**, 349 (1962).
7. Lipson, H. S., *Optical Transforms* (Academic Press, New York, 1973).
8. Cornell, R. M., Goodwin, J. W., and Ottewill, R. H., *J. Colloid Interface Sci.* **71**, 254 (1979).
9. Ise, N., Ito, K., Okubo, T., Dosho, S., and Sogami, I., *J. Am. Chem. Soc.* **107**, 8074 (1985).
10. Overbeek, J. T. G., *Faraday Discuss.* **90**, 153 (1990).
11. Brenner, S. L., *J. Phys. Chem.* **80**, 1473 (1976); Chan, D. Y. C. et al., *J. Chem. Soc., Faraday Trans. II*, **74**, 136 (1978); Hachisu, S., and Takano, K., *Adv. Colloid Interface Sci.* **16**, 233 (1982): Furusawa, K., and Yamashita, S., *J. Colloid Interface Sci.* **89**, 574 (1982); Okubo, T., *J. Chem. Phys.* **86**, 2394, 5182, 5528 (1988); **88**, 2083, 6581 (1988); *Colloid Polym. Sci.* **265**, 522, 597 (1987); **266**, 1042, 1049 (1988); *J. Phys. Chem.* **98**, 1472 (1994).
12. Hirtzel, C. S., and Rajagopalan, R., *Colloidal Phenomena, Advanced Topics* (Noyes Publications, Park Ridge, NJ, 1985), Ch. 6.
13. Ito, K., Ieki, T., and Ise, N., *Langmuir* **8**, 2952 (1992).
14. Robinson, R. A., and Stokes, R. H., *Electrolyte Solutions* (Butterworths, London, 1959), Ch. 7.
15. Ito, K., Nakamura, H., Yoshida, H., and Ise, N., *J. Am. Chem. Soc.* **110**, 6955 (1988).
16. Ise, N., Matsuoka, H., Ito, K., and Yoshida, H., *Faraday Discuss.* **90**, 153 (1990).
17. Kittel, C., *Introduction to Solid State Physics* (John Wiley, New York, 1986), 4th Edn., Ch. 4.
18. Ise, N., Matsuoka, H., and Ito. K., *Z. Phys. Chem. (Leipzig)* **269**, 345 (1988).
19. Dosho, S., et al., *Langmuir* **9**, 394 (1993).
20. Henderson, B., *Defects in Crystalline Solids* (Edward Arnold, London, 1972), Ch. 1.
21. Ise, N., et al. *J. Am. Chem. Soc.* **102**, 7901 (1980).
22. Matsuoka, H., and Ise, N., *Adv. Polymer Sci.* **114**, 187 (1994).
23. Lin, S.-C., Lee, W. I., and Schurr, J. M., *Biopolymers*, **17**, 1041 (1978).
24. Schmitz, K. S., Lu, M., Singh, N., and Ramsay, D. J., *Biopolymers* **23**, 1637 (1984).
25. Schmitz, K. S., *Macroions in Solution and Colloidal Suspension* (VCH, New York, 1993).
26. Yoshida, H., Ito, K., and Ise, N., *J. Am. Chem. Soc.* **112**, 592 (1990).
27. Marcus, Y., *Introduction to Liquid State Chemistry* (John Wiley, New York, 1977), Chap. 2.

28. Matsuoka, H., Tanaka, H., Hashimoto, T., and Ise, N., *Phys. Rev.* **B 36**, 1754 (1987).
29. Matsuoka, H., Tanaka, H., Iizuka, N., Hashimoto, T., and Ise N., *Phys. Rev.* **B 41**, 3854 (1990).
30. Yoshida, H., Ito, K., and Ise, N., *J. Chem. Soc. Faraday Trans.* **87**, 371 (1991).
31. Ostwald, W., *Z. phys. Chem.* **34**, 495 (1900).
32. Ise, N., in *Proc. 19th Yamada Conference on Ordering and Organization in Ionic Solutions*, Eds. N. Ise and I. Sogami (World Scientific, Singapore, 1988), p. 624.
33. Kesavamoorthy, R., Rajalakshimi, M., and Rao, C. B., *J. Phys. Condens. Matter* **1**, 7149 (1989).
34. Ito, K., Yoshida, H., and Ise, N., *Chem. Letters* **1992** 2081 (1992).
35. Ito, K., Yoshida, H., and Ise, N., *Science* **263**, 66 (1994).
36. Ito, K., et al. unpublished.
37. Uyeda, N., Tanaka, K., Aoyama, K., Matsumoto, M., and Fujiyoshi, Y., *Nature* **327**, 319 (1987).
38. Chi, L. F., Anders, M., Fuchs, H., Johnston, R. R., and Ringsdorf, H., *Science* **259**, 213 (1993).
39. Sogami, I. S., and Yoshiyama, T., *Phase Transitions* **21**, 171 (1990).
40. Lu, J. R., Simister, E. A., Thomas, R. K., and Penfold, J., *J. Phys. Chem.* **97**, 13907 (1993).
41. Ito, K., Kuramoto, K., and Kitano, H., *J. Am. Chem. Soc.* **117**, 5005 (1995).
42. Skelton, E. F., et al., *Science* **263**, 1416 (1994).
43. Yoshiyama, T., and Sogami, I., *Langmuir* **3**, 851 (1987).
44. Matsuoka, H., Schwahn, D., and Ise, N., *Macromolecules* **24**, 4227 (1991).
45. Boue, F., Daoud, M., Nierlich, M., Jannink, G., Benoit, H., Dupplessix, R., and Picot, C., *Proc. Symp. Neutron Inelastic Scattering* **1**, 563 (1977).
46. van der Maarel, J. R. C. et al., *Macromolecules* **26**, 7295 (1993).
47. Matsuoka, H., Schwahn, D., and Ise, N., in *Macro-Ion Characterization from Dilute Solutions to Complex Fluids*, Ed. K. Schmitz (American Chemical Society, Washington, DC, 1994), Ch. 27.
48. Brown, J. C., Pusey, P. N., Goodwin, J. W., and Ottewill, R. H., *J. Phys.* **A8**, 664 (1975).
49. Matsuoka, H., Kakigami, K., Ise, N., Kobayashi, Y., Machitani, Y., Kikuchi, T., and Kato, T., *Proc. Natl. Acad. Sci. USA* **88**, 6618 (1991).
50. Konishi, T., Ise, N., Matsuoka, H., Yamaoka, H., Sogami, I. S., and Yoshiyama, T., *Phys. Rev.* **B51**, 3914 (1995).
51. Bonse, U., and Hart, M., *Z. Phys.* **189**, 151 (1966).
52. Matsuoka, H., Kakigami, K., and Ise, N., *The Rigaku Journal* **8**, 21 (1991).
53. Matsuoka, H., Kakigami, K., and Ise, N., *Proc. Japan Acad.* **B67**, 170 (1991).
54. Matsuoka, H., and Ise, N., *Chemtracts-Macromol. Chem.* **4**, 59 (1993).
55. Verwey, E. J. W., and Overbeek, J. Th. G., *Theory of the Stability of Lyophobic Colloids* (Elsevier, Amsterdam, 1948).
56. Matsuoka, H., Nakatani, Y., and Ise, N., *Polymer Preprints, Japan*, **42**, 4695 (1993).
57. Arora, A. K., Tata, B. V. R., Sood, A. K., and Kesavamoorthy, R., *Phys. Rev. Lett.* **60**, 2438 (1988).
58. Tata, B. V. R., Arora, A. K., and Valsakumar, M. C., *Phys. Rev.* **E 47**, 3404 (1993).
59. Sogami, I., and Ise, N., *J. Chem. Phys.* **81**, 6320 (1984).
60. Tata, B. V. R., Rajalakshmi, M., and Arora, A. K., *Phys. Rev. Lett.* **69**, 3778 (1992).
61. Palberg, T., and Würth, M., *Phys. Rev. Lett.* **72**, 786 (1994).
62. Tata, B. V. R., and Arora, A. K., *Phys. Rev. Lett.* **72**, 787 (1994).
63. Huizenga, J. R., Grieger, P. F., and Wall, F. T., *J. Am. Chem. Soc.* **72**, 2636 (1950).
64. Ito, K., Ise, N., and Okubo, T., *J. Chem. Phys.* **82**, 5732 (1985)
65. Yamanaka, J., Matsuoka, H., Kitano, H., and Ise, N., *J. Coll. Interface Sci.* **134**, 92 (1990).
66. Wall, F. T., and Eitel, M. J., *J. Am. Chem. Soc.* **79**, 1550 (1957).
67. Wall, F. T., and Eitel, M. J., *J. Am. Chem. Soc.* **79**, 1556 (1957).
68. Ise, N., et al., *J. Am. Chem. Soc.* **107**, 8074 (1985).
69. Ito, K., Nakamura, H., and Ise, N., *J. Chem. Phys.* **85**, 6143 (1986).
70. Matsuoka, H., et al., *J. Chem. Phys.* **83**, 378 (1985).
71. Ito, K., Okumura, H., Yoshida, H., Ueno, Y., and Ise, N., *Phys. Rev.* **B 38**, 10852 (1988).
72. Asher, S. A., et al. *J. Chem. Phys.* **94**, 711 (1991).

73. Ise, N., and Smalley, M. V., *Phys. Rev.* **B 50**, 16722 (1994).
74. Tata, B. V. R., Sood, A. K., and Kesavamoorthy, R., *Pramana - J. Phys.* **34**, 23 (1990).
75. Lindsay, H. M., and Chaikin, P. M., *J. Chem. Phys.* **76**, 3774 (1982).
76. Ito, K., Sumaru, K., and Ise. N., *Phys. Rev.* **B 46**, 3105 (1992).
77. Eigen, M., and Winkler, R., *Law of the Game: How the Principles of Nature Govern Chance* (Penguin Books, New York, 1981), Ch. 17: Original German version *Das Spiel: Naturgestze steuern den Zufall* (Piper, Muñchen, 1976).
78. Overbeek, J. Th. G., *J. Chem. Phys.* **87**, 4406 (1987).
79. Ise, N., Matsuoka, H., Ito, K., Yoshida, H., and Yamanaka, J., *Langmuir* **6**, 296 (1990).
80. Smalley, M. V., *Molec. Phys.* **71**, 1521 (1990).
81. Larson, I., Drummond, C. J., Chan, D. Y. C., and Grieser, F., *J. Am. Chem. Soc.* **115**, 11885 (1993).
82. Vondermassen, K., Bongers, J., Müller, A., and Versmold, H., *Langmuir* **10**, 1351 (1994).
83. Mahanty, J., and Ninham, B., *Dispersion Forces* (Academic Press, New York, 1976).
84. Langmuir, I., *J. Chem. Phys.* **6**, 873 (1938).
85. Mori, S., Okamoto, H., Hara, T., and Aso, K., in *Fine Particle Processing*, ed. P. Somasundaram (American Institute of Mining, Metallurgical and Petroleum Engineerings, New York, 1980), Ch. 33.
86. Hunter, R., *Zeta Potential in Colloid Science* (Academic Press), 1981, Ch. 7.
87. Konishi, T., and Ise, N., *J. Am. Chem. Soc.* **117**, 8422 (1995).
88. Yoshiyama, T., and Sogami, I. S., unpublished.

6

Vapor-Liquid Condensation and Reentrant Transition in Charged Colloids

B. V. R. Tata and Akhilesh K. Arora

A phase transition analogous to the well-known vapor-liquid (VL) condensation in atomic systems has been recently observed in dilute aqueous charged colloidal suspensions of polystyrene particles. A noninteracting homogeneous suspension, when deionized, is found to phase separate into dense (liquidlike) and rare (vaporlike) phases. The structure factors of these phases are obtained from angle-resolved polarized light scattering measurements. Upon further deionization the dense phase is found to occupy the full volume; (that is, homogeneous again) suggesting the occurrence of a reentrant transition, which is unique to these systems. The VL transition observed in these suspensions unambiguously demonstrates the existence of an attractive interaction between charged polystyrene spheres. Monte Carlo (MC) simulations with effective interparticle interaction having an attractive component are carried out over a wide range of impurity ion concentration. The simulations indeed show the occurrence of VL and reentrant transitions upon decreasing impurity ion concentration. The free energy calculations provide an understanding for the occurrence of the above transitions in these suspensions. The phase diagram obtained from these calculations is also discussed.

6.1 Introduction

Vapor-liquid (VL) condensation is a well-known phenomenon in atomic systems. The pioneering works of van der Waals [1] and Maxwell [2] have provided a good understanding about this phenomenon occuring in atomic and molecular systems. It is now a well known fact that the attractive term in the effective interatomic interaction [3] is responsible for vapor-liquid condensation below the critical temperature. An analogous phase transition has been recently [4] observed in dilute charge stabilized colloidal suspension of polystyrene spheres when the impurity ion concentration n_i is reduced. In addition to the VL condensation, these suspensions are found to exhibit another subtle phase transition [5] upon further reduction of n_i across which the suspension again becomes homogeneous; that is, the suspension undergoes phase separation only in a limited range of n_i and remained homogeneous otherwise.

In addition to these phase transitions, there have been several other observations [6-16] reported in charged colloids, which show that these suspensions can be inhomogeneous under the appropriate conditions. Some of these have already been discussed in earlier chapters. These observations cannot be understood on the basis of a repulsive screened Coulomb potential [17, 18], which until recently was believed to be the dominant interaction between the colloidal particles and has been extensively used to understand the ordering phenomena in these suspensions. A system of particles interacting via a purely repulsive potential, when restricted to finite volume, is expected to remain homogeneous except for the fluid-solid transition. The experimental observations mentioned suggest that there exists an attraction that dominates at large interparticle separation r, whereas the conventional screened Coulomb repulsion dominates at small r. The recent theory of interactions in macroionic solutions derived initially by Sogami [19] and later by Sogami and Ise [20], indeed show the existence of a screened Coulomb repulsive term and a long-range attractive term in the effective interparticle interaction.

The expermental and theoretical investigations carried out on aqueous suspensions of polystyrene colloids that exhibit the phenomena of vapor-liquid condensation and the reentrant phase transition reported recently are reviewed in this chapter. The implications of these observations on the current understanding of the effecive pair potential are also highlighted. The organization of the chapter is as follows: In the next section the experimental details and the sample preparation are discussed. Section 6.3 deals with experimental results on VL condensation, and reentrant transitions. Section 6.4 describes the Monte Carlo investigations carried out to simulate the experimentally observed phenomena. MC simulation results along with a discussion are given in Section 6.5. A comparison of simulation results with experiments is discussed in detail in Section 6.6. The free energy calculations and the phase diagram obtained using these calculations are discussed in Section 6.7. Conclusions are given in Section 6.8.

6.2 Structural Ordering Using Angle-Resolved Polarized Light Scattering

Since the interparticle separation is of the order of the wavelength of light, light scattering is the appropriate technique to investigate the structural order in colloidal suspensions. Angle-resolved polarized light scattering can be analyzed to obtain the static structure factor $S(Q)$, as will be discussed in the following.

Consider light scattering from a scattering volume V_s containing N spherical colloidal particles. The intensity of light scattered by the solvent is negligible as compared with that scattered quasielastically by the particles. The time-averaged intensity in the Rayleigh-Gans approximation is given as [21, 22]

$$I_s(Q) = A_c P(Q) S(Q) \tag{6.1}$$

where the scattering wave vector Q is $4\pi \mu_m \sin(\theta/2)/\lambda$, μ_m is the refractive index of the medium, and λ is the incident wavelength. The particle scattering form factor $P(Q)$ for a spherical particle of radius a is given as

$$P(Q) = \left(3\frac{\sin(Qa) - (Qa)\cos(Qa)}{(Qa)^3} \right)^2 \tag{6.2}$$

The interparticle structure factor $S(Q)$ in Eq. (6.1) is given by

$$S(Q) = 1 + \frac{1}{N} \sum_{i>j=1}^{N} \exp[i\mathbf{Q} \cdot (\mathbf{r}_i - \mathbf{r}_j)] \tag{6.3}$$

where $\mathbf{r}_i$ is the position of the center of mass of the ith particle and A_c is a constant, which for a vertically polarized scattered light is given by

$$A_c = \frac{9\pi^2 \mu_m (m_r - 1)^2 V_s n_p v_p^2}{\lambda^4 (m_r + 2)^2} \frac{I_0}{R_d^2} \tag{6.4}$$

Here $m_r = \mu_p/\mu_m$, μ_p is the refractive index of the particle, v_p is the volume of the particle, I_0 is the intensity of the incident radiation, and R_d is the distance between the scattering volume and the detector. $S(Q)$ can be obtained by measuring $I_s(Q)$ as a function of Q and then correcting for $P(Q)$ using Eq. (6.2). The $S(Q)$ thus obtained using Eqs. (6.1) and (6.3) has been used to identify the structural ordering.

The details of the home-built angle resolved polarized laser light scattering setup used for measuring $S(Q)$ is described elsewhere [23]. The schematic diagram of the photon counting system is shown in Fig. 6.1. Using a microprocessor-based data acquisition system [24] and a personal computer the $S(Q)$ is obtained from measured incident and scattered intensities in the vertical-vertical geometry (the first vertical stands for the polarization of the incident light and the latter for that of scattered light) for various scattering angles. The multiple scattering

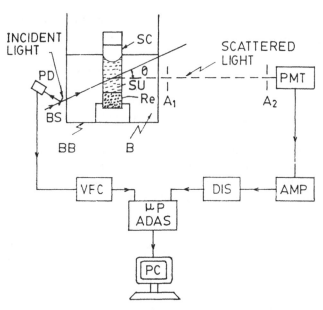

Figure 6.1   Schematic diagram of the light scattering setup along with sample cell and photon-counting system. BB, benzene bath; BS, beam splitter; PD, photodiode; SC, sample cell; B, benzene, SU, suspension; O, scattering angle; Re, mixed-bed ion exchange resin; PMT, photomultiplier tube; AMP, amplifier; DIS, discriminator; μP ADAS, microprocessor-based automation and data acquisition system; VFC, voltage-to-frequency converter; PC, personal computer.

correction is known to be small ($< 4\%$) for an aqueous suspension of polystyrene spheres of diameter 110 nm with $n_p \leq 5 \times 10^{12}$ cm^{-3} [25]; hence no multiple scattering correction is applied.

6.2.1 Sample Preparation

Monodisperse suspensions of polystyrene spheres of diameter $d = 110$ nm, obtained from Polysciences, Inc., USA, are prepared by diluting the stock suspension in deionized water to the desired concentration. Suspensions thus prepared are homogeneous and have a conductivity of the order of a few tens of μS cm^{-1}. Suspensions with particle concentrations n_p ranging over an order of magnitude (samples S1: 4.20, S2: 4.11, S3: 3.26, S4: 1.61, S5: 1.58, and S6: 0.46×10^{12} cm^{-3}) are investigated. Suspensions are taken in cylindrical cells of internal diameter 1 cm. In order to deionize the samples, a mixed bed of ion-exchange resins is prepared by mixing equal volumes of activated cation- and anion-exchange resins. A small quantity of this is then added to the suspensions and gently shaken and left for observation. The particle concentration in various samples has been obtained by carefully drying the known amount of the suspension followed by accurate weighting [26] after the completion of the light scattering measurements. The conductivity of the deionized samples also has been measured after deionization equilibrium is reached (the lowest conductivity is reached after 10 days with occasional shaking) and is found be 0.4 to 0.6 μS cm^{-1}, which corresponds to $n_i \sim 2$ to 3×10^{15} cm^{-3}. In order to measure the transmittance of the samples a 2 mW He-Ne laser beam was focused into a sample cell placed in an index-matching bath. The transmitted intensity is measured using an EG&G photodiode (SGD-100A). The height dependence of the transmitted intensity is measured by scanning the sample using the laser beam.

6.3 Experimental Results and their Implications

The experimental results obtained from angle-resolved polarized scattering and transmission measurements are discussed in this section.

6.3.1 Reentrant Phase Transition

The suspensions prepared by the process of dilution are always homogeneous; that is, the particles occupy the full volume of the solvent, having a uniform n_p. Since these suspensions are not treated with a mixed bed of ion-exchange resins, and their measured conductivity is of the order of few tens of μS cm^{-1}, these are expected to be noninteracting. In order to confirm this, the structure factor is measured and is shown in Fig. 6.2. Note that except for a small initial rise at low Q, which is due to the scattering from dust particles, $S(Q)$ is featureless, indicating that no spatial correlations exist in these suspensions and hence these can be characterized as gaslike (noninteracting) ordered.

In order to reduce the impurity ion concentration, the mixed bed of ion-exchange resins is added to the suspension. The resins settle down at the bottom of the cell. As n_i reduces due to the action of ion-exchange resins, the homogeneous suspensions are found to phase separate. A dense phase with particle concentration n_d much higher than n_p settles below due to gravity, leaving the rest of the volume occupied by a rare phase with particle concentration n_r ($n_r < n_p$). This occurs because the density of the polystyrene particle (1.05 g/cm^{-3}) is higher than that of water. The transmitted intensity through the sample cell, which is related to the particle concentration, is shown in Fig. 6.3(a) as a function of height at various times after adding the resins. The horizontal sections of the curves *a-d* represent the rare phase, and the sedimented dense phase is identified by the sudden decrease of the transmitted intensity. Angle-resolved polarized light scattering measurements from the dense and the rare phases, show that the former has strong liquidlike order, whereas the latter does not have any structural order (gaslike). The coexistence of crystalline and liquidlike orders observed in the dense phase is due to the concentration gradient arising due to gravitational effects [27]. The transmittance through the crystalline order is very small as most of the light is scattered away due to the Bragg diffraction [28]. The attenuation

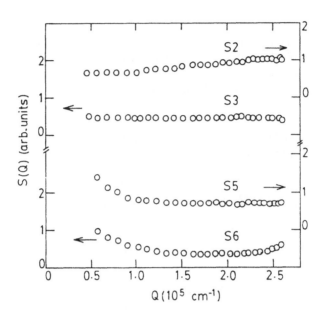

Figure 6.2 $S(Q)$ versus Q from light scattering measurements for suspensions S2, S3, S5, and S6 before adding ion-exchange resins.

of the light is scattered away due to the Bragg diffraction [28]. The attenuation of the laser beam through the dense phase does not faithfully represent the particle concentration profile because of multiple scattering; however, in the rare phase it can be used to estimate n_r. The attenuation in the rare phase decreases as the grains (domains) of dense phase settle down under gravity. Note from Fig. 6.3(b) that the settling of dense phase grains is nearly complete in less than 10 hours.

There is another interesting observation that the volume occupied by the dense phase continues to increase (interface between the dense and the rare phase continues to move upward) as impurities are further reduced by the action of the resins. Figure 6.3(c) shows the position of the interface as a function of time. In approximately 100 hours the dense phase occupies the full volume of the solvent and again becomes a single phase except for the concentration gradient due to gravitational effects. This suggests that the average interparticle separation in the dense phase continues to change (increase) as n_i decreases. This is confirmed from the observed shift in the first peak of the $S(Q)$ measured within the dense phase. Similar behavior is also observed in suspensions S2 and S3. Thus the suspensions S1-S3, are homogeneous to start with and noninteracting, while exhibit a phase separation for a restricted range of n_i, have reentered into a homogeneous interacting state (liquidlike order) upon further reduction of n_i.

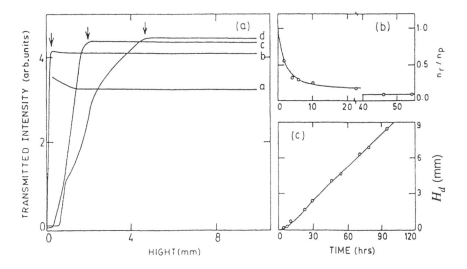

Figure 6.3   (a) Transmitted intensity as a function of height in the suspension S1 at different times after the addition of the ion-exchange resins: a, 0.1 h; b, 4.2 h; c, 22 h; d, 54 h. Vertical arrows indicate the positioning of the interface between the dense and the rare phases. The shoulder observed in the curve d within the dense phase is due to the interface between crystalline and liquid orders. (b) Concentration in the rare phase relative to n_p as a function of time. (c) Height of the dense phase (H_d) as a function of time [5].

6.3.2 Vapor-Liquid Condensation

While the suspensions S1-S3 exhibited a reentrant interacting homogeneous state, suspensions S4-S6 are found to remain in the phase-separated state even after 10 days. In other words, the dense phase is not found to occupy the full volume of the solvent. As it is known that the deionization equilibrium is reached within 10 days [27], the phase separated state can be identified as the equilibrium state of these suspensions. In order to remove the possible n_i gradient and with an aim to see whether the phase separation reappears or not, the suspensions are shaken occasionally and left again undisturbed. After approximately 50 hours, the samples are found to show phase separation once again with same volume of dense phase, separated by a rare phase, thus confirming that the phase-separated state observed is indeed an equilibrium state of the suspensions. Figure 6.4 shows photographs of some of these sample cells.

The dense phase, which is in the lower part of the cell, appears very turbid or milky because its large particle concentration, whereas the rare phase in the upper part of the cell is much more transparent, suggesting a much lower n_p in this region. Figure 6.4 also shows the sample cells S2 and S3 exhibiting uniform turbidity because these are in the reentered homogeneous state, as mentioned earlier. The structural ordering in the dense and rare phase regions has been

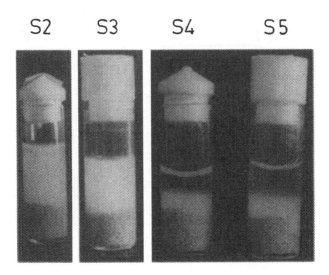

Figure 6.4 Photograph of sample cells showing the suspensions S2, S3, S4, and S5 several days after reaching the deionization equilibrium. Suspensions S4 and S5 exhibit the dense phase (lower part of the cell) macroscopally separated from the dilute phase upper part of the cell). Suspensions S2 and S3 became homogeneous after an initial phase separation. Ion-exchange resin is at the bottom of the cell.

investigated using angle resolved polarized light scattering, and the structure factor is shown in Fig. 6.5. Note that $S(Q)$ in the rare phase region shows no structural ordering indicating gaslike behavior, while that measured in the dense phase exhibits a clear liquidlike order. These observations clearly indicate vapor-liquid coexistence in these suspensions. On the other hand, the $S(Q)$ of samples S1-S3, which exhibited reentrant behavior, shows liquidlike order over the full height of the suspension, as shown in Fig. 6.6. This indicates that the suspensions S1-S3 are homogeneous and strongly interacting.

One can get some qualitative idea about the particle concentration profile as a function of height if the light scattering is measured at sufficiently large Q, at which the structure factor is asymptotically close to 1. The scattered intensity is then proportional to the particle concentration except for the factors governed by mulitiple scattering and attenuation. The scattered intensity $I(Q_M)$ as function of height, which is measured with respect to the top of the resin bed, is shown in Fig. 6.7 for several sample cells. The Q in the measurement corresponds to the largest possible scattering vector $Q_M = 2.56 \times 10^5$ cm^{-1}. Note from Fig. 6.7 the sharp fall in $I(Q_M)$ occurring at different heights for samples S3-S6, whereas no such decrease is seen for suspension S3. Samples S1 and S2 also show a behavior similar to that of S3. This clearly shows the existence of a boundary between the dense and rare phases in samples S4-S6 and the homogeneous nature of samples S1-S3. The rise in $I(Q_M)$ seen for sample S3 at large heights is due to the extra scattering from the meniscus. One can notice in Fig. 6.7 that $I(Q_M)$ across the

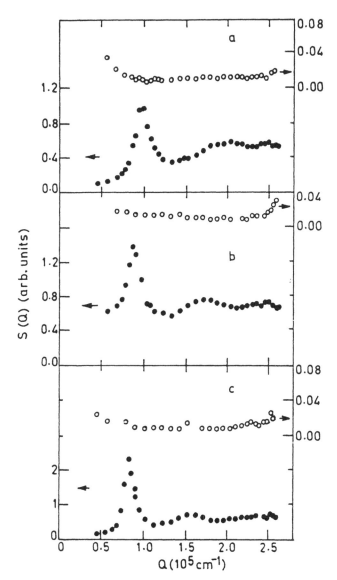

Figure 6.5 $S(Q)$ for different heights for suspensions that remained in the phase-separated state after deionization equilibrium is reached. (a) Sample S4, (b) sample S5 and (c) sample S6. ($\bullet$): Lower region of the cell; (o) top region of the cell.

phase boundary changes by more than an order of magnitude, indicating the ratio n_d/n_r is of similar magnitude. This ratio can vary from sample to sample, as n_d is expected to depend on the inverse Debye screening length κ, which in turn depends on n_p and n_i.

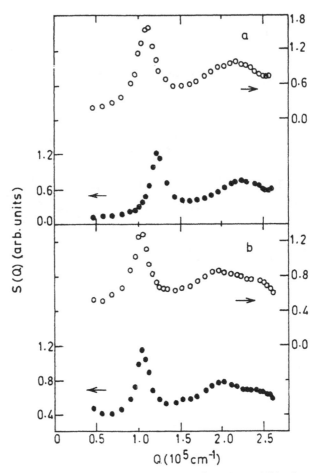

Figure 6.6 $S(Q)$ for different heights for suspensions that exhibited a reentered homogeneous state after deionization equilibrium is reached. (a) Sample S2, (b) sample S3. (●): Lower region of the cell, (o) top region of the cell.

To summarize, all suspensions exhibit phase separation upon deionization in the form of a dense phase coexisting with a rare phase, which is analogous to vapor-liquid condensation. On further reduction of n_i, concentrated samples S1-S3 reentered into an interacting homogeneous state that has liquidlike order, while the dilute samples S4-S6 remained in a phase-separated state with a dense phase exhibiting a liquidlike structure and a rare phase exhibiting a gaslike ordering indicating the coexistence of vapor-liquid phases with widely differing particle concentrations. It is important to point out that in suspensions exhibiting a reentrant phase transition both gaslike as well as liquidlike states have the same n_p. Such a situation never occurs in atomic systems, and hence it is unique to these suspensions.

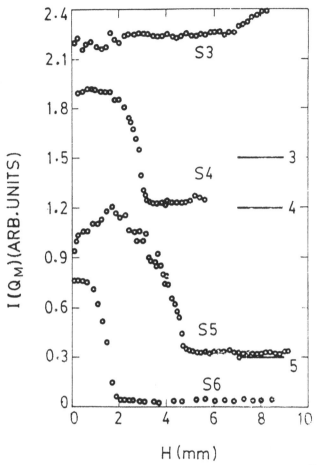

Figure 6.7   Scattered intensity at a fixed scattering vector ($Q_M = 2.6 \times 10^5$ cm^{-1}) as a function of height in various suspensions. The decrease in the scattering beyond certain heights in the suspensions S4 to S6 is because of the phase boundary between the dense and rare phases. Horizontal lines marked 3 to 5 are the zeros of the ordinates for sample S3 to S5 [4].

It is of interest to estimate the average nearest-neighbour (nn) distance d_{nn} or the particle concentration in the condensed phase. The position Q_{max} of the first peak in $S(Q)$ can be used for this purpose. As suspensions at low volume fractions crystallize in the body centered cubic (BCC) structure, the liquidlike order is also assumed to have a BCC-like local short-range order [7]. Then d_{nn} is $\sqrt{6}\pi\, Q_{max}^{-1}$ and is shown in Fig. 6.8. The average interparticle separation R_0 [for BCC-like coordination, $R_0 = \sqrt{3}(4n_p)^{-1/3}$] expected for a homogeneous dispersion is also shown as a continuous curve. Note that d_{nn} is significantly smaller than R_0 for suspensions exhibiting condensation. This implies that the particle concentration in the condensed phase is several times higher than the average concentration

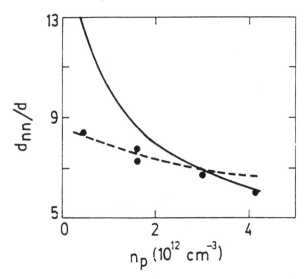

Figure 6.8   Estimated nn distance from first peak height of $S(Q)$ as a function of n_p for suspensions S2 to S6 (•). R_0 and R_m (calculated for $n_i = 3 \times 10^{15}$ cm^{-3}) are also shown as full and dashed curves, respectively [4].

n_p. On the other hand, for the suspensions that become homogeneous (S1-S3) $d_{nn} \simeq R_0$ as expected.

Vapor-liquid condensation can occur, in principle, if there exists a sufficiently deep minimum in the effective interaction potential at a distance smaller than R_0. One also has to understand why concentrated suspensions S1 to S3 remain homogeneous. According to the recent theory of electrostatic interaction by Sogami and Ise [19, 20] in charge-stabilized colloids, the effective interparticle potential $U_S(r)$ has the form

$$U_S(r) = 2 \frac{(Ze)^2}{\epsilon} \left(\frac{\sinh(\kappa d/2)}{\kappa d} \right)^2 \left(\frac{A}{r} - \kappa \right) \exp(-\kappa r) \tag{6.5}$$

where $A = 2 + (\kappa d) \coth(\kappa d/2)$. The inverse Debye screening length κ is given by $\kappa^2 = 4\pi e^2 (n_p Z + n_i)/\epsilon k_B T$, where k_B is the Boltzmann constant, ϵ is the dielectric constant of medium, T is the temparature (298 K), and Ze is the charge on the particle. This potential has a minimum at $R_m = \{A + [A(A + 4)]^{\frac{1}{2}}\}/2\kappa$. The dependence of R_m on n_p is also shown in Fig. 6.8 for $Z = 500$ [29] and $n_i = 3 \times 10^{15}$ cm^{-3}. Note the excellent agreement between d_{nn} and R_m for suspensions S4 to S6. The impurity ion concentration n_i is chosen as 3×10^{15} cm^{-3} because it is close to the estimated equilibrium value in these samples (see Section 6.2). It may be mentioned that the structure factor $S(Q)$ obtained from computer simulation [30, 31] using this potential is found to agree well with that measured experimentally for a homogeneous liquidlike ordered suspension. Although the DLVO potential can also explain the structural ordering in homogenous suspensions, it cannot explain the VL condensation and the reentrant transition discussed here.

6.4 Monte Carlo Simulation

Encouraged by the success of the potential $U_S(r)$ in explaining a number of phenomena, Monte Carlo simulations have also been carried out to investigate the consequences of an attractive term in $U_S(r)$. MC simulations are done [32] in a canonical ensemble (constant NVT, where N, V and T are, respectively, the number of particles, volume, and the temperature). For the required particle concentration n_p, the length L_s of the MC cell is fixed from the relation $L_s^3 = N/n_p$. Particles of diameter 109nm are assumed to interact via an effective pair potential given by Eq. (6.5). Suspensions with n_i ranging between zero and $1000 n_p Z$ have been investigated. Table 6.1 gives the important parameters of $U_S(r)$, that is, the position and depth of the potential minimum for the values of n_i considered in the simulations.

The simulations are carried out using the well-known Metropolis algorithm [32, 33] with periodic boundary conditions. The initial configuration is that of 432 particles arranged on a BCC lattice. As the particles are far separated from each other in the initial configuration, an initial step size of $4d$ is chosen to make the system evolve rapidly. Usually the step size is chosen in such away that the trial acceptence ratio (T_r), the ratio of the number of moves that are

Table 6.1 Position of the potential minimum R_m, the depth of the potential well U_{min}, first peak position R_p of $g(r)$, and equilibrium trial acceptance ratio T_r for different impurity ion concentrations n_i [32]

$n_i/n_p Z$	κd	R_m/d	$U_{min} = -U(R_m)/k_B T$	R_p/d	T_r (%)
0	0.27	18.12	0.30	9.1	50
1.5	0.42	11.50	0.47	8.7	68
2.0	0.46	10.46	0.51	8.5	72
2.6	0.51	9.57	0.56	8.3	75
3.0	0.53	9.13	0.58	8.1	76
5.0	0.65	7.49	0.71	7.1	82
7.0	0.76	6.52	0.80	6.1	80
10.0	0.89	5.60	0.92	5.2	75
30.0	1.49	3.49	1.32	3.3	55
78.0	2.38	2.40	1.51	2.2	40
100.0	2.69	2.21	1.49	2.2	92
150.0	3.28	1.95	1.39	1.95	93
400.0	5.35	1.55	0.86	1.55	96
1000.0	8.45	1.35	0.42	1.33	99

accepted to the number of moves that are attempted, is approximately 0.5 [34]. This ensures sampling of the relevent region of the configuration space as well as optimization of equilibration run time. Initially a step size of $4d$ is chosen, and when T_r reached some steady state value, the step size was lowered to $0.5d$ and the systems evolved further until equilibrium is reached. With this step size T_r ranged between 40 and 75% for all $n_i \leq 78n_pZ$. For suspensions with $n_i \geq 100n_pZ$, either of the step sizes gave $T_r > 93\%$, exhibiting an insensitivity of the weakly interacting system to the step size. Values of T_r after reaching equilibrium are also shown in Table 6.1. Most of the simulations away from the vapor-liquid transition take $\sim 9 \times 10^5$ configurations to reach equilibrium, whereas those close to the transition take approximately 5×10^6 configurations to reach equilibrium. Constancy of the average total interaction energy U_T $[\frac{1}{2} \sum_{i=1}^{N} U_S(r_{ij})]$ is taken to be the criterion for reaching equilibrium.

After reaching equilibrium, $g(r)$ is calculated using the standard method [35]. The interval Δr is taken to be $0.1d$, and averaging was performed over configurations generated from 550 Monte Carlo steps. An MC step (MCS) is defined as a set of N configurations, during which, on the average, each particle gets a chance to move. The osmotic pressure p^*, expressed in the units of $n_p k_B T$ ($p^* = p/n_p k_B T$), is calculated using the virial equation [34, 36]. A cluster of dense phase is defined in a conventional manner [37]. Here constraining volumes of radius R_c are marked around each macroion. A given macroion belongs to a cluster if the neighborhood of that macroion has nonzero overlap with the neighbourhood of at least another one belonging to that cluster. R_c is taken to be the distance at which the minimum of $g(r)$, after the first peak, occurs. This is reasonable, as it takes care of the spread in the nearest-neighbor distance, and hence all the first neighbors are included. Having identified all the clusters, their size distribution $P_c(n)$ and the total fraction of particles participating in the clustering F_c are obtained.

6.5 Simulation Results and Discussion

Figure 6.9 shows the the pair potential $U_S(r)$ for different n_i values for suspensions with $n_p = 1.33 \times 10^{12}$ cm^{-3}. The vertical line in Fig. 6.9 denotes the average interparticle spacing $\ell = n_p^{-1/3}$. Note from Table 6.1 that the depth U_{min} and the position R_m of the well depend strongly on n_i. It is clear from Fig. 6.9 that the dispersion will be homogeneous when $R_m > \ell$. The extent of structural ordering in this homogeneous state depends upon the strength of the potential at $r = \ell$. However, when $R_m < \ell$, the system can, in principle, phase separate into dense (particle concentration higher than n_p) and rare (particle concentration lower than n_p) phase regions, provided U_{min} is sufficiently large.

The simulation results can be grouped into three ranges of n_i. Figures 6.10, 6.11, and 6.12 show the pair correlation functions for various impurity ion concentrations ranging between zero and 1000 n_pZ. Note from Fig. 6.10 that for

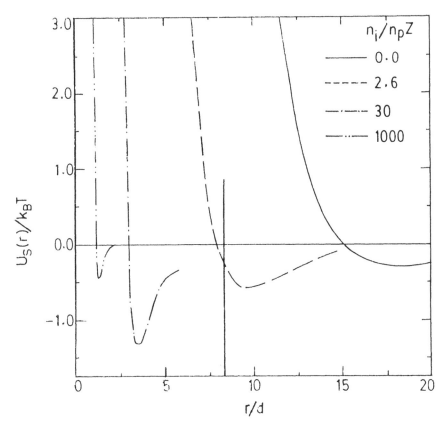

Figure 6.9   The pair potential $U_S(r)/k_B T$ for different impurity ion concentrations. The vertical line corresponds to the average interparticle separation ℓ [32].

$1000 \geq n_i/n_p Z \geq 100$ there is a single peak in $g(r)$ at a position R_p that is close to the value of R_m. Also the pair correlation function saturates to a value 1 immediately after the first peak, suggesting that there are no short-range structural correlations. As seen in Fig. 6.11, the $g(r)$ values obtained for suspensions with impurity ion concentration in the range $78 \geq n_i/n_p Z \geq 7$ are markedly different from those at high n_i. These have several peaks in $g(r)$, which suggest structural correlation up to several neighbor distances. However, the first peak position still corresponds to the minimum of the potential well. Further, the heights of the first peak are very large; for example, it has a value 80 in the case of suspensions with $78\, n_p Z$, and $g(r)$ gradually decays to 1 as r increases. This type of behavior is expected only when dispersion is inhomogeneous. Suspensions with $n_i \leq 5 n_p Z$ exhibit typical liquidlike behavior except at $n_i=0$, which shows BCC crystalline order.

The appearance of peaks in pair correlation functions at distances much shorter than ℓ suggests the possibility of either clustering or formation of a condensed

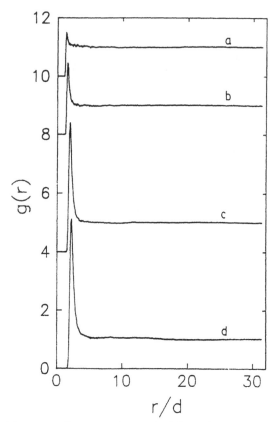

Figure 6.10 $g(r)$ versus r for suspensions at high impurity ion concentrations. Curves a, b, c, and d correspond to $n_i/n_p Z = 1000, 400, 150$, and 100, respectively. Curves a, b, and c are shifted vertically for the sake of clarity by amounts 10, 8, and 4, respectively [32].

phase where $R_p \simeq R_m$. Clusters can remain homogeneously distributed over all the volume of the MC cell whereas condensation would result in the formation of large compact droplets of the dense phase, leaving the remaining volume practically empty. The cluster size distributions are shown in Figs. 6.13 and 6.14 for various values of n_i. Note that for $n_i = 1000 n_p Z$ only a few pairs are seen, and as n_i reduces, larger clusters are observed and the fraction F_c of particles participating in clustering also increases.

Table 6.2 gives values of various equilibrium parameters like the total interaction energy U_T, height g_{max} of the first peak in $g(r)$, F_c, and the osmotic pressure p^*. Note that F_c increases significantly as n_i reduces. The clustering behavior in suspensions with $78 \geq n_i/n_p Z \geq 7$ is also markedly different from that in systems with $n_i/n_p Z \geq 100$. Note from Fig. 6.14 that apart from a few small clusters, a single large cluster encompassing most of the particles is seen. This

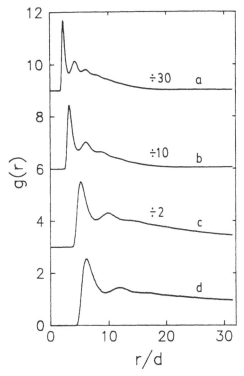

Figure 6.11 $g(r)$ versus r for suspensions at intermediate impurity ion concentrations. Curves a, b, c, and d correspond to $n_i/n_p Z = 78$, 30, 10, and 7, respectively. Curves a, b, and c are compressed by factors 30, 10, and 2, respectively and are shifted vertically for the sake of clarity. The vertical shift for curves a, b, and c correspond to 9, 6, and 3, respectively [32].

cluster, which exhibits pronounced short-range order, as seen in the corresponding $g(r)$ values, can be identified with a droplet of condensed phase. Figure 6.15 shows the projection of the coordinates of the particles in the MC cell onto the XY plane for suspensions with $n_i = 1000, 78$, and $2.6n_p Z$. From these projections one can see that the suspensions at high n_i and at low n_i appear homogeneous, whereas those at $78n_p Z$ show a droplet of condensed phase consistent with the results shown in Fig. 6.14. Often a few particles are found to disscociate and reassociate from the dense phase droplets during the MC evolution. Similar dissociation and reassociation behavior has been noted in the case of smaller clusters as well. However, the total fraction F_c of particles constituting the clusters/dense phase remains constant during further MC evolution. Figure 6.16 shows F_c as a function of MCSs for suspensions with different n_i. The constancy of F_c for a system with $n_i = 1000n_p Z$ suggests a dynamical equilibrium between clusters and single particles. A similar inference can be made about the dense phase droplet and single particles from the behavior of F_c for suspensions at $78n_p Z$.

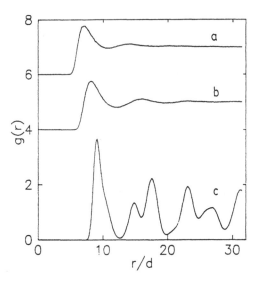

Figure 6.12 $g(r)$ versus r for suspensions at low impurity ion concentrations. Curves a, b, and c correspond to $n_i/n_p Z$ = 5, 3, and 0. Curves a and b are shifted vertically for the sake of clarity by amounts 6 and 4, respectively [32].

Table 6.2   Monte Carlo simulation results after reaching equilibrium for suspensions at different n_i. The numbers in parentheses represent the error in the least significant digits. The abbreviations HG, HL, HC, and PS represent the homogeneous gas, homogeneous liquid, homogeneous crystalline, and phase-separated states, respectively [32].

$n_i/n_p Z$	$U_T/Nk_B T$	g_{max}	F_c (%)	p^*-1	State
0	57.18(9)	3.68(2)	100	33.70(4)	HC
1.5	−1.08(5)	2.23(2)	100	−6.39(4)	HL
2.0	−2.87(4)	2.03(2)	100	−7.42(5)	HL
2.6	−3.30(4)	1.81(2)	100	−7.39(4)	HL
3.0	−3.28(4)	1.75(2)	100	−7.10(6)	HL
5.0	−2.65(5)	1.76(2)	100	−5.58(5)	HL
7.0	−2.70(5)	2.54(2)	99.6	−5.58(6)	PS
10.0	−3.46(5)	5.00(3)	97.4	−7.08(4)	PS
30.0	−4.14(8)	24.38(12)	88.8	−9.86(7)	PS
78.0	−5.14(2)	80.71(35)	81.0	−11.52(9)	PS
100.0	−2.75(3)	5.14(3)	49.6	−0.60(4)	HG
150.0	−0.130(9)	4.40(4)	31.2	−0.35(5)	HG
400.0	−0.021(18)	2.43(4)	12.0	−0.065(6)	HG
1000.0	−0.0033(20)	1.48(4)	3.1	−0.014(4)	HG

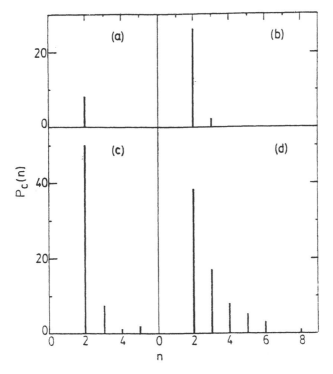

Figure 6.13   Cluster size distribution for suspensions at large n_i. (a), (b), (c), and (d) correspond to $n_i/n_p Z$ = 1000, 400, 150, and 100, respectively [32].

Based on the behavior of the pair correlation function and the manner in which clustering occurs for suspensions with $n_i \geq 7n_p Z$, the systems can be grouped into two classes. (1) Suspension with $n_i \geq 100 n_p Z$ exhibit small clusters without any structural correlation, and the cell volume is homogeneously occupied. One can see from Table 6.2 that for these systems p^* and g_{max} are low. Based on these features the suspension can be characterized to be in a homogeneous vapor (gas) phase. (2) Suspensions with impurity ion concentrations in the range $78 \geq n_i/n_p Z \geq 7$ exhibit dense phase droplets that have short-range structural correlations extending up to several neighbors and occupy only small regions of the simulation cell. The osmotic pressure and g_{max} are high for these suspensions. The gradual decay of $g(r)$ to a value 1 as r increases also reflects the boundary effects arising due to a compact droplet. These properties suggest that these suspensions can be considered to be a mixture of a condensed phase (liquidlike) in equilibrium with the vapor phase constituted by small clusters and free particles. Thus the colloidal suspension clearly exhibits a vapor-liquid condensation as the impurity ion concentration is reduced. These simulation results are analogous to condensation of a gas in a container of fixed volume. It may be mentioned that

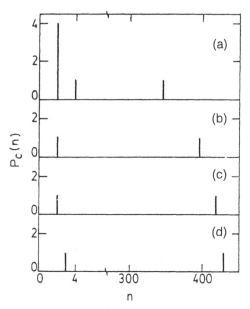

Figure 6.14 Cluster size distribution for suspensions at intermediate n_i. (a), (b), (c), and (d) correspond to n_i/n_pZ = 78, 30, 10, and 7, respectively [32].

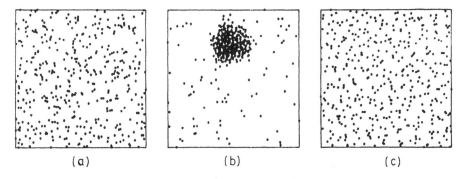

Figure 6.15 Projection of coordinates of the particles in the MC cell onto the XY plane for suspensions with (a) n_i = 1000 (b) 78, and (c) 2.6 n_pZ [32].

this phenomenon occurs in atomic systems (e.g., in rare gases) when the temperature is lowered. The magnitude of the potential minimum relative to thermal energy k_BT decides the transition temperature [34]. In the case of colloids one can change the magnitude of the interaction energy at a fixed temperature simply by changing the impurity ion concentration. Note from Table 6.1 that the depth of the potential minimum increases when n_i is reduced from 1000 to 78 n_pZ. Thus a reduction of n_i is analogous to reduction of the equivalent temperature T^*

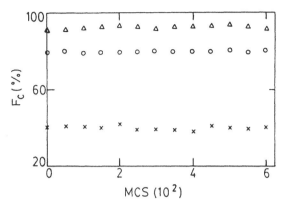

Figure 6.16   Total fraction F_c of particles constituting the clusters/dense phase as function of MCS. The symbols $\triangle$, O, and x correspond to $n_i/n_pZ = 30$, 78, and 100, respectively [32].

$= U_{\min}^{-1}$. Hence the occurrence of a vapor-liquid condensation as a consequence of a reduction of n_i is understandable.

Having identified the phases for suspensions with different n_i values, it is meaningful to identify the value of n_i at which the transition takes place. Figures 6.17 and 6.18 show the behavior of osmotic pressure and $g_{\max}$, respectively, as a function of n_i. Note the sudden increase of both these parameters around 90

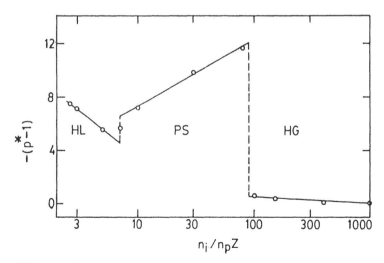

Figure 6.17 $-(p^* - 1)$ versus impurity ion concentration n_i. HG, PS, and HL represent the homogeneous gaseous state, phase-separated state, and homogeneous liquid state, respectively. Lines drawn through the simulation points are guides to the eye [32].

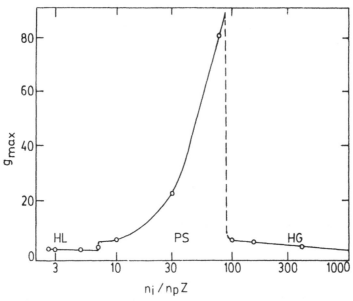

Figure 6.18   Height g_{max} of the first peak of $g(r)$ as a function of n_i. The abbreviations HG, PS, and HL are the same as those in Fig. 6.17. Lines drawn through the simulation points are guides to the eye [32].

$n_p Z$ as n_i is reduced. This change is associated with the vapor-liquid condensation. The large values of these parameters arise due to scaling with respect to particle concentration n_p corresponding to the homogeneous dispersion instead of the particle concentration in the condensed phase, n_d, which is several times higher than n_p. The behavior of p^* and g_{max} as a function of n_i exhibits another discontinuous change at $n_i = 7 n_p Z$, and these quantities increase when n_i is further reduced. One can also see from Fig. 6.12 that the suspensions exhibit either liquidlike or crystalline structural ordering and the suspensions again become homogeneous (see Fig. 6.15). Hence the simulations predict a "subtle" transition from a phase-separated state once again to a homogeneous state. This can happen if the droplet of the condensed phase, which exhibits strong structural correlations, now expands to occupy the full volume of the simulation cell. This can be understood if one examines the behavior of R_m, which increases monotonically as n_i is reduced. Hence in the phase-separated state the volume occupied by the dense phase droplets increases as n_i is decreased. Eventually the droplet occupies the full volume of the MC cell, as is evident from Fig. 6.9, which shows that $R_m \geq \ell$ at low impurity ion concentrations. The behaviors of g_{max} and p^* suggest that the transition to this reentrant homogeneous state occurs at $7 n_p Z$. Thus the MC simulations carried out using the potential $U_S(r)$ predict a vapor-liquid condensation and a reentrant transition. It is important to point out that the reentrant transition at low n_i is unique to colloidal systems as it arises purely due to the specific dependence of the potential minimum on n_i. Atomic systems that have a fixed R_m do not exhibit this transition. As mentioned earlier, it is found that the suspension at $n_i = 0$ is crystalline. Note from Table 6.2 that in contrast to other suspensions U_T and $p^* - 1$ are positive in this case. This is because $R_m \geq \ell$ and the particles essentially see the repulsive edge of the potential. p^* being negative at finite n_i is due to attractive part dominating over the repulsive part [34].

6.6 Interpretation of Experimental Results and Comparison with Simulations

The prediction of the simulations discussed in the previous section, that a weakly interacting suspension at $n_i \sim 100 n_p Z$ or equivalently at $n_i \sim 10^{17}$ cm^{-3} goes to a phase-separated state upon deionization only over a limited range of n_i and subsequently to a homogeneous and strongly interacting state at $n_i \sim 10^{15}$ cm^{-3} having liquidlike order, is essentially the same as what is exhibited by samples S1-S3 (see Section 6.3). The agreement is even quantitative because, as mentioned earlier, the samples S1-S6 contain impurity ion concentration $\sim 10^{17}$ cm^{-3} before deionization, and after reaching deionization equilibrium these have $n_i \sim 3 \times 10^{15}$ cm^{-3}. Thus one can see a complete agreement between simulation results [32] and experimental observations. The reason for not observing a reentrant phase transition in samples S4-S6 is due to the inability to reduce n_i further such that $R_m > R_0$. However, if one uses advanced gradient-free method [38], it is possible to achieve a very low value of n_i. It is worth pointing out that recently Palberg and Würth [39] have carried out experiments similar to those of Tata et. al. [4] under different sealing conditions of the sample cells. They also observe phase separation in cells with relatively large n_i, whereas the suspensions in the cells with low n_i remain homogeneous, consistent with Ref. 4; however, the phenomenon is interpreted as a nonequilibrium one arising due to the possible gradient in n_i. These authors argue that the gradient causes the particles to move toward the region of low n_i, that is, toward the ion-exchange resins leading to a phase separation, and further claim that the reported phenomenon [4] does not experimentally justify a discussion of the validity of the DLVO potential. In order to study the possible effects of the gradient in n_i on the VL condensation, Tata and Arora [40] have carried out further experiments where the ion-exchange resin is confined to the upper part of the cells. This would reverse the direction of the gradient of n_i, if any. These cells again exhibit phase separation, and the dense phase is found to settle to the bottom of the cell, that is, away from the resins as shown in Fig. 6.19. This result clearly demonstrates that the n_i gradient has little effect on the VL condensation and also disproves the hypothesis of Palberg and Würth [39].

It is worth mentioning the important difference between the phase-separated state seen in the MC simulations and that observed in experiments. In the experimental situation, the dense phase appeared as a macroscopic phase with clear

Figure 6.19   Photograph of the cell exhibiting VL condensation in an aqueous suspension of polystyrene particles of diameter 110 nm at $n_p = 2 \times 10^{12}$ cm^{-3} with the ion-exchange resin in a bag in the upper part of the cell [40].

phase boundary separating the rare phase, whereas simulations show the dense phase in the form of liquid droplets coexisting with a rare phase constituted by isolated particles. The appearance of macroscopic dense phase in the experiments is due to the action of gravity on dense phase clusters, which is not considered in the simulations. In experiments the dense phase clusters formed during phase separation settle due to the action of gravity (sedimentation of clusters) and constitute the macroscopic phase. This is shown schematically in Fig. 6.20. In a density-matched (in $H_2O + D_2O$ mixture as the background medium instead of water) experiment dense phase clusters are expected to float in the sea of rare phase. In such a situation it would be possible to observe the existence of dense phase clusters if proper scattering experiments are performed.

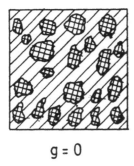
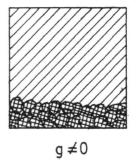

$$g = 0 \qquad\qquad g \neq 0$$

Figure 6.20   Schematic representation of effect of gravity on dense phase clusters.

It may be pointed out that in addition to VL condensation and reentrant transitions, a number of other experimental observations can also be understood in light of the MC simulation results. The recent observation of isolated bound pairs in dilute aqueous polystyrene suspensions [14] is also consistent with the prediction of MC results as evident from the cluster size distribution (see Figs. 6.13 and 6.14). One can see that the suspensions in the homogeneous vapor phase as well as in the phase-separated state have bound pairs. Only the fraction of particles forming the bound pairs changes as a function of n_i. The microscopic observation of the coexistence of ordered and disordered regions in colloidal suspensions, which is known as the two-state structure [8–10, 15,16], is similar to the dense phase coexisting with the rare phase, seen in the present simulation results. The dense phase observed microscopically is an ordered phase, in contrast to the liquid droplet seen in MC simulations. This is because of the high surface charge density on polystyrene spheres, which leads to a well depth several times larger than $k_B T$, where as in the present simulations the well depth is $\simeq k_B T$. In view of the present simulation results and the experimental observations, it is possible to conjecture a vapor-solid transition in colloidal systems with particles having high charge density for appropriate values of n_i.

6.7 Phase Diagram Based on the Free Energy Calculations

Having seen that MC simulations based on $U_S(r)$ reproduce the experimental observations, it is of interest to obtain the phase diagram in the n_p versus n_i plane. As performing simulations over a wide range of combinations of n_p and n_i could be time consuming, an alternate approach based on free energy calculations of homogeneous and phase separated states has also been attempted [5, 31] and is briefly described here.

Consider first a homogeneous suspension of N particles, each of radius a in a volume V ($n_p = N/V$) interacting via $U_S(r)$. The free energy F_h of the homogeneous state can then be written as

$$F_h = \tfrac{1}{2} p_n N U_S(R_0) - T k_B N S(\phi_e) \tag{6.6}$$

where p_n is the number of nearest neighbors and the second term in Eq. (6.6) is the contribution of the entropy. Entropy is obtained in a straightforward manner [41] from the number of possible ways in which N particles can be distributed in N_c cells, each of an equivalent hard-sphere volume v_h, such that $N_c = V/v_h$. The effective volume fraction is then $f = n_p v_h$, and the function $S(\phi_e)$ is given as [41]

$$S(\phi_e) = \phi_e^{-1} \ln(\phi_e^{-1}) - (\phi_e^{-1} - 1) \ln(\phi_e^{-1} - 1) \tag{6.7}$$

As the Coulomb barrier [the repulsive part of $U_S(r)$] does not allow the particles to come close to each other, it is meaningful to visualize the particles as hard spheres of equivalent diameter d_h ($v_h = \pi d_h^3/6$) for the purpose of calculating the entropy. The hard sphere diameter is obtained as [36]

$$d_h = d + \int_d^{r_0} \{1 - \exp[-U_S(r)/k_B T]\} \, dr \tag{6.8}$$

where r_0 is the distance at which $U_S(r)$ goes to zero.

In the phase-separated state the particles in the dense phase are assumed to stay in the potential minimum. The particle concentration in the dense phase is $n_d = (3\sqrt{3}/4) R_m^{-3}$. If xN particles out of the total N are present in the dense phase, the volume occupied by them is xN/n_d and the remaining $(1-x)N$ particles are assumed to be uniformly distributed in the rest of the volume to constitute the rare phase. The particle concentration in the rare phase is then $n_r = (1-x)[1/n_p - x/n_d]^{-1}$ and the average interparticle separation R_r in the rare phase becomes $(4n_r)^{-1/3}/\sqrt{3}$. The free energy of the phase-separated state, $F_{PS}(x)$, which depends on the fraction x, can now be written as

$$F_{PS}(x) = E_{PS}(x) - T S_{PS}(x) \tag{6.9}$$

where the interaction energy is

$$E_{PS}(x) = N[x p_1 U_S(R_m) + (1-x) p_2 U_S(R_r)] \tag{6.10}$$

where p_1 and p_2 are the number of nearest-neighbors in the dense and the rare phases, respectively. The entropy is

$$S_{PS}(x) = K_B N[x S(\phi_1) + (1-x) S(\phi_2)] \tag{6.11}$$

where $\phi_1 = n_d v_h$ and $\phi_2 = n_r v_h$ are effective volume fractions in the dense and the rare phases, respectively. The phase separation will occur if the energy of the phase-separated state is lower than that of the homogeneous state; that is, $\Delta F(x) = F_{PS}(x) - F_h < 0$. Figure 6.21 shows $\Delta F(x)$ in units of the thermal energy $Nk_B T$ as a function of x for various n_i values for $n_p = 4.2 \times 10^{12}$ cm^{-3} or volume fraction $\phi = 0.284\%$. The number of nearest-neighbors that could be different for different phases also enters into the calculations as a parameter. In order to keep the calculations free from the arbitrariness of parameters to the extent possible, we take $p_n = p_1 = p_2 = 8$, which corresponds to a BCC-like coordination. This is reasonable as dilute suspensions are known to crystallize into BCC structures [42]. One can see from Fig. 6.21(a) that as n_i decreases, $\Delta F(x)$ becomes negative and again becomes positive for smaller values of n_i. The equilibrium fraction x corresponds to the minimum in these curves. Figure 6.21(b) shows the potential $U_S(r)$ for different values of n_i. Note that the depth and the position of the well depends strongly on n_i.

The phase diagram (ϕ versus n_i) obtained from the condition $\Delta F(x) < 0$ is shown in Fig. 6.22 and can be physically understood as follows. The suspension is homogeneous on the high-impurity side because the depth of the well is not large enough to trap the particles, as seen in curve I_1 of Fig. 6.21(b), and entropy forces the system to remain homogeneous. As one moves across the

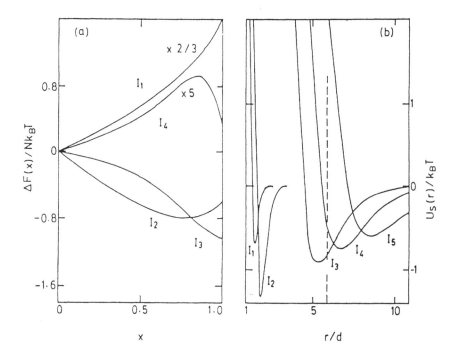

Figure 6.21 The difference in the free energy $\Delta F(x)$ relative to the thermal energy NK_BT as a function of the fraction x for various n_i values expressed relative to counterion concentration $n_p Z$ ($I = n_i/n_p Z$). $I_1 = 170$, $I_2 = 60$, $I_3 = 3$, and $I_4 = 1.5$. (b) $U_S(r)$ for impurity-ion concentration I_1 to I_4 same as those in (a); $I_5 = 0.5$. The vertical dashed line shows the position of average interparticle separation ℓ. The parameters of the suspension are the volume fraction $\phi = 0.284\%$, $d = 109$ nm, and $Z = 500$ [5].

phase boundary PB1, the system phase separates and the particle concentration in the dense phase is dictated by the position of the potential well [see curves I_2 and I_3 of Fig. 6.21(b)]. Experimentally n_d can be estimated from a knowledge of n_r and the height of the dense phase [see Fig. 6.3(c)] using the conservation of the total number of particles, as $n_d/n_p = [H - (H - H_d)n_r/n_p]/H_d$. H is the total height of the suspension and H_d that of the dense phase. The average n_d estimated in this manner is found to be as high as $45n_p$ at $t = 1.75$ h and reduces monotonically to a value close to n_p as system approaches the phase boundary PB2. Note similar behavior of n_d from MC simulations (see Table 6.2). As mentioned earlier for the values of n_i for which $\ell < R_m$, the particles see the repulsive edge of $U_S(r)$; hence the system is expected to be homogeneous again [see curve I_5 in Fig. 6.21(b)]. The phase transition across the boundary PB1 is analogous to the condensation of monatomic gas where one observes the dense (liquid droplet) and the rare (vapor) phases coexisting; however, the phase boundary PB1 is entirely due to the strong dependence of R_m on n_i.

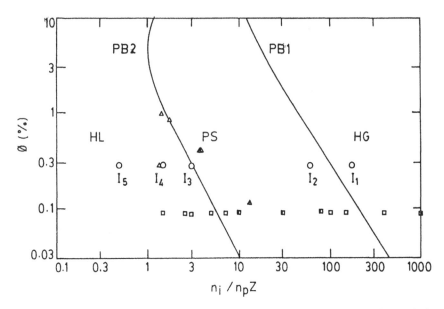

Figure 6.22 Calculated phase diagram as a function of volume fraction ϕ. Open circles correspond to $\phi = 0.284\%$ and n_i values I_1 to I_5 of Fig. 6.21. Open sqares represent the homogeneous state, and half-filled squares represent the phase-separated state predicted by MC simulations. Triangles represent samples S1 to S6, and n_i is taken to be 3×10^{15} cm^{-3}.

In the present theoretical formalism, though the charge on the colloidal particle, and the number of nearest neighbors in different states enter as parameters, it may be mentioned that the reentrant phase transition is predicted for a wide range of the values of Z and p_n that are physically reasonable. The effect of increasing Z or p_n is to increase the separation between the boundaries PB1 and PB2. For very large values of Z it may not be possible to encounter the reentrant boundary PB2 even for zero n_i on the high ϕ side of the phase diagram. This seems to be the cause of the existence of two-state structures in high-charge-density latex suspensions [8–10]. The value of κ has been taken to be the same in the rare and the dense phases. It is also possible to generalize the model to have κ different in the two phases in a self-consistent manner.

6.7.1 Comparison of Phase Diagram with Experimental and Simulation Results

Since MC simulations as well as the free energy calculations have been carried out using the same pair potential, a comparison of the results of the two calculations provides a valuable information on the validity and reasonability of the assumptions involved in free energy calculations, as the simulations do not involve any assumptions. The simulation results are shown in Fig. 6.22 for comparison. Note that except for n_i values close to the phase boundary PB2, free energy calculations agree well with simulation results. Simulation results for n_i = $100n_pZ$ and $150n_pZ$ predict that the suspensions are in a homogeneous and noninteracting state. However, the free energy calculations indicate that these belong to a phase-separated state. This discrepancy is due to (1) the assumption that the number of nearest neighbors are the same in the homogeneous and phase-separated states and (2) approximations involved in estimating entropy in rare phase, for example, the simple procedure to estimate the hard-shere diameter. Thus the free energy calculations as well as MC simulations based on $U_S(r)$ predict the colloidal suspension to exhibit either a homogeneous (interacting or noninteracting) or a phase-separated state depending on the impurity ion concentration present in the suspension. Since for the samples S1-S6 n_p, n_i, and the equilibrium states of the system are known, the experimental data are also shown in Fig. 6.22. Note the good agreement between experimental results and that obtained from free energy calculations.

Thus the report of the experimental observation of vapor-liquid condensation and reentrant phase transitions in dilute charged colloidal suspensions *unambiguously* suggest the existence of an attraction in the effective pair potential between two charged colloidal particles. Since $U_S(r)$ could reasonably well explain all the experimental observations, the effective pair potential in these systems might be of the form of $U_S(r)$. The widely used screened Coulomb repulsive potential cannot explain these experimental observations and also other experimental results [6-16], which suggest the existence of attraction in these systems. However, in view of the success of the interparticle potential $U_S(r)$ in explaining these phase transitions and also in explaining [43, 30] the elastic constant data of Lindsay and Chaikin [29] as well as the measured $S(Q)$ of liquidlike ordered suspensions [26], one can conclude that it is a good model potential for describing the charge-stabilized colloidal suspensions.

6.8 Conclusions

In conclusion, the charge-stabilized colloidal suspensions exhibit vapor-liquid condensation and a subtle reentrant phase transition upon deionization. Angle-resloved polarized light scattering is shown to be an appropriate tool for investigating these phenomena in colloids. Suspensions having a finite n_i exhibit reentrant behavior only when their particle concentration is above a critical value. The present experimental observations unambiguously suggest the existence of an attraction at large interparticle separations between colloidal particles in dilute charge-stabilized colloids. Hence the widely used repulsive screened Coulomb potential or DLVO potential needs reconsideration. MC simulations carried out for different n_i values, with an effective pair potential having an attractive term, reproduces all the experimental observations. The free energy calculations also lead to two phase boundaries. The strong dependence of the potential minimum on κ leads to the above phase transitions. The theoretical as well as MC simulation results suggest that Sogami's potential appears to be a better choice than DLVO potential in explaining the ordering and associated phase transitions between different phases.

Acknowledgements

The authors wish to thank M. C. Valsakumar for discussions and Dr. K. Krishan for encouragement.

References

1. van der Waals, J. D., *Thesis Leiden* (1873); also see *Physica* **73**, (1974) (this volume contains both technical and historical articles on this subject).
2. Maxwell, J. C., *Nature* **10**, 477 (1874).
3. Lebowitz, J. L., and Waisman, E. M., *Physics Today* **33**, 24 (1980).
4. Tata, B. V. R., Rajalakshmi, M., and Arora, A. K., *Phys. Rev. Lett.* **69**, 3778 (1993).
5. Arora, A. K., Tata, B. V. R., Sood, A. K., and Kesavamoorthy, R., *Phys. Rev. Lett.* **60**, 2438 (1988).
6. Daly, C. G., and Hasting, R., *J. Phys. Chem.* **85**, 294 (1981).
7. Arora, A. K., and Kesavamoorthy, R., *Solid State Commun.* **54**, 1047 (1985).
8. Yoshiyama, T., Sogami, I., and Ise, N., *Phys. Rev. Lett.* **53**, 2153 (1984).
9. Ise, N., Matsuoka, H., and Ito, K., *Macromolecules* **22**, 1 (1989).
10. Okubo, T., *J. Chem. Soc. Faraday Trans. I* **82**, 3163 (1986).
11. Ise, N., Ito, K., and Yoshida, H., *Polymer Preprints* **33**, 769 (1992).
12. Ito, K., Yoshida, H., and Ise, N., *Science* **1**, 1 (1993).
13. Kesavamoorthy, R., Rajalakshmi, M., and Baburao, C., *J. Phys.: Condens. Matter* **1**, 7149 (1989).
14. Yoshino, S., in *Ordering and Organization in Ionic Solutions*, Eds. N. Ise and I. Sogami (World Scientific, Singapore, 1988) p. 449.
15. Ise, N., Matsuoka, H., Ito, K., and Yoshida. H., *Faraday Discuss. Chem. Soc.* **90**, 153 (1990).
16. Ito, K., Okumura, H., Yoshida, H., and Ise, N., *J. Am. Chem. Soc.* **111**, 2347 (1989).
17. Verwey, E. J. W., and Overbeek, J. Th. G., *Theory of the Stability of Lyophobic Colloids* (Elseveir, Amsterdam, 1948).
18. Sood, A. K., in *Solid State Physics*, Eds. H. Ehrenreich and D. Turnbull (Academic, New York, 1991) Vol. 45, p. 1.
19. Sogami, I., *Phys. Lett.* **96A**, 199 (1983).
20. Sogami, I., and Ise, N., *J. Chem. Phys.*. **81**, 6320 (1984).
21. Kerker, M., *The Scattering of Light and Other Electromagnetic Radiation* (Academic Press, New York, 1969).
22. Berne, B. J., and Pecora, R., *Dynamic Light Scattering* (Wiley, New York, 1976).
23. Arora, A. K., *J. Phys. E: Sci. Instruments* **17**, 1119 (1984).
24. Kesavamoorthy, R., Arora, A. K., and Vasumathi, D., *A Microprocessor Based Automation System for Raman and Rayleigh Spectrometers*, IGC Report No. IGC-103 (1988) unpublished.
25. Kesavamoorthy, R., Sood, A. K., Tata, B. V. R., and Arora, A. K., *J. Phys. C: Solid State Phys.* **21**, 4737 (1988).
26. Tata, B. V. R., Kesavamoorthy, R., and Sood, A. K., Molec. Phys. **61**, 943 (1987).
27. Kesavamoorthy, R., and Arora, A. K., *J. Phys. A: Math. Gen.* **18**, 3389 (1985).
28. Asher, S. A., Flaugh, P. L., and Washinger, G., *Spectroscopy* **1**, 26 (1986).
29. Lindsay, H. M., and Chaikin, P. M., *J. Chem. Phys.* **76**, 3774 (1982).
30. Tata, B. V. R., Sood, A. K., and Kesavamoorthy, R., *Pramana - J. Phys.* **34**, 23 (1990).
31. Tata, B. V. R., Ph.D. Thesis, University of Madras, 1992 (unpublished).
32. Tata, B. V. R., Arora, A. K., and Valsakumar, M. C., *Phys. Rev.* **E 47**, 3404 (1993).

33. Binder, K., in *Monte Carlo Methods in Statistical Physics*, Ed. K. Binder (Springer Verlag, New York, 1979), p. 1.

34. Hansen, J. P., and McDonald, I. R., *Theory of Simple Liquids*, 2nd edn. (Academic, London, 1986).

35. Hirtzel, C. S., and Rajagopalan, R., in *Micellar Solutions and Microemulsions: Structure, Dynamics and Statistical Thermodynamics*, Eds. S.-H. Chen and R. Rajagopalan (Springer Verlag, New York 1990), p. 111.

36. Castillo, A., Rajagopalan, R., and Hirtzel, C. S., *Rev. Chem. Engg.* **2**, 237 (1984).

37. Müller-Krumbhaar, H., in *Monte Carlo Methods in Statistical Physics*, Ed. K. Binder (Springer Verlag, New York, 1979) p. 195.

38. Palberg, T., Hartl, W., Wittig, U., Versmold, H., Würth, M., and Simnacher, E., *J. Phys. Chem.* **96**, 8180 (1992).

39. Palberg, T., and Würth, M., *Phys. Rev. Lett.* **72**, 786 (1994).

40. Tata, B. V. R., and Arora, A. K., *Phys. Rev. Lett.* **72**, 787 (1994).

41. Williams, R., Crandall, R. S., and Wojtowicz, P. J., *Phys. Rev. Lett.* **37**, 348 (1976).

42. Williams, R., and Crandall, R. S., *Phys. Lett.* **48A**, 225 (1974).

43. Ito, K., Sumaru, K., and Ise, N., *Phys. Rev.* **B 46**, 3105 (1992).

7

Order-Disorder Transition in Charge-Polydisperse Colloids

Akhilesh K. Arora and B. V. R. Tata

Size and charge polydispersities are inherent to colloidal suspensions. At large polydispersities colloidal crystals can become disordered due to size-mismatch frustration. Charge polydispersity (CPD) plays an important role in determining the structural ordering in dilute charged colloids. Monte Carlo simulations show that beyond a critical CPD the long-range order of the colloidal crystal is destroyed and an amorphous (disordered) structure emerges. The amorphous state is characterized by the shape of the pair correlation function and significantly larger mean-square displacement representing enhanced thermal diffusion. The equilibrium positions of the particles in the time-averaged structure also exhibit a sudden increase of spatial disorder across the crystalline (c) to amorphous (a) transition and can be used to estimate configurational entropy. The critical polydispersity of effective hard-sphere diameters corresponding to the $c \rightarrow a$ transition, estimated using a simple model based on distance of closest approach, turns out to be comparable to that in size-polydisperse systems, suggesting the equivalence of the two systems. MC simulations are also used to probe the relative stability of random and charge-ordered states. These results are discussed in the light of those on random binary atomic systems. Further, upon increasing the ionic strength of the suspension, the charge-polydisperse colloidal crystals are found to be less stable against melting than the monodisperse counterparts. The reasons for the reduced stability are also pointed out.

7.1 Introduction

In contrast to the atomic systems, particles in a colloidal suspension are usually not identical. The variations of size, shape, and charge from particle to particle, which can be described by suitable distributions, can influence the structural ordering [1, 2], particle diffusion [3], and also the thermodynamic properties [4] of the suspension. Whereas the shape polydispersity can be avoided, the size and the charge polydispersities are unavoidable. It is now possible to synthesize colloidal particles with desired diameter and surface charge density [5]; hence one can, in principle, control the size and the charge on the particle rather independently and also their polydispersities. The knowledge of the effect of various types of polydispersities on the stability of the colloidal dispersions is also of importance from the point of view of their applications such as in optical Bragg filters [6].

7.1.1 Conseqences of Size Polydispersity: The Present Understanding

The effect of size polydispersity on the structure of colloidal fluids and colloidal crystals has been a subject of considerable current interest over the past several years. Qualitatively one expects a *departure from perfect crystalline order* to occur in randomly substituted binary colloidal crystals similar to that found in atomic alloys [7]. The larger the number of components, the greater will be this departure from perfect order or equivalently the *disorder* [8]. Although one may choose a variety of distributions with appropriate widths to describe the size-polydisperse system, the size polydispersity (SPD) is best quantified in terms of the standard deviation of the size relative to its mean $< a >$

$$\text{SPD} = (< a^2 > - < a >^2)^{1/2} / < a > \tag{7.1}$$

There may arise a situation where in a system with sufficient SPD the long-range order may cease to exist. The system may then exhibit only short-range order similar to that found in amorphous or glassy materials. In hard-sphere systems this can be understood to arise due to the size-mismatch frustration. In charged colloids, electrostatic interactions also play a role in determining the polydispersity at which crystalline order disappears.

Dickinson and co-workers [1, 9, 10, 11] carried out molecular dynamics (MD) simulations of size-polydisperse colloidal crystals interacting via a DLVO potential. The disapparence of the volume change across melting is used as the criterion to identify the order-disorder transition. These results are in qualitative agreement with those obtained from the comparison of the Helmholtz free energy as a function of SPD calculated using the polydisperse Evans-Naper model [12] in the ordered and disordered states [10]. The osmotic pressure is also found to increase with SPD [1]. The disordered state has been assigned to a fluid. The charge- and the size-polydisperse fluids have been investigated in much more detail because of the availability of analytic solutions to the hard-sphere systems under the Percus-Yevick (PY) approximation. Vrij and co-workers [13, 14] obtained the structure factor $S(Q)$ of a polydisperse colloidal system from the Baxter solution of the PY approximation for hard-sphere fluids. The structure factor is found to smear out as SPD is increased, suggesting a loss of structural correlations. An alternate approach [15] to obtain $S(Q)$ also yields similar results. Barat and Hansen [16] applied the density-functional theory of freezing to polydisperse hard-sphere systems and obtained its phase diagram. In polydisperse systems freezing is predicted to occur at a volume fraction ϕ higher than that corresponding to the monodisperse system. Investigations on the polydisperse Yukawa systems have begun only recently [17]. The mean spherical approximation [18] (MSA) has been used to obtain the structure of binary mixtures [19]. The difficulty in selecting acceptable solutions from a set of solutions has prevented the application of MSA to suspensions with continuous polydispersity. In order to solve the Ornstein-Zernike equation under the Rogers-Young (RY) approximation for a continuous polydisperse Yukawa system, a discretization method has been recently proposed [20]. The charge on the particle is modeled to vary as the square of its diameter, keeping the surface charge density constant. Discretization of the continuous polydispersity into three components has been claimed to be sufficient to obtain good agreement between the RY structure factor and the Monte Carlo data.

Pusey [21] presented a very simple geometrical argument to obtain the critical polydispersity for a hard-sphere system that is sufficient to disrupt the crystalline order. A set of particles with average radius $< a >$ and SPD δ can remain ordered if their diameters are less than the nearest-neighbor separation between their centers R

$$2 < a > (1 + \delta) < R \tag{7.2}$$

R is related to the particle concentration n_p, and for the face-centered cubic lattice, which can have the maximum packing fraction ≈ 0.74, it has a value $(6 \times 0.74/\pi n_p)^{1/3}$. Similarily, the diameter $2 < a >$ can also be expressed as $(6\phi/\pi n_p)^{1/3}$. Substituting these into Eq. (7.2), one gets

$$1 + \delta < (0.74/\phi)^{1/3} \tag{7.3}$$

Choosing $\phi = 0.545$, a value corresponding to the melting of hard sphere crystals [22], one gets $\delta < 0.11$. The MD results for a triangular size distribution [11] and those of density-functional theory for triangular and rectangular distributions [16] are in good agreement with this estimate when polydispersity is calculated using Eq. (7.1).

In charged colloids, the crystalline order arises due to strong electrostatic interactions. Colloidal crystals can be made to melt by increasing the ionic strength [2], which reduces the range of interaction represented by the Debye screening length. Investigations of the effect of charge polydispersity on the structural and other properties of colloidal suspensions have been started only recently [2–4, 8]. In some of these, both the charge and the size have been made to vary simultaneously, according to some hypothesis, such as linear [4] or quadratic [20]. Using the variational method based on the Gibbs-Bogoliubov inequality, the equivalence of the charge-polydisperse Yukawa fluid and a reference polydisperse hard-sphere fluid has been obtained. In contrast to the hard sphere systems, the excess free energy and the osmotic pressure are found to decrease as the charge polydispersity is increased [4]. Computer simulations are also extensively used in obtaining detailed information on the structure, diffusion, and thermodynamic properties [2, 3, 8, 23]. Beyond a critical charge polydispersity the colloidal crystals are found to turn amorphous [3]. The behavior of the melting of the colloidal crystals by increasing the ionic strength has also been investigated, and the resulting phase diagram has been reported [2]. This chapter reviews the results of recent Monte Carlo (MC) simulations of the crystalline-to-amorphous ($c \rightarrow a$) transition and the melting of the charge-polydisperse colloidal crystals.

The organization of the chapter is as follows. Section 7.2 briefly discusses the MC simulation technique applied to charge-polydisperse colloidal suspensions. In Section 7.3 the $c \rightarrow a$ transition is identified and an effective hard-sphere model discussed. Section 7.4 presents the results of charge-polydisperse crystalline and amorphous colloidal suspensions. The phase diagram exhibiting regions of stability of crystalline, amorphous, and liquidlike order is discussed in Section 7.5. The possibilities of phase separation of charge-polydisperse colloidal suspension and those of charge ordering are examined in Section 7.6. The chapter ends with Section 7.7, giving a summary and conclusions.

7.2 Monte Carlo Simulations

In order to simulate a charge-polydisperse colloidal suspension, one takes N particles of diameter d in a cubic simulation cell of volume V, and the charges Z_i are assigned on the particles randomly from a symmetric rectangular distribution $P(Z)$ with mean Z_0 and width 2Δ given as

$$P(Z) = \begin{cases} (2\Delta)^{-1}, & \text{for} Z_0 - \Delta \leq Z \leq Z_0 + \Delta \\ 0, & \text{otherwise} \end{cases} \tag{7.4}$$

Charge polydispersity is defined in the same way as was done in the case of SPD

$$\text{CPD} = (< Z^2 > - < Z >^2)^{1/2} / < Z > \tag{7.5}$$

which in this case is $\Delta/\sqrt{3}Z_0$. Metropolis sampling [24] is used to obtain the equilibrium at temperature T in a canonical ensemble [25] (constant NVT ensemble). The use of periodic boundary conditions eliminates the surface effects. The simulation cell length is fixed according to $L_s^3 = N/n_p$ to get the required particle concentration n_p. Although it may be more appropriate to use the Sogami potential as the interparticle interaction, the size-corrected Yukawa potential is often used in investigations of the consequences of charge polydispersity because this allows one to compare these results with those already reported on size-polydisperse systems. $U(r_{ij})$ between particles i and j with charges $Z_i e$ and $Z_j e$ (e being the electronic charge) along with a hard-sphere repulsion is then written as

$$U(r_{ij}) = \begin{cases} \infty, & r_{ij} \le d \\ \dfrac{Z_i Z_j e^2}{\epsilon} \dfrac{4 \exp(\kappa d)}{(2 + \kappa d)^2} \dfrac{\exp(-\kappa r_{ij})}{r_{ij}}, & r_{ij} > d \end{cases} \tag{7.6}$$

where r_{ij} is the center-to-center distance between particles i and j. κ, the inverse Debye screening length, arising from the counterion concentration $n_p Z_0$ and the impurity ion concentration n_i in the aqueous suspension, is given as

$$\kappa^2 = 4\pi e^2 (n_p Z_0 + n_i)/\epsilon k_B T \tag{7.7}$$

where k_B is the Boltzmann constant and ϵ is the dielectric constant of water. At low volume fractions, $\phi \sim 10^{-3}$, the average interparticle separation is several times the particle diameter; hence a step size of $0.5d$ is normally optimum [2], which gives the trial rejection ratio around 50% after equilibrium is reached. Further the choice of the initial configuration as body-centerd cubic (BCC) is more appropriate at low ϕ because the BCC structure has been found to be stable there [26].

The system is allowed to evolve, and the total energy $U_T = U_T(r_1, \ldots, r_N)$, obtained by assuming pairwise additivity of interactions, is monitored. Approximately a million configurations are initially discarded during which the system evolves rapidly toward equilibrium. An MC step (MCS) is defined as a set of N configurations during which, on the average, each particle gets a chance to move. U_T and a few other structural parameters like the pair correlation function $g(r)$ and structure factor $S(Q)$ are monitored to ascertain whether equilibrium has been reached or not. $g(r)$ is calculated with a small interval $\Delta r = 0.1d$ using the standard method [27]. The virial equation is used to obtain the osmotic pressure P [25]. Because the colloidal system is polydisperse only in charge and not in size, it is sufficient to describe the system in terms of a number-number structure factor $S(Q)$ and $g(r)$ [2, 28]. In the case of crystalline order, $S(\mathbf{Q})$ at selected wave vectors $\mathbf{Q}$ can be calculated using the expression

$$S(\mathbf{Q}) = < \rho_{\mathbf{Q}} \rho_{-\mathbf{Q}} > /N \tag{7.8}$$

where

$$\rho_{\mathbf{Q}} = \sum_{j=1}^{N} \exp(-i\mathbf{Q} \cdot \mathbf{r}_j)$$

and $< \cdot >$ represents the ensemble average. However, in the case of liquidlike order or amorphous solids, it is sufficient to obtain $S(Q)$ by Fourier transforming $g(r)$ because the system is then isotropic.

Diffusion of the particles is widely different in suspensions with crystalline, amorphous, or liquidlike orders. In charge-polydisperse systems different particles would diffuse differently because of the charges on each being different; however, the mean-square displacement (MSD) averaged over all the particles still represents the overall behavior of the system. MSD is defined as

$$< r^2(m) >= \frac{1}{N} \overline{\sum_{i=1}^{N} [r_i(m+n) - r_i(n)]^2} \qquad (7.9)$$

where $r_i(m)$ is the position of the ith particle after m MCSs and the overbar denotes averaging over the initial configurations n.

7.3 Effect of Charge Polydispersity

The polydispersity of charge leads to a polydispersity of the interaction energy U_{ij} between various neighbors. This interaction polydispersity causes the particles to relax to new equilibrium positions such that $U_{ij}(r_{ij}) \sim U_0$, where U_0 is the average interaction energy. This causes the nearest-neighbor distance to have a distribution leading to the broadening of the peaks in $g(r)$ as compared to those of a monodisperse colloidal crystal. Figure 7.1 shows the pair correlation

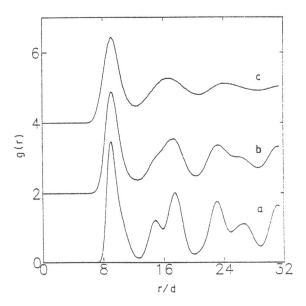

Figure 7.1   Pair correlation function $g(r)$ of an aqueous colloidal suspension of polystyrene particles at different CPDs. (a) 0% (monodisperse), (b) 24%, and (c) 28% CPD. The curves are vertically displaced for the sake of clarity. Note the absence of features characteristic of crystalline order in the suspension with 28% CPD [3].

function $g(r)$ obtained at several CPDs for a system of 432 particles representing an aqueous colloidal suspension of polystyrene particles of diameter $0.109\ \mu$m at a concentration of 1.33×10^{12} cm^{-3} ($\phi \sim 0.001$) and $n_i = 0$. The average charge on the particle is 600 [29], and the temperature of the simulation is taken to be 298 K. Broadening of the peaks in $g(r)$ is clearly seen in Fig. 7.1 as CPD is increased. Note that the suspension with 24% CPD exhibits peaks characteristic of the crystalline order, whereas that with 28% CPD does not, suggesting that it is probably disordered. The widths of the peaks in $g(r)$ also have a contribution from the thermal motion of the particles. The equilibrium positions of the particles are obtained by averaging the particle coordinates for large enough time or equivalently MCSs, such that the thermal motions are averaged out [30]. This procedure is equivalent to cooling the system to low temperatures. The pair correlation function obtained from such coordinate-averaged frames, $g_c(r)$, is shown in Fig. 7.2. 5×10^5 configurations have been found to be adequate for averaging. Sharp peaks corresponding to the twelve neighbor shells of BCC structure for zero CPD gradually broaden as CPD is increased. Note the abrupt qualitative change in $g_c(r)$ when CPD is increased from 24 to 28%. The characterstic crystalline peaks are absent in the suspension with 28% CPD.

7.3.1 Chacterization of Structural Ordering and Disorder

The disappearence of the characteristic crystalline peaks in $g(r)$ and in $S(Q)$ is often taken to indicate the onset of the disordered state, which could either be amorphous, arising from the positional disorder due to polydispersity, or a liquid

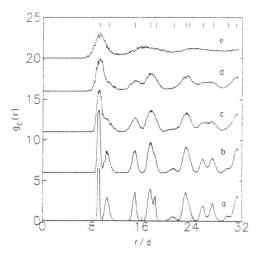

Figure 7.2   Pair correlation function $g_c(r)$ obtained from time-averaged particle coordinates for various CPDs. (a) 0%, (b) 9.4%, (c) 17%, (d) 24%, and (e) 28% CPD. Vertical bars correspond to the position of different neighbor shells for ideal BCC structure [3].

state, arising due to the melting of colloidal crystals resulting from an increase of the impurity ion concentration n_i. No distinctions between the amorphous or the liquidlike states were made in the early investigations of the size-polydisperse systems [11]; however, these can be distinguished from each other if the behavior of the diffusion is examined [31]. Commonly used criteria for the identfication of different states are given in Table 7.1.

Diffusion in the amorphous solid is known to be more than that in the corresponding crystalline state [32], whereas the liquid is expected to exhibit nonsaturating linear behavior. Figure 7.3 shows the mean-square displacement (MSD) as a function of MCSs for suspensions with different parameters. Although the MC dynamics lacks a strict or rigorous sense of time scale, and the true dynamics is obtained using the Brownian dynamics [23], the long-time slope of the MSD as a function of MCSs is related to the diffusion constant of the particles [33]. The difference between the behaviors of the three suspensions is distinctly seen in Fig. 7.3. Thus the pair correlation functions $g(r)$ and $g_c(r)$ and the behavior of the particle diffusion can completely identify the structural ordering.

As mentioned earlier, in a system with finite CPD, the particles are displaced from the ideal lattice positions, and one can treat the system as disordered; however, at low CPD it may still be possible to correlate the new positions of the particles to the reference lattice points and describe the relaxed structure on the basis of the reference lattice as a *marginally disordered crystalline structure* [8]. On the other hand, at large CPD the displacement of the particles from the ideal sites may be large enough to destroy the long-range order. A quantity that can be used to describe the extent of disorder is the magnitude of the displacement Δr_s of the equilibrium positions of the particles (*static displacement*) relative to the reference lattice.

Table 7.1   Criterion for identifying crystalline, amorphous, and liquidlike orders. R_0 is the nearest-neighbor distance

State	Structural ordering	Behavior of MSD with MCS
Crystalline	Long-range order	Saturation of MSD and root MSD $\ll R_0$
Amorphous	Short/medium-range order	Tendency towards saturation of MSD and root MSD $\ll R_0$
Liquidlike	Short-range order	MSD varies linearly with time and root MSD may exceed R_0 depending on time

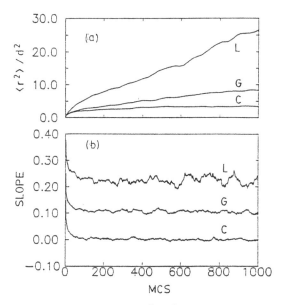

Figure 7.3   (a) Mean square displacement $\langle r^2 \rangle / d^2$ as a function of MCSs for suspensions with 0% (labeled C) and 28% CPD (labeled G). MSD for a liquidlike ordered suspension (labeled L) at an impurity concentration $1.4 n_p Z_0$ is also shown for comparison. (b) Slopes of the curves shown in (a) representing the diffusion constant. The asymptotic values of the slopes are 0.0011(35), 0.0060(55), and 0.024(14) for curves C, G, and L, respectively. The curves are vertically shifted by 0.1 and 0.2 for clarity [3].

7.3.2 $c \rightarrow a$ Transition

Having established the existence of an amorphous state at high CPD, the critical CPD at which the order-disorder or equivalently $c \rightarrow a$ transition occurs can be obtained from the behavior of the root-mean-square (rms) Δr_s, the MSD and other structural parameters obtained from $g(r)$ and $S(Q)$. The structural parameter R_g defined as the ratio g_{min}/g_{max}, where g_{min} is the value of first minimum and g_{max} is the height of the first peak in $g(r)$, exhibits a discontinuity across melting/freezing and has been extensively used to identify these transitions in Lennard-Jones systems [34] and colloidal crystals [35]. A comparison of the structural parameters across the melting transition with those across the $c \rightarrow a$ transition is appropriate because both these transitions lead to a loss of long-range order.

The other parameter that is very sensitive to disorder is the height of the first peak, S_{max}, in the structure factor $S(Q)$. In the case of BCC structures, this corresponds to the (110) Bragg reflection, corresponding to the wave vector $Q_0 = (2\pi/l_a)(110)$. One must calculate $S(Q_0)$ for all possible (six) orientations of Q_0 and then average over them to get S_{max} which represents the Bragg intensity in the powder diffraction. Figure 7.4 shows the $(\Delta r_s)_{rms}$ as a function of

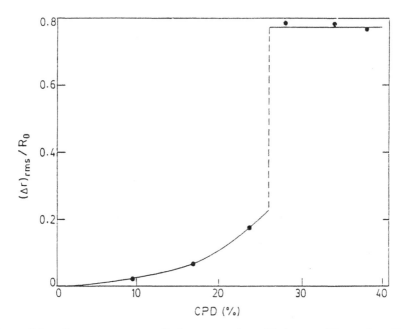

Figure 7.4   Root-mean-square displacement of equilibrium positions of particles $(\Delta r_s)_{rms}$ relative to the reference lattice as a function of CPD. The sudden rise of $(\Delta r_s)_{rms}$ around 26% CPD corresponds to the $c \rightarrow a$ transition [8].

CPD. The sudden rise of this parameter is associated with the $c \rightarrow a$ transition occurring around 26% CPD. One expects the crystalline order to be destroyed if the displacement of the particles is more than half of the nearest-neighbor (nn) distance R_0. Under such a condition the correlation between the relaxed positions of the particles and the corresponding lattice sites ceases to exist. The value of $(\Delta r_s)_{rms} \sim 0.8R_0$ in the amorphous state is consistent with the above argument. Further, the value of $(\Delta r_s)_{rms}$ at which *marginally disordered crystalline structure* becomes grossly disordered is $\sim 0.23R_0$, which is close to the largest root-mean-square displacement (thermal motion) that a particle can have in a colloidal crystal just before melting [2, 35]. This is analogous to the Lindemann criterion of melting [36] of atomic solids, according to which a crystalline solid melts, when the atomic motion is a fraction W of the nn distance. In atomic solids W has a value ~ 0.1, while in colloidal crystals it is ~ 0.23. This implies that the colloidal crystals can sustain much larger thermal motion than the atomic solids, probably because of their dilute nature and the softness of the interaction potential [2]

The dynamics of the particles in charge-polydisperse colloidal crystals also depend on CPD. The behavior of the diffusion as a function of CPD is shown in Fig. 7.5. The MSD increases slowly in the crystalline state, while it is independent of CPD in the amorphous state. The broadening of the pair correlation function due to finite polydispersity causes g_{max} to reduce and g_{min} to increase, resulting

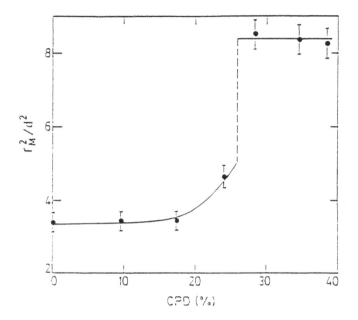

Figure 7.5   MSD at the end of M ($M = 1000$) MCSs as a function of CPD. Note the sudden increase in diffusion in the amorphous state [3].

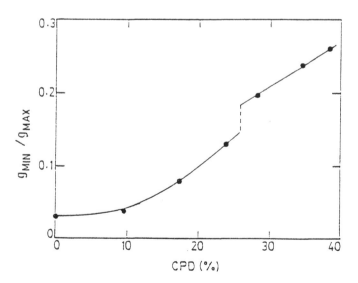

Figure 7.6   The structural parameter R_g as a function of CPD. The jump in this parameter at around 26% CPD is associated with $c \rightarrow a$ transition [3].

in an increase of the structural parameter R_g. This implies a loss of structural correlations. From the dependence of R_g on CPD, shown in Fig. 7.6, one can note the discontinuity arising due to $c \rightarrow a$ transition. This parameter continues to increase even in the amorphous state, because even in a system with only short-range order the structural correlations further diminish as CPD is increased. The value of R_g at the order-disorder transition is around 0.18, which is close to the value of 0.2 reported for the melting of Lennard-Jones systems [34, 35]. It may be mentioned that R_g has also been used in the identification of the glass transition during the quenching of atomic liquids [37], which occurs at $R_g = 0.14$. The value of R_g at the glass transition being lower than that at freezing into crystalline order is understandable, as the glass transition temperature is always lower than the freezing temperature.

The other structural parameter S_{max}, which is related to the first Bragg peak intensity, exhibits a dramatic decrease, the moment the long-range order is lost. For a perfect single crystal having N particles, it has a value N, whereas in a liquid it drops to a value less than 2.85 [38] and is independent of N. In the case of amorphous solids, it has a value slightly higher than that corresponding to the liquids. Figure 7.7 shows the behavior of S_{max} as a function of CPD. The sharp fall in the value of S_{max} clearly identifies the order-disorder transition. A value

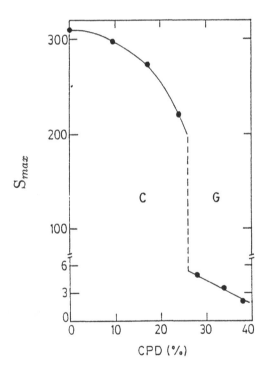

Figure 7.7 S_{max} as a function of CPD. The sharp decrease in its value is due to a loss of long-range order [3].

of S_{max} as high as 5 in the disordered state rules it out as a liquid state. Also from the magnitude of MSD one can clearly assign it to be a solid.

As mentioned earlier, by keeping the average charge fixed while CPD is varied, the average strength of interaction and hence U_T is expected to remain unchanged. U_T is indeed the same for the initial frames; however, its value after reaching equilibrium is found to be slightly lower than that of monodisperse crystals. This is shown in Fig. 7.8. This is probably because of the relaxation of the particles to new equilibrium positions. Similar behavior has been predicted theoretically in the case of charge-polydisperse colloidal fluids [4]. A decrease in U_T could also occur if large charges make small charges as neighbors. The interaction energy of such a pair would be lower than that of a pair each with charge Z_0; that is, $(Z_0 + \Delta)(Z_0 - \Delta) < Z_0^2$. Hence the decrease in U_T as a function of CPD could also be due to the charge ordering. This aspect is further discussed in Section 7.6.

The magnitude of disorder $(\Delta r_s)_{rms}$ can also be used to estimate the configurational entropy S of the disordered crystal. In a system with finite CPD, as any particle resides within a region $(\Delta r_s)_{rms}$ around each lattice point, the volume of such regions will be proportional to $[2(\Delta r_s)_{rms} + d]^3$, where d is the particle diameter. Hence the number of possible ways W_d in which the particle can

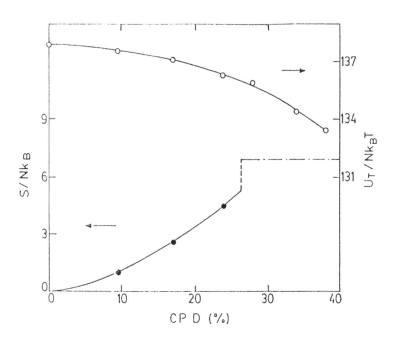

Figure 7.8   The total interaction energy U_T and the configurational entropy as a function of CPD. U_T:(○); Entropy of disordered colloidal crystal: (●). Entropy of an a-colloid is shown as a dot-dash line. Curves through the points are guides to the eye [8].

be placed in this volume is $[1 + 2(\Delta r_s)_{rms}/d]^3$. This allows one to estimate the configurational entropy as

$$S = Nk_B \ln(W_d) \tag{7.10}$$

which is also shown in Fig. 7.8. The configurational entropy of the disordered crystal increases with CPD as expected. The entropy of the amorphous state can be estimated from Eq. (7.10) if one uses it in the limit of maximum possible disorder [i.e., $(\Delta r_s)_{rms} + d/2 \rightarrow R_0/2$], half the nn distance. This is shown as dot-dash line in Fig. 7.8. The two curves corresponding to the entropies of the disordered crystal and the a-colloid, respectively, when extrapolated toward the center from either side, exhibit a mismatch. The increase in the entropy is thus associated with the $c \rightarrow a$ transition. It is important to point out that the entropy of the a-colloid obtained from Eq. (7.10) agrees well [8] with that obtained from the number of ways of arranging N particles of volume v_p in a volume V at a given volume fraction ϕ [39].

7.3.3 Effective Hard Sphere Model of Charge-Polydisperse Colloids

Because of the electrostatic interaction, the particles in a suspension can be considered to have an effective diameter d_h that is larger than the actual diameter. The fact that $g(r)$ remains zero up to a fairly large value of r is a clear manifestation of the existence of an effective hard-sphere diameter; that is, the particles are unable to go closer than this distance because of the presence of a Coulomb barrier. The largest r until which $g(r)$ remains zero in a monodisperse suspension represents the distance of closest approach for two particles and is taken to be $\bar{d}_h$ for a particle with charge Z_0. The effective hard-sphere diameter $d_h(Z)$ is expected to depend on its charge, and hence charge polydispersity would lead to a polydispersity of $d_h(Z)$, which can be estimated in the following manner [3]. Figure 7.9 schematically shows the distance D_Z of closest approach between particles of charges Z and Z_0. $d_h(Z)$ is then obtained as

$$h(Z) = 2(D_Z - \bar{d}_h/2) \tag{7.11}$$

where D_Z is obtained by solving

$$U(D_Z, Z_0, Z) = U(\bar{d}_h, Z_0, Z_0) \tag{7.12}$$

This is reasonable as the interaction energy between two particles of charge Z and Z_0 at the distance of closest approach is expected to be the same as that for two particles each with charge Z_0, which is $\sim 40k_B T$ at $\bar{d}_h = 7.15d$. Figure 7.10 shows the polydispersity ESPD of $d_h(Z)$ for the CPDs used in the investigation. ESPD turns out to be significantly lower than the corresponding CPD. Similar conclusions are drawn from the investigation on dense charge-polydisperse Yukawa fluids [4]. It may be mentioned that although $d_h(Z)$ depends on the magnitude of $U(\bar{d}_h, Z_0, Z_0)$, the ESPD is relatively insensitive to it. The

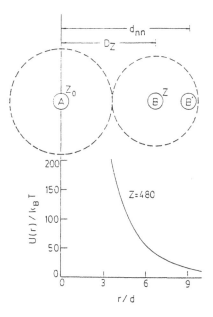

Figure 7.9   Schematic representation of the effective hard-sphere diameters (broken circles) and the distance of closest approach of particle A with charge Z_0 to another particle B with $Z = 480$. B' represents the equilibrium position of particle B. The lower part of the figure gives the interaction energy between these two particles as a function of interparticle separation [3].

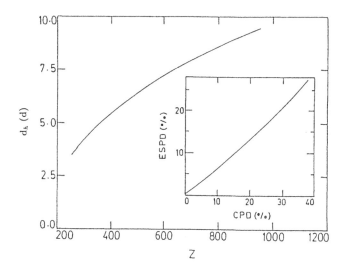

Figure 7.10   Dependence of the effective hard-sphere diameter d_h on the charge on the particle. The inset shows the estimated polydispersity of effective hard-sphere diameter ESPD as a function of CPD [3].

critical CPD of 26±2% corresponding to the $c \rightarrow a$ transition corresponds to an ESPD of 17±1%. This may be compared with that of 11% predicted for hard spheres by Pusey [21]. The difference between these may be due to the simple procedure used in estimating $d_h(Z)$. In fact the presence of the interaction potential beyond h makes the actual system qualitatively different from a true hard-sphere system.

7.3.4 The Distribution of Nearest-Neighbor Distance: The nn Model

The broadening of the first and the subsequent peaks in the pair correlation functions $g(r)$ and $g_c(r)$ of charge-polydisperse colloidal crystals arises basically due to the nearest-neighbor distance R being distributed resulting from the relaxation of particles to new equilibrium positions. In equilibrium one expects the nn distance R_{ij} to be such that all the pair interaction energies are nearly the same within the thermal energy $k_B T$

$$U_{ij}(R_{ij}) = Z_i Z_j A \exp(-\kappa R_{ij})/R_{ij} = U_0 \pm k_B T \tag{7.13}$$

where U_0 is the average pair interaction energy and A is a constant that contains terms independent of R. This nn model, based on the equipartition of energy, can be used to obtain the distribution of nn distance R_{ij}. In order to get the distribution from the equilibrium positions, one can set $k_B T = 0$, and then one has

$$U_0 = A Z_i Z_j \exp(-\kappa R_{ij})/R_{ij} \tag{7.14}$$

and after dropping the subscripts ij on R, Eq. (7.14) can be rewritten as

$$Z_i Z_j = (U_0/A) R \exp(\kappa R) \tag{7.15}$$

The distribution $P(R)$ of R can now be obtained from that of the product y of the random variables Z_i and Z_j: $y = Z_i Z_j$. Using the method of cumulative probability, the distribution $f(y)$ can be obtained as [8]

$$f(y) = \begin{cases} (4\Delta^2)^{-1} \ln(y/y_-), & y_- \leq y \leq y_m \\ (4\Delta^2)^{-1} \ln(y_+/y), & y_m \leq y \leq y_+ \\ 0, & \text{otherwise} \end{cases} \tag{7.16}$$

where $y_{\pm} = (Z_0 \pm \Delta)^2$ and $y_m = Z_0^2 - \Delta^2$.

It is now straightforward to obtain the distribution $P(R)$ from $f(y)$ using Eq. (7.15). After a few algebraic simplifications, $P(R)$ is given as [8]

$$P(R) = \begin{cases} \frac{Z_0}{4\Delta^2 R_0}[\kappa(R - R_-) + \ln\frac{R}{R_-}](1 + \kappa R)e^{[\kappa(R-R_0)]}, & \text{for } R_- \leq R \leq R_m \\ \\ \frac{Z_0}{4\Delta^2 R_0}[\kappa(R_+ - R) + \ln\frac{R_+}{R}](1 + \kappa R)e^{[\kappa(R-R_0)]}, & \text{for } R_m \leq R \leq R_+ \\ \\ 0, & \text{otherwise} \end{cases} \tag{7.17}$$

where $R_\pm$ are the solutions of $Ay_\pm = U_0 R_\pm \exp(\kappa R_\pm)$ and R_m is that of $Ay_m = U_0 R_m \exp(\kappa R_m)$. $R_\pm$ and R_m are obtained by solving the corresponding equations numerically for various CPDs. The distributions $P(R)$ thus obtained, shown in Fig. 7.11, exhibit broadening similar to that of the first peak of $g_c(r)$ as CPD is increased.

Figure 7.12 shows the full width at half maximum (FWHM) of the distribution $P(R)$ as the continuous curve. The results of the model can be compared with those of simulations if one compares these with the width of the first peak of $g_c(r)$, which represents the distribution of first neighbors. The data obtained from simulations are also shown in Fig. 7.12. The agreement between the two is excellent except for a scale factor of 1.6 by which the width predicted by the model is more than that obtained from simulations [8]. This is understandable as in the model one considers the relaxation of isolated $Z_i Z_j$ pairs, whereas in a real system every such pair is surrounded by its other first and subsequent neighbors. Hence the extent of actual relaxation is small, and its distribution is narrower.

7.4 Melting of Charge-Polydisperse Colloidal Crystals

As mentioned in the introduction, the formation of colloidal crystals occurs due to the competition between interaction energy and thermal energy. The melting of a colloidal crystal occurs when the dimensionless parameter $\Gamma = U(R_0)/k_B T$ is less than a critical value. Unlike atomic systems, where one varies Γ by

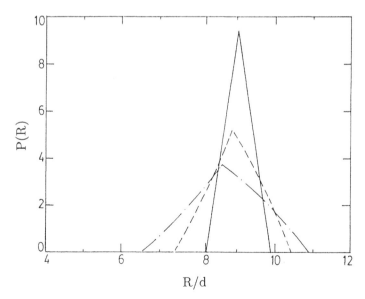

Figure 7.11   The probability distribution function $P(R)$ calculated from the nn model [Eq. (7.17)] for different CPDs. The full curve, dashed curve, and the dot-dash curve correspond to 9, 17, and 24% CPD [42].

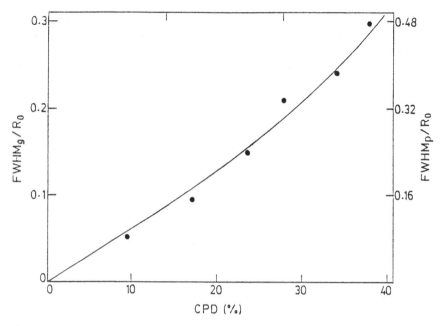

Figure 7.12   The full width at half-maximum $FWHM_p$ of the nn distance distribution $P(R)$ as a function of CPD (full curve). The points correspond to the full width at half-maximum $FWHM_g$ of the first peak of $g_c(r)$ obtained in the MC simulation [8].

varying the temperature, in aqueous colloidal suspension it is possible to vary Γ at fixed T by changing n_i or n_p. The change in $U(R_0)$ arises from the variation of the inverse Debye screening length κ, which depends on n_i and n_p. As the range of temperature variation is rather limited in aqueous colloidal suspensions, the relevant parameter for investigating melting/freezing experimentally is the impurity ion concentration [40, 41]. This aspect has been duly considered in a recent MC investigation of the melting of monodisperse and charge-polydisperse colloidal crystals [2, 42].

In order to identify the melting transition uniquely, MSD and the structural parameters R_g and S_{max} can again be analyzed as a function of impurity ion concentration. The behavior of R_g and S_{max} as a function of n_i is shown in Fig. 7.13 for a monodisperse suspension as well as those with finite CPD. The sharp decrease in S_{max} and a corresponding increase in R_g at $1.4n_pZ_0$ in the case of monodisperse suspensions can unambiguously be identified with the melting of the colloidal crystal. In the case of the charge-polydisperse colloidal crystal with 24% CPD, the melting occurs at $\sim 0.75n_pZ_0$. Note that at melting $R_g \simeq 0.2$, which is the same as that obtained from a computer simulation of Lennard-Jones systems [34]. Subsequently $R_g = 0.2$ has also been used as a criterion for identifying the melting/freezing of colloidal crystals [35]. Figure 7.14 shows the value of MSD at the end of M MCSs as a function of n_i. Note the sudden increase

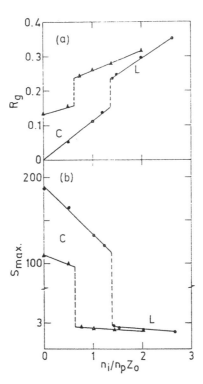

Figure 7.13   The structural parameters (a) R_g and (b) S_{max} as a function of n_i for monodisperse (•) and 24% polydisperse (▲) suspensions. The discontinuous changes in these parameters correspond to the melting transition of colloidal crystals [42].

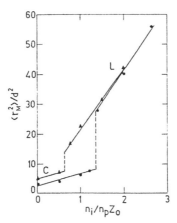

Figure 7.14   Mean square displacement at the end of M ($M = 1000$) MCSs as a function of n_i for monodisperse (•) and 24% polydisperse (▲) suspensions. The sudden increase in the diffusion of the system is associated with the melting of colloidal crystals.

in $< r_M^2 >$ in both the suspensions when n_i is increased beyond some critical values. The value of n_i at which $< r_M^2 >$ exhibits discontinuities is the same as that observed in Fig. 7.13 for R_g and S_{max}. This increase is hence associated with the melting of colloidal crystals. It is important to point out that melting in charge-polydisperse colloidal crystals occurs at lower n_i. The possible physical reasons for this will be discussed in Section 7.5.

7.4.1 Melting of Charge-Polydisperse a-Colloid

An amorphous-colloidal suspension would also melt if the impurity ion concentration were increased. As the amorphous as well as the liquidlike states are disordered, the structural parameters are not expected to show sharp discontinuities, as was observed in the case of the melting of charge-polydisperse colloidal crystals. The behaviors of R_g and $< r_M^2 >$ as a function of n_i are shown in Fig. 7.15 for an a-colloid with 34% CPD. The change in slope of R_g and $< r_M^2 >$ around $n_i = 1.5 n_p Z_0$ is associated with the melting of a-colloid. A similar change in slope of R_g has been found in the case of Lennard-Jones systems as a function of temperature across the glass transition. This occurred at $R_g = 0.14$, whereas in charge-polydisperse a-colloids the value of R_g is around 0.3. The largest value of R_g up to which an a-colloid can exist being considerably different from that in glassy monatomic Lennard-Jones systems has suggested investigators to propose [2, 42] the nature of the disordered state in a polydisperse system to be different from that of atomic glasses; that is, an amorphous polydisperse system may be stable while the monatomic quenched glasses are always metastable. The fact that a 34% charge-polydisperse colloid with BCC structure as an initial configuration evolves into an amorphous state also supports this hypothesis. A binary Lennard-Jones system with sufficiently different atomic diameters and a random initial configuration has also been recently found to turn amorphous [43]. As different charges on the particles lead to a variety of effective hard-sphere diameters, the incompatibility packing (i.e., size-mismatch frustration) into an ordered state leads to amorphous order. Thus the charge-polydisperse amorphous state can be considered somewhat analogous to a spin-glass, which also arises due to frustration in spin orientations due to random interactions.

7.5 The Phase Diagram

The phase diagram of a charge-polydisperse colloid in the CPD versus n_i plane will thus have crystalline, amorphous, and liquidlike ordered regions. The $c \rightarrow a$ transition and melting of colloidal crystals and a-colloids will constitute the boundaries between different phases. One such phase diagram obtained recently for a dilute charge-polydisperse colloidal suspension [42] is shown in

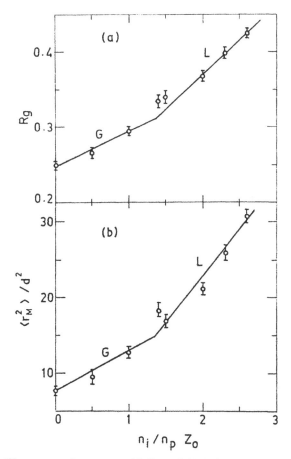

Figure 7.15   The structural parameter (a) R_g and (b) MSD at the end of M ($M = 500$) MCSs as a function of n_i for an a-colloid at 34% CPD. Note the change in the slope in these parameters that occurs due to the melting of the a-colloid [2].

Fig. 7.16. Note the decrease of the impurity ion concentration at which a charge-polydisperse colloidal crystal melts as CPD is increased. This behavior can be explained if one uses the concept of reduced temperature T^* and invokes a criterion of melting based on critical disorder. As a variation in n_i causes the nn interaction energy U_0 to change, reduced temperature can be defined as $T^* = k_B T / U_0$, expressing the thermal energy in units of interaction energy. Some other ways of defining the reduced temperature have also been frequently used in the literature [23, 44]. Figure 7.16 also shows the reduced temperature thus obtained for the colloidal system. Monodisperse crystals melt at $T^* = 0.16$ [2]. Crystals with finite CPD melt at lower T^*. As noted earlier from Fig. 7.14, the value of R_g at melting being independent of CPD implies that a colloidal crystal melts when structural correlations become weaker than a certain value irrespective of CPD.

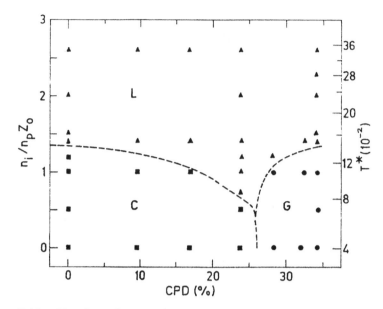

Figure 7.16   The phase diagram of a charge-polydisperse colloidal suspension with $n_p = 1.33 \times 10^{12}$ cm^{-3}. The symbols (■), (●), and (▲) represent the crystalline, amorphous, and liquidlike regions. Broken curves representing phase boundaries are guides to the eye. The reduced temperature T^* corresponding to n_i is also marked on the y axis.

In a suspension with finite CPD, apart from the thermal fluctuations, the static randomness in positions due to charge polydispersity also contribute to R_g, and hence at finite CPD melting can occur when the thermal contribution to randomness is smaller in magnitude, that is, at lower T^*. The increase in the value of n_i or equivalently T^* at which the a-colloid melts as a function of CPD is because of the better stability of the amorphous structure (against crystallization) away from the $c \rightarrow a$ transition boundary.

It may be mentioned that these phase boundaries [42] only represent the points of instability of the crystalline phase to the formation of an amorphous or liquid phase, and there may be hysteresis for the reverse transitions. Further, it may be of interest to explore any possible regions of coexistence and investigate the order of the $c \rightarrow a$ transition. These aspects can be examined only with large system sizes, typically a thousand particles or more.

7.6 Possibilities of Phase Separation and Charge Ordering

The results discussed so far are for charge-polydisperse colloidal systems where in the initial configuration the charges are randomly placed on BCC lattice points. As mentioned in Section 7.3.2, it is likely that a system may try to reduce its interaction energy by choosing small charges as the neighbors of large charges and vice versa. If such a thing occurs in a colloidal crystal, it would lead to a

CsCl-like structure. Another possibility which exists is of a phase separation or decomposition of a random a-colloid into ordered (crystalline) phases of large and small charges, each with CPD less than the critical CPD for the $c \rightarrow a$ transition. Such order-disorder transitions and phase separations are well known in binary atomic systems [45] and arise due to competition between interaction energy and the entropy. A substitutionally disordered binary system upon cooling may either go to a chemically or charge-ordered state or to a phase-separated state depending on the relative magnitudes of pair interaction energies U_{aa}, U_{bb}, and U_{ab}. Table 7.2 gives the interaction energies and entropies for different states using a lattice model and assuming nearest-neighbor interaction.

A charge-polydisperse system with continuous polydispersity can also be divided into a pseudobinary system with charges $Z1 < Z_0$ and $Z2 > Z_0$ having narrower distributions, as shown in Fig. 7.17. In order to get a charge-ordered (CO) state, one can place charges from one distributions on the cation sublattices of the CsCl structure and on the anion sublattice from the other distributions and then allow the system to reach equilibrium. Similarly a phase-separated (PS) state is obtained by filling one half of the simulation cell with charges from one distribution and the other half with the other distribution. Preliminary investigations show that charge-polydisperse colloidal suspensions with 34% CPD in these states, when equilibrated, remain crystalline and have interaction energies that are not very different from each other [46]; for example, these, when expressed in units of Nk_BT, are 133.9 for CO and 134.3 for PS. These may be compared with $U_T = 134.5$ of the amorphous state obtained from random initial conditions (see Fig. 7.8). Hence the entropies in the CO, amorphous, and PS states will be the deciding factor in determining their relative stability.

7.7 Conclusions and Some Outstanding Problems

The polydispersity of charge on colloidal particles in a colloidal crystal makes it marginally disordered if its value is low, whereas at high values it drives the colloidal crystal into an amorphous state. The $c \rightarrow a$ transition occurs abruptly

Table 7.2 Interaction energies, entropies, and order parameters for the disordered and the ordered phases of a binary system using lattice models

	Charge ordered	Disordered	Phase separated
Interaction energy	U_{ab}	$\frac{1}{4}(U_{aa} + 2U_{ab} + U_{bb})$	$\frac{1}{2}(U_{aa} + U_{bb})$
Entropy	0	ln 2	0
Warren-Cowley short-range order parameter	-1	0	1

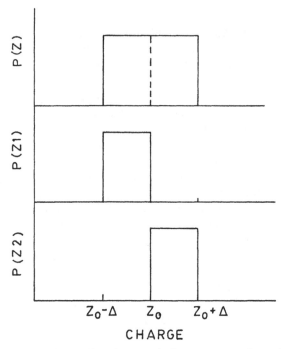

Figure 7.17 Decomposition of a charge-polydisperse suspension with a symmetric rectangular distribution of width 2Δ into two distributions $P(Z1)$ and $P(Z2)$, each of width Δ, leading to an equivalent pseudobinary system.

at a critical CPD. Further, the charge-polydisperse colloidal crystals melt at lower impurities or equivalently at lower reduced temperatures than the monodisperse counterparts. These results suggest a critical disorder model for the melting as well as for the amorphization, that is, the the order-disorder transition.

As compared to the phase-separated state, a charge-ordered state appears more plausible. This opens up a number of possibilities that remain to be examined; for example, a charge-polydisperse random colloidal suspension becomes amorphous at 26% CPD, whereas that in the charge-ordered state remains crystalline even at 34% CPD. This suggests that the $c \rightarrow a$ transition in a CO system would occur at even higher CPD, and even the phase diagram would be different. Further, would a CO colloidal suspension exhibit CO (CsCl structure) $\rightarrow$ random (BCC structure) $\rightarrow$ liquidlike transitions similar to those observed in binary atomic systems such as β-brass? In order to obtain the phase diagram of charge-polydisperse colloidal systems theoretically, the possibility of generalizing the density functional theory of freezing to these systems remains to be explored.

References

1. Dickinson, E., Parker, R., and Lal, M., *Chem. Phys. Lett.* **79**, 578 (1981).
2. Tata, B. V. R., and Arora, A. K., *J. Phys.: Condens. Matter* **3**, 7983 (1991).
3. Tata, B. V. R., and Arora, A. K., *J. Phys.: Condens. Matter* **4**, 7699 (1992).
4. Lowen, H., Roux, J. N., and Hansen, J. P., *J. Phys.: Condens. Matter* **3**, 997 (1991).
5. Ito, K., Nakamura, H., and Ise, N., *J. Chem. Phys.* **85**, 6136 (1986).
6. Asher, S. A., Flaugh, P. L., and Washinger, G., *Spectroscopy* **1**, 26 (1986).
7. Hume-Rothery, W., and Raynor, B. R., *Proc. R. Soc. London A* **174**, 471 (1954).
8. Arora, A. K., and Tata, B. V. R., *J. Phys. Chem. Solids* **55** 377 (1994).
9. Dickinson, E., *Chem. Phys. Lett.* **57**, 148 (1978).
10. Dickinson, E., *J. Chem Soc. Faraday Trans. II* **75** 466 (1979).
11. Dickinson, E., and Parker, R., *J. Physique Lett.* **46**, L229 (1985).
12. Evans, R., and Napper, D. H., *J. Colloid Interface Sci.* **63**, 43 (1978).
13. Vrij, A., *J. Chem. Phys.* **71**, 3267 (1979).
14. van Beurten, P., and Vrij, A., *J. Chem. Phys.* **74**, 2744 (1981).
15. Blum, L., and Stell, G., *J. Chem. Phys.* **71**, 42 (1979).
16. Barrat, J. L., and Hansen, J. P., *J. Physique* **47**, 1547 (1986).
17. Ginoza, M., *Mol. Phys.* **71**, 145 (1990).
18. Hansen, J. P., and Hayter, J. B., *Mol. Phys.* **46**, 651 (1982).
19. Ruiz-Estrada, H., Medina-Noyola, M., and Nägele, G., *Physica A* **168**, 919 (1990).
20. D'Aguanno, B., and Klein, R., *Phys. Rev. A* **46**, 7652 (1992).
21. Pusey, P. N., *J. Physique* **48**, 709 (1987).
22. Hoover, W. G., and Ree, F. H., *J. Chem. Phys.* **49**, 3609 (1968).
23. Lowen, H., Hansen, J. P., and Roux, J. N., *Phys. Rev. A* **44**, 1169 (1991).
24. Binder, K., in *Monte Carlo Methods in Statistical Physics*, Ed. K. Binder (Springer, New York, 1979), p. 1.
25. Hansen, J. P., and McDonald, I. R., *Theory of Simple Liquids*, 2nd edn. (Academic, London, 1986).
26. Williams, R., and Crandall, R. S., *Phys. Lett. A* **48**, 225 (1974).
27. Rahman, A., *Phys. Rev.* **136**, A405 (1964).
28. Waseda, Y., *The Structure of Non-Crystalline Materials* (McGraw-Hill, New York, 1980).
29. Lindsay, H. M., and Chaikin, P. M., *J. Chem. Phys.* **76**, 3774 (1982).
30. Nose, S., and Yonezawa, F., *J. Chem Phys.* **84**, 1803 (1986).
31. Rosenberg, R. O., Thirumalai, D., and Mountain, R. D., *J. Phys.: Condens. Matter* **1**, 2109 (1989).
32. Cantor, B., and Cahn, R. W., in *Amorphous Metallic Alloys*, Ed. F. E. Luborsky (Butterworths, London, 1983), p. 487.
33. Kehr, K. W., and Binder, K., in *Applications of Monte Carlo Methods in Statistical Physics*, Ed. K. Binder (Springer, Berlin, 1984), p. 181.
34. Raveche, H. J., Mountain, R. D., and Street, W. B., *J. Chem. Phys.* **61**, 1970 (1974).
35. Rosenberg, R. O., and Thirumalai, D., *Phys. Rev.* **A36** 5690 (1987).
36. Lindemann, F. A., *Z. Phys.* **11**, 609 (1934).
37. Wendt, H. R., and Abraham, F. F., *Phys. Rev. Lett.* **41**, 1244 (1978).
38. Hansen, J. P., and Verlet, L., *Phys. Rev.* **184**, 151 (1969).
39. Williams, R., Crandall, R. S., and Wojtowicz, P. J., *Phys. Rev. Lett*, **37**, 348 (1976).
40. Hachisu, S., Kobayashi, Y., and Kose, A., *J. Colloid Interface Sci.* **42**, 342 (1973).
41. Sirota, E. B., Ou-Yang, H. D., Sinha, S. K., Chaikin, P. M., Axe, J. D., and Fujii, Y., *Phys. Rev. Lett.* **62**, 1524 (1989).
42. Tata, B. V. R., Ph.D. thesis, Madras University (1992) (unpublished).
43. Li, M., and Johnson, W. L., *Phys. Rev. Lett.* **70**, 1120 (1993).
44. Robbins, M. O., Kremer, K., and Grest, G. S., *J. Chem. Phys.* **88**, 3286 (1988).
45. Kittel, C., *Introduction to Solid State Physics*, 5th edn. (John Wiley, New York, 1976).
46. Tata, B. V. R., and Arora, A. K., *J. Phys.: Condens. Matter* **7**, 3817 (1995).

Dynamics of Charged Colloidal Suspensions Across the Freezing and Glass Transition

Hartmut Löwen

Dynamical correlations of charge-stabilized colloidal suspensions are investigated using Brownian dynamics simulations. Special emphasis is put on the kinetic glass transition in a charge-polydisperse suspension, and on long-time self-diffusion, particularly on a dynamical freezing rule for colloidal fluids. First, structural slowing down near the kinetic glass transition that shows up as a plateaulike behavior in the time-dependent density autocorrelation function is discussed for a supercooled polydisperse Yukawa fluid. Brownian dynamics results are compared with those of molecular dynamics, which ignores solvent effects. It is found that only the intermediate time region is affected by the different types of short-time dynamics, but the long-time behavior is at least qualitatively similar. Second, a dynamical scaling law at the freezing line of the fluid is empirically found stating that the ratio of long-time and short-time self-diffusion has a universal value close to 0.1. This constitutes a dynamical phenomenological freezing rule for colloidal suspensions, similar to the Lindemann melting criterion. Third, long-time translational and orientational diffusion is discussed for a system of Brownian hard spherocylinders. Along the fluid-crystal and fluid-nematic coexistence line, both long-time self-diffusion coefficients measured in terms of their short-time limits are nonmonotonic as a function of the length-to-width ratio of the spherocylinders. The ratio of long-time and short-time orientational self-diffusion is roughly 0.1, constituting a simple dynamical freezing rule for anisotropic fluids. For all topics, the connection to mode-coupling theories on the one hand, and to experiments on the other hand, is also briefly discussed.

8.1 Introduction

Suspensions of colloidal particles embedded in aqueous or some other organic solvent exhibit a local structure and a phase behavior that are highly reminiscent of those observed in simple atomic systems [1–4]. In contrast to atomic systems, however, the length scale of structural ordering is *mesoscopic* roughly determined by the mean interparticle spacing $\ell = n_p^{-1/3}$, where n_p is the particle concentration. This permits a direct visualization of the particle positions using direct image processing methods or video microscopy and opens a fascinating way to see and measure structural and dynamical correlations in real space; see for example, [5]. If the colloidal particles are practically monodisperse spheres, they serve as ideal realizations of simple model liquids from classical statistical mechanics. Two major interaction models have frequently been applied to describe the interaction between index-matched colloidal particles: the hard-sphere model and the Yukawa model. For *sterically stabilized* suspensions, a simple hard-sphere pair interaction simply incorporating the excluded volume of the spheres can be employed. For an interparticle separation r the potential energy between two particles is then given by

$$U(r) = \begin{cases} 0, & \text{for } r \geq d \\ \infty, & \text{for } r < d \end{cases} \tag{8.1}$$

where d is the diameter of the particles. For *charge-stabilized* suspensions, on the other hand, Derjaguin, Landau, Verwey, and Overbeek (DLVO) [6, 7] have calculated that at infinite dilution the effective interaction is pairwise and of the Yukawa type

$$U(r) = \frac{(Z_{\text{eff}}e)^2}{4\pi\epsilon_0\epsilon r} \exp(-\kappa r) \tag{8.2}$$

with

$$Z_{\text{eff}} = Z \exp(\kappa d/2)/(1 + \kappa d/2) \tag{8.3}$$

Ze denoting the bare charge of the colloidal "macroions" and

$$\kappa = \left(\sum_i n_i z_i^2 / k_B T \epsilon_0 \epsilon\right)^{1/2} \tag{8.4}$$

being the inverse Debye-Hückel screening length. Here n_i and z_i are the concentration and charge of the ith type of impurity ion (including the counterions) in the solvent and T is the temperature. A Poisson-Boltzmann-cell model designed for strongly interacting macroions also results in an effective Yukawa interaction [8]. A more refined ab initio study of the "primitive model," which includes counter- and impurity-induced effective many-body forces between the macroions [9, 10] indeed reveals that the Yukawa-pair-interaction remains a satisfying description of the effective macroionic forces [11]. These calculations show that the Yukawa picture is justified even for high concentrations provided the packing fraction of the macroions is not too high. However, the actual values of the parameters Z_{eff} and κ entering into Eq. (8.2) have to be renormalized with respect to the DLVO predictions Eqs. (8.3) and (8.4) valid at infinite dilution. In order to check the charge renormalization, it is highly motivated to study structural and dynamical quantities for Yukawa systems by theory and computer simulations and to compare with the experimental data gained from charge-stabilized colloidal suspensions.

The special purpose of this chapter is to study the *dynamics* of Yukawa systems, particularly in the neighborhood of *phase transitions*. In contrast to atomic and molecular fluids whose dynamics are Newtonian, the complete time-scale separation between solvent and colloidal particle relaxation leads to irreversible Brownian motion of the macroparticles in the solvent. For dilute systems, these dynamics can explicitly be obtained by writing the finite-difference version of the particle displacements as follows: We consider N particles confined to a large volume V with positions $\{r_i : i = 1, \ldots, N\}$. The particle positions after a small time step Δt are gained from those at time t by the formula [12, 13]:

$$\mathbf{r}_i(t + \Delta t) = \mathbf{r}_i(t) + \frac{1}{\xi}\mathbf{F}_i(t)\Delta t + (\Delta \mathbf{r})_R + O(\Delta t^2) \tag{8.5}$$

where the random displacement $(\Delta \mathbf{r})_R$ is sampled from a Gaussian distribution of zero mean, $\overline{(\Delta \mathbf{r})_R} = 0$, and variance $\overline{(\Delta \mathbf{r})_R^2} = 6k_B T \Delta t/\xi$. Here $F_i(t)$ are the total forces on the particles derived from the pairwise interaction Eq. (8.1) or (8.2), and $\xi = 3\pi\eta d$ is the solvent friction coefficient (η denoting the solvent shear viscosity), which is related to the short-time diffusion constant as follows:

$$D_0 = k_B T/\xi \tag{8.6}$$

D_0 provides a natural scale to measure the long-time diffusion coefficient D_L defined by

$$D_L = \lim_{t \to \infty} \left(\frac{1}{6t} \Big\langle \sum_{j=1}^{N} \frac{1}{N}[\mathbf{r}_i(t) - \mathbf{r}_i(0)]^2 \Big\rangle \right) \tag{8.7}$$

where $< \cdots >$ denotes a canonical average. We remark that Eq. (8.5) also constitutes a direct algorithm for a Brownian dynamics computer simulation. In concentrated colloidal systems, *hydrodynamic forces* induced by the solvent are relevant. In principle, these could be approximately included by replacing ξ by a $3N \times 3N$ matrix depending parametrically on the positions $\{r_i\}$. In the following, however, also for simplicity, we take ξ to be diagonal and constant, thus neglecting any hydrodynamic interactions, which is a reasonable assumption for dilute though highly interacting charged suspensions.

As regards phase transformations, several types of phase transitions are conceivable. The best known is the *freezing transition* [4]. The hard-sphere fluid exhibits a strong first-order freezing transition into a dense-packed crystal with a large density jump of about 10%. Also the Yukawa fluid freezes into a BCC or FCC crystal depending on whether the interaction is soft (small κ) or harsh.

Another transition is the kinetic *glass transition* of an undercooled or compressed fluid. This is not a true thermodynamic phase transition with a non analyticity in the free energy but a smooth transition of dynamical origin where time-dependent correlations decay only for very long times.

This chapter is concerned with dynamical signatures at the freezing and fluid-to glass transition of a charged colloidal suspension. Near the glass transition the long-time self-diffusion coefficient D_L practically drops to zero, exhibiting a power law as a function of a typical parameter measuring the distance to the glass transition point. One main point of this chapter is to check explicitly the validity of this power law by performing extensive Brownian dynamics computer simulations for a charge-polydisperse colloidal suspension [14]. Also the relaxation of the density autocorrelation function is studied in order to detect the kinetic glass transition by simple dynamical diagnostics. Particularly, the scenario of the kinetic glass transition is compared to that of a system governed by Newtonian dynamics. It is found that the long-time relaxation is very similar but the short-time and intermediate-time relaxation is different.

At the freezing transition, the long-time self-diffusion coefficient D_L jumps from a finite value at the fluid side of the phase coexistence line to a very small value corresponding to diffusion of grain boundaries and vacancies in the crystalline phase. Interestingly enough, Löwen et al. [15] found that the ratio D_L/D_0 exhibits a *universality* at the fluid coexistence line. It always equals 0.098 at the freezing transition of the fluid regardless of the interaction between the colloidal particles. This constitutes a dynamical phenomenological freezing rule, similar in spirit to the Lindemann rule [16] of melting or the Hansen-Verlet [17] freezing rule which have proved to be very helpful in estimating fluid-solid coexistence lines. As early as 1910 Lindemann [16] put forward the empirical fact that the ratio of the root-mean-square displacement and the average interparticle distance at the solid melting line has a value of roughly 0.15. The Hansen-Verlet criterion [17] states that the amplitude of the first maximum of the liquid structure factor has a universal value of $\simeq 2.85$ along the liquid freezing line. The Hansen-Verlet and the dynamical freezing criterion of Löwen et al. [15] are universal in the sense that they do not depend on the detailed nature of the spherical interaction potential.

It is interesting to check whether the concept of dynamical universality at the freezing line of the fluid also holds for *anisotropic fluids* where one has coexisting crystalline or nematic phases. In order to clear that up, we report Brownian dynamics simulations of hard spherocylinders for different total length-to-width ratios p_c. As a result the orientational self-diffusion coefficient measured in terms of it short-time limit also drops one order of magnitude at the fluid-liquid-crystalline transition line. The actual value, however, depends a bit on p_c; hence the self-diffusion is not that universal as in the case of isotropic interactions.

The chapter is organized as follows: First we discuss Brownian dynamics of a supercooled charge-polydisperse Yukawa fluid near the kinetic glass transition in Section 8.2. Then we turn to self-diffusion at the fluid freezing line in Section 8.3. For both topics, the experimental and theoretical situation is first briefly reviewed and then computer simulation results are discussed. Section 8.4 is devoted to self-diffusion in anisotropic fluids. We state some open problems in Section 8.5.

8.2 Kinetic Glass Transition in Colloidal Suspensions

In this section we review some recent experimental theoretical and simulational results for the kinetic glass transition in colloidal suspensions. The main emphasis is placed on results from a Brownian dynamics computer simulation.

8.2.1 Light Scattering Experiments

Pusey, van Megen, and co-workers [18-22] measured the time-dependent density autocorrelation function over a broad time window for a sterically stabilized colloidal suspension as a function of the packing fraction of the particles. The experimental method they used was dynamical light scattering. Despite the enormous differences in time scales between atomic and mesoscopic glass formers, the supercooled colloidal liquid exhibits qualitative features very similar to that of an atomic liquid at the kinetic glass transition. The advantage for interpretation is that the experimental system is a rather simple: It represents a hard-sphere-like system with a small polydispersity.

Also the relaxation of spherical polystyrene micronetwork particles of mesoscopic size, swollen in a good solvent, was recently measured over a very broad time window by Bartsch and co-workers [23-25], representing another type of colloidal suspension. The samples are a bit more polydisperse ($p_\sigma = 0.16$) than that used by van Megen and co-workers; consequently the glass transition occurred at higher volume fractions of the colloidal spheres. Again the long-time relaxation was found to be very similar to that of simple atomic liquids. Charge-stabilized colloidal suspensions also form glasses; experimental studies at the glass transition were done by Sirota et al. [26] and Meller and Stavans [27].

It would be interesting to apply real-space methods by tagging a single particle and following its way using video microscopy. Then one could link the results more easily to computer simulations. Also most of the concepts and jargon concerning the relaxation of the kinetic glass transition (like particle cage relaxation and thermally activated hopping) are borrowed from real-space pictures and should be tested experimentally.

8.2.2 Theory

The most prominent theory of the glass transition capable of making nontrivial predictions for the relaxation scenario is the mode-coupling theory (MCT) developed by Götze and co-workers [28]. The experimental results of van Megen and Pusey [19] were compared with predictions of mode-coupling theory for a hard-sphere system by Götze and Sjögren [29] and by Fuchs et al. [30]; good agreement was found between mode-coupling theory and the experimental data.

The MCT was originally derived for molecular dynamics. It was shown explicitly by Szamel and Löwen [31] that the asymptotic predictions of MCT do not change for Brownian dynamics. Hence, within MCT, the asymptotics of the density relaxation are universal with respect to the short-time dynamics. For example, the ideal glass transition occurs at the same temperature for MD and BD. There have only been a few attempts to incorporate hydrodynamic interactions into MCT, a first step having been done by Fuchs [32]. Again the asymptotic scenario remains unaffected by the explicit form of the hydrodynamic interactions, while there are changes for finite times.

Recently, Kawasaki [33] proposed a stochastic model particularly designed for the Brownian dynamics of colloidal suspensions. Here also activated hopping is incorporated in some sense into the theory.

8.2.3 Brownian Dynamics Simulations

In order to make a direct comparison between undercooled atomic and colloidal fluids, Löwen et al. [34] performed an extensive simulation for a charge-stabilized polydisperse colloidal suspension near the kinetic glass transition for both Brownian and Newtonian dynamics. To date this is the only simulation for the kinetic glass transition which takes solvent friction into account. It may be mentioned that there are other Monte Carlo simulations of the glass transition which interpret the fictitious Monte Carlo move dynamics as a real dynamics; see for example [35]. If the long-time behavior is independent with respect to short-time dynamics, then one can chose one suitable fictitious dynamics to extract the long-time dynamics such that the actual computational time to explore sufficient statistics is smaller. An idea of this kind was used by Kob et al. [36, 37].

As a model system, a charge-polydisperse colloidal fluid, described by the potential

$$U_{ij}(r) = U_0 \frac{\ell}{r} \frac{Z_i Z_j}{\bar{Z}^2} \exp\left(-\kappa^* \frac{r - \ell}{\ell}\right) \tag{8.8}$$

was chosen in Ref. [34] where $\kappa^* \equiv \kappa\ell \equiv 7$ and $n_p \equiv \ell^{-3}$ are fixed. The system is then cooled from a temperature $T^* \equiv k_B T/U_0 = 0.45$ down to $T^* = 0.10$. The macroionic charges $\{Z_i : i = 1, \ldots, N\}$ are continuously distributed according to a Schultz distribution with a relative charge polydispersity of 0.5. The characteristic time scales for BD and MD are $\tau_B = \xi\ell^2/U_0$ for Brownian and $\tau_N = \sqrt{m\ell^2/U_0}$ for Newtonian dynamics, where m is the mass of the particle. The high value of κ^* chosen implies that the system behaves similar to a polydisperse hard-sphere system. In order to perform an explicit mapping from the charge-polydisperse onto a size-polydisperse system by using the Gibbs-Bogoliubov inequality, an effective relative size polydispersity of about 0.13 was obtained [34]. Thus the data correspond roughly to those of the experiments of Bartsch et al. [23-25]. Although more complicated than a monodisperse model, the motivation for the charge-polydisperse model is twofold: First, crystallization is suppressed, and one is sure to encounter a glassy state in the simulations. Second, polydispersity is an intrinsic property of any colloidal suspension, and it is thus natural to incorporate it into the theoretical model directly.

The dynamical key quantities characterizing the glass transition are the density autocorrelation functions in real space as well as in Fourier space. We define the general density autocorrelation function as follows

$$C_\rho(t) \equiv C_\rho(\mathbf{r}, \mathbf{r}', t) = \left\langle \sum_{i,j=1}^{N} \delta(\mathbf{r} - \mathbf{r}_j(0)) \, \delta(\mathbf{r}' - \mathbf{r}_i(t)) \right\rangle \tag{8.9}$$

This can be split into a self (s) and distinct (d) part

$$C_\rho(\mathbf{r}, \mathbf{r}', t) = C_\rho^{(s)}(\mathbf{r}, \mathbf{r}', t) + C_\rho^{(d)}(\mathbf{r}, \mathbf{r}', t) \tag{8.10}$$

with

$$C_\rho^{(s)}(\mathbf{r}, \mathbf{r}', t) = \left\langle \sum_{j=1}^{N} \delta(\mathbf{r} - \mathbf{r}_j(0)) \, \delta(\mathbf{r}' - \mathbf{r}_j(t)) \right\rangle \tag{8.11}$$

$$C_\rho^{(d)}(\mathbf{r}, \mathbf{r}', t) = \left\langle \sum_{i\neq j=1}^{N} \delta(\mathbf{r} - \mathbf{r}_j(0)) \, \delta(\mathbf{r}' - \mathbf{r}_i(t)) \right\rangle \tag{8.12}$$

$C_\rho^{(s)}(\mathbf{r}, \mathbf{r}', t)$ [resp., $C_\rho^{(d)}(\mathbf{r}, \mathbf{r}', t)$] give the joint probability density to find a particle at position $\mathbf{r}'$ after a time t and the same (resp. another) particle at position $\mathbf{r}$ for zero time.

By normalization, we obtain the *van Hove correlation function*

$$G(\mathbf{r}, \mathbf{r}', t) = C_\rho(\mathbf{r}, \mathbf{r}', t)/n_p^2 \tag{8.13}$$

that also naturally splits into a self- and distinct part

$$G(\mathbf{r}, \mathbf{r}', t) = G_s(\mathbf{r}, \mathbf{r}', t) + G_d(\mathbf{r}, \mathbf{r}', t) \tag{8.14}$$

The distinct part of the van Hove correlation function is the time-dependent generalization of the pair distribution function $g(r)$

$$G_d(r, t) = \frac{1}{n_p N} \left\langle \sum_{i\neq j=1}^{N} \delta(\mathbf{r} - \mathbf{r}_i(0) + \mathbf{r}_j(t)) \right\rangle \tag{8.15}$$

Of course, $G_d(r, 0) = g(r)$ and $\lim_{t\to\infty} G_d(r, t) = 1$ in a fluid, whereas the van Hove function has frozen-in components for large times in a glass. The self-part reads:

$$G_s(r, t) = \frac{1}{n_p N} \left\langle \sum_{j=1}^{N} \delta(\mathbf{r} - \mathbf{r}_j(0) + \mathbf{r}_j(t)) \right\rangle \tag{8.16}$$

For $t = 0$, we get $G_s(r, 0) = \delta(\mathbf{r})/n_p$, and the long-time limit is given by the hydrodynamic behavior

$$G_s(r, t) \cong \frac{1}{n_p}(4\pi D_L t)^{3/2} \exp(-r^2/4D_L t) \tag{8.17}$$

Furthermore, one can take the Fourier transform of $G_s(r, t)$ and $G_d(r, t)$ with respect to r to gain corresponding Q-dependent structure factors $S_s(Q, t)$, $S_d(Q, t)$ that are defined as

$$S_s(Q, t) = \frac{1}{n_p N} \sum_{j=1}^{N} \left\langle \exp\{i\mathbf{Q} \cdot [\mathbf{r}_j(t) - \mathbf{r}_j(0)]\} \right\rangle \tag{8.18}$$

$$S_d(Q, t) = \frac{1}{n_p N} \sum_{\ell \neq j=1}^{N} \left\langle \exp\{i\mathbf{Q} \cdot [\mathbf{r}_j(t) - \mathbf{r}_\ell(0)]\} \right\rangle \tag{8.19}$$

It is clear that $S_d(Q, 0) = S(Q)$; that is, $S_d(Q, t)$ equals the static structure factor for $t = 0$. Further Fourier transformation with respect to time t then leads to the dynamical structure factors $S_s(Q, \omega)$, $S_d(Q, \omega)$. The latter quantity is directly accessible in dynamical scattering experiments.

As a first dynamical diagnostic for the kinetic glass transition, only the relaxation of the self-part of the van Hove function, $G_s(r, t)$, can be used. As the system is gently cooled down further, there is a sudden drastic change in the relaxation. The function $r^2 G_s(r, t)$ now shows the buildup of a secondary peak roughly at a mean particle distance ℓ, whereas the position of the first peak remains frozen over "long" (i.e., $100\tau_B$, τ_N) times. This is shown in Fig. 8.1, where $r^2 G_s(r, t)$ is plotted above and near the kinetic glass transition for Brownian dynamics (BD). Hence we have a first indication that hopping processes do occur. Of course this qualitative change occurs gradually in a smooth manner but still in a relatively narrow temperature interval, and it can be used to determine an estimate for the kinetic glass transition, which is now microscopically connected with a change in the relaxation behavior from hydrodynamic relaxation to relaxation by thermally activated jumps. Remarkably, the temperature interval in which this dynamical crossover occurs is the same for BD and molecular dynamics (MD). Also the buildup of the secondary peak is present in BD, indicating that there is the *same* crossover to thermally activated jumps in the Brownian case although there are no phonons present in the Brownian case. By this diagnostics one may estimate the temperature for the kinetic glass transition for both MD and BD to be within

$$0.115 < T_{\text{glass}}^* < 0.12. \tag{8.20}$$

The long-time diffusion constant drops to very small values near T_{glass}^* and a power law with a small residual contribution ΔD due to jumps fits well the data of the supercooled liquid

$$D_L = \Delta D + A(T^* - T_{\text{glass}}^*)^\gamma \tag{8.21}$$

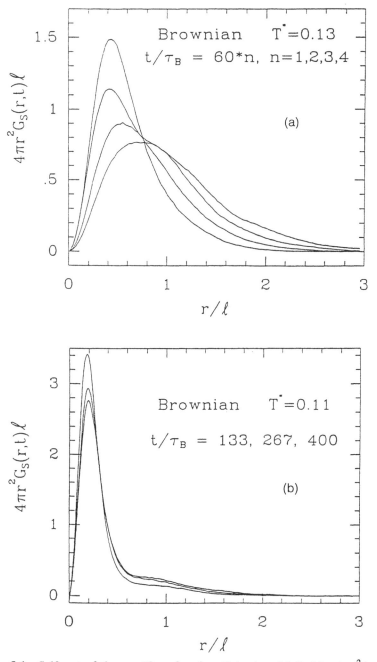

Figure 8.1   Self-part of the van Hove function $G_s(r, t)$ multiplied by $4\pi r^2 \ell$ versus reduced distance r/ℓ calculated with Brownian dynamics; the curves from left to right (or top to bottom) are for increasing time arguments. (a) Results for $T^* = 0.13$ and $t^* = t/\tau_B = 60, 120, 180, 240$. (b) Results for $T^* = 0.11$ and $t^* = 133, 267, 400$. From Ref. [34].

with $\gamma \approx 1.4$ for *both* BD and MD. Plots of the long-time self-diffusion constant as a function of $T^* - T^*_{glass}$ are shown in Fig. 8.2. On a double logarithmic plot the values D_L fall on a straight line, indicating the validity of the power law [Eq. (8.21)].

Another interesting quantity is the distinct part of the van Hove function, $G_d(r, t)$, defined in Eq. (8.15). At the kinetic glass transition, it turns out, again both for BD and MD, that a peak at $r = 0$ is built up, giving again strong evidence for particle-exchange hopping processes. In a dense supercooled liquid, however, these processes are more complicated than simple pair exchanges. In general, more than two particles (small clusters of particles) participate to a real position-exchange process; see for example, Migayawa et al. [38].

The other interesting quantities are the spatial and time Fourier transformations of $G_s(r, t)$ and $G_d(r, t)$. In Fig. 8.3, $S_s(Q, t)$ is plotted as a function of time t on a logarithmic scale for a fixed wave vector Q near the first peak of the static structure factor $S(Q)$. From this figure it becomes evident that there are qualitative different relaxations for BD and MD. For short times t/τ_N, t/τ_B, the decay of the density

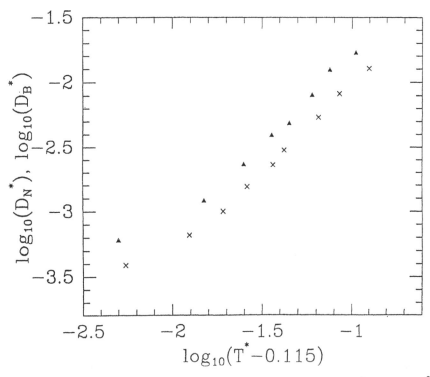

Figure 8.2   Double logaritmic plot of the reduced diffusion constants $D^*_N = D_L\tau_N/\ell^2$ (Newtonian dynamics, crosses) and $D^*_B = D_L\tau_B/\ell^2$ (Brownian dynamics, triangles) versus distance $T^* - 0.115$ to the kinetic glass transition. From Ref. [34].

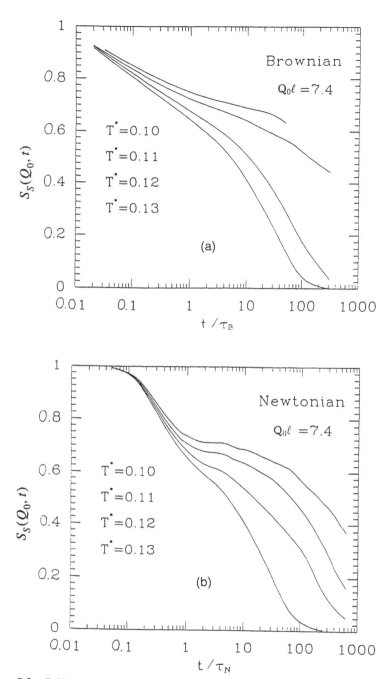

Figure 8.3 Self-part of the density autocorrelation function $S_s(Q, t)$ versus reduced time $t^* = t/\tau_N, t/\tau_B$ (on a logarithmic scale) for $Q = Q_0 = 7.4/\ell$ and (from bottom to top) $T^* = 0.13, 0.12, 0.11$, and 0.10. (a) Brownian dynamics; (b) molecular dynamics. From Ref. [34].

autocorrelation function is different. It starts with $1 - \mathcal{O}(t^2)$ in the MD and with $1 - \mathcal{O}(t)$ in the BD case. For very long times, $S_s(Q, t)$ and $S_d(Q, t)$ can be fitted by a stretched exponential law $\sim \exp[-(t/t_0)^\nu]$, where t_0 is strongly temperature dependent and, as ν, also depends on Q. The exponent ν is found to be practically the same for BD as for MD. So the α-relaxation scenario is quite similar. From Fig. 8.3 it becomes clear that, for times small compared to α-relaxation but still larger than τ_B, τ_N, there are qualitative differences; that is, the different short-time behavior also induces a different crossover to the long-time behavior. The decay is smoother in the BD case, and there is no clear indication for a buildup of a plateau near T_{glass}^*. (Of course, for smaller temperatures there must be a quasiplateau.) If one looks at the time Fourier transforms of $S_d(Q, t)$ or $S_s(Q, t)$ (at $\kappa^* = 7$), one finds that there is a shoulder at the corresponding frequency for MD that is missing for BD. It is tempting to call this a β relaxation, and one main conclusion is that the dynamical onset of β-relaxation is qualitatively different for BD and MD. In mode-coupling theory, β-relaxation is defined in a different way, namely, by an additional scaling law near T_{glass}^* whose asymptotic behavior can be studied analytically. Again, mode coupling theory predicts no difference for MD and BD in the asymptotic case. The difference obtained in the simulation, however, occurs for smaller times that are not yet in the asymptotic regime.

Summarizing, the kinetic glass transition manifests itself microscopically as a crossover from hydrodynamic relaxation to relaxation by thermally activated jumps. This crossover is not completely sharp but occurs on a very narrow temperature interval upon cooling. The transition temperature is the same for BD and MD. The dynamical onset of β-relaxation, however, is different. A shoulder in the dynamical structure factor at intermediate frequencies is present for MD but missing for BD. On the other hand, α-relaxation is similar, supporting the prediction of simple mode-coupling theory.

We finally note that a detailed analysis has also been made distinguishing between low-charge and high-charge particles [39]. As expected, the high-charge particles freeze first and then the low-charge particles follow, but everything happens in a smooth manner. The kinetic glass transition is connected with a considerable change in the ratio of the long-time self-diffusion coefficients for the low-charge and high-charge particles.

8.3 Long-Time Self-Diffusion Across the Freezing Transition

In this section we review some recent experiments, theories, and simulations of the long-time self-diffusion coefficient for a Brownian Yukawa system viz. charge-stabilized colloidal suspension and also sterically stabilized suspension.

8.3.1 Measurements of the Long-Time Self-Diffusion Coefficient

Different experimental techniques have been used to measure the long-time self-diffusion coefficient D_L. Most of the studies have used dynamical light scattering [40-49]. This technique, however, may have problems with multiple scattering for concentrated colloidal suspensions. Next also more refined methods where some particle are dyed have been applied as forced Rayleigh scattering [50, 51, 15] and fluorescence recovery after photobleaching [52, 53]. Finally fiber-optical quasielastic light scattering, where multiple scattering effects are negligible [54], and pulsed-field NMR [55] experiments were performed. It has to be pointed out that the error obtained in using the different experimental methods is still large, lying on the 10% level for concentrated or strongly interacting suspensions.

In Refs. [15] and [51] D_L/D_0 was measured close to the freezing transitions of a charged colloidal suspension for different salt concentrations. It was found that the ratio D_L/D_0 equals 0.1 on the fluid freezing line. Also in Ref. [51] D_L/D_0 was systematically investigated across the freezing transition involving diffusion in the compressed fluid, which is finite and very similar to the usual fluid diffusion and diffusion in the solid phase which is much slower and results from vacancy hopping and grain boundary diffusion.

8.3.2 Theories for the Long-Time Self-Diffusion Coefficient in Strongly Interacting Colloidal Suspensions

In most of the theoretical approaches to the long-time self-diffusion coefficient in concentrated suspensions hydrodynamic interactions are neglected completely. Historically the first approach is the mode-coupling theory, which was developed by Mori and Zwanzig [56] and applied to colloids by Hess and Klein [57]. Its starting point is the exact expression for the tagged particle intermediate scattering function. This expression is then analyzed and approximated using ideas borrowed from the mode-coupling theories of simple liquids. The final result is a nonlinear self-consistent equation for the scattering function. This equation also involves the collective scattering function $S_d(Q, t)$, for which another self-consistent equation is written down. As nonlinear self-consistent equations are difficult to solve, one usually approximates them further, introducing short-time limits of the scattering functions into friction kernels (for details, see [57-59] and references cited therein). In this way one obtains the following expression for D_L:

$$\frac{D_L}{D_0} = \left(1 + \frac{1}{6\pi^2 n_p} \int_0^\infty dQ\, Q^2 \frac{[S(Q) - 1]^2}{1 + S(Q)}\right)^{-1} \tag{8.22}$$

Another theory has been proposed by Medina-Noyola [60]. It is based on an analysis of the generalized Langevin equation describing the coupled motion of the tagged particle and the surrounding colloidal suspension. The final result for D_L is:

$$D_L/D_0 = \left(1 + \frac{n_p}{6} \int dr \, [g(r) - 1]^2\right)^{-1} \tag{8.23}$$

It turns out that numerically the Eqs. (8.22) and (8.23) are very close. These also provide a very simple explanation for the observed universality at the fluid freezing line: Since $S(Q)$ is universal at freezing provided Q is scaled by $n_p^{1/3}$ according to the Hansen-Verlet rule (see, e.g., [61]), D_L/D_0 also has to be invariant at the fluid freezing transition. This theoretical justification is, however, too simple since the theories do only work at high dilution. Indeed the actual value of D_L/D_0 is 0.3; hence it is strongly overestimated by theory.

More sophisticated theories also designed for concentrated suspensions are the modified Enskog approach proposed by Cichocki [62] and investigated by Löwen and Szamel [63] and the Enskog approach first used by Szamel and Leegwater [64, 65] for hard spheres, which was then generalized to the soft Yukawa potential [63]. These approaches involve the particle interaction potential explicitly and do not give universality at freezing. However, these actually give better absolute values than the former theories for high concentrations. In Figs. 8.4 to 8.6 we

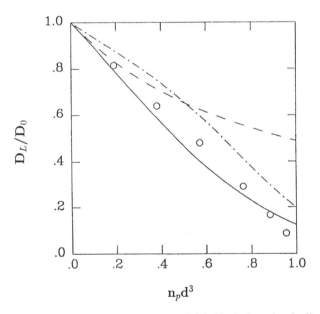

Figure 8.4 Long-time self-diffusion coefficient divided by its low-density limit, D_L/D_0, as a function of the reduced density $n_p d^3$ for a hard-sphere suspension. Circles: Brownian dynamics data of Ref. [68]; solid line: Enskog theory; dashed line: modified Enskog theory; dot-dashed line: theory of Medina-Noyola, Eq.(23). From Ref. [63].

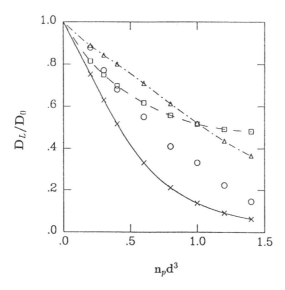

Figure 8.5 Self-diffusion coefficient divided by its low-density limit, D_L/D_0, as a function of the reduced density $n_p \ell^3$ for a Yukawa suspension where the Yukawa potential is written as $U(r) = U_0(\ell/r)\exp[-\kappa^*(r-\ell)/\ell]$. The parameters of the Yukawa potential are $\kappa^* = 8$, $k_B T/U_0 = 1$. Circles: Brownian dynamics results; crosses: Enskog theory; squares: modified Enskog theory; triangles: theory of Medina-Noyola, Eq. (8.23). Lines serve only as guides to the eyes; the actual calculations were performed at the data points. From Ref. [63].

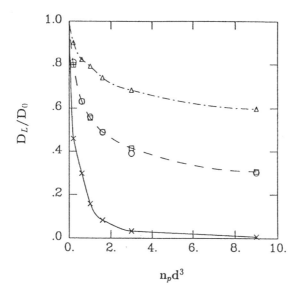

Figure 8.6 Same as Fig. 8.5; but now for a softer Yukawa potential with parameters $\kappa^* = 3$, $k_B T/U_0 = 0.8$. From Ref. [63].

have shown the results for D_L/D_0 for different theories and the "exact" computer simulational results for varying particle concentration from the dilute limit to high concentrations near the fluid freezing point. A hard-sphere system was taken in Fig. 8.4, while two different Yukawa interactions were assumed for Figs. 8.5 and 8.6. For hard spheres the Enskog description gives satisfactory results, while the modified Enskog theory fails. For a soft Yukawa interaction, on the other hand, the Enskog theory fails, and its modification works well. Medina-Noyola's expression always fails for large concentrations. Consequently, at the moment, there is no satisfactory theory that works for both soft and hard interactions at low and high concentrations. Recently, Indrani and Ramaswamy [66] used a self-consistent mode-coupling approach to check the universality of D_L/D_0 at freezing. Their values of ≈ 0.05 also seem to be too low as compared to the computer simulation and experimental value of 0.1.

8.3.3 Computer Simulations of the Long-Time Self-Diffusion Coefficient

In the fluid phase, the ratio D_L/D_0 of the hard-sphere system with simple Brownian dynamics given by Eq. (8.5) has been "exactly" calculated by Brownian dynamics simulations by Cichocki and Hinsen [67, 68] and for the Yukawa system by Löwen et al. [63, 69]. The main problem is the finite time window, where the extrapolation to infinite times is somewhat arbitrary. One can use the exact solution for hard spheres in the semidilute limit to find an explicit long-time extrapolation [68] or fit to a stretched exponential or some other function. The resulting values of D_L/D_0 change by about 5% and are thus fortunately rather insensitive to the extrapolation procedure, while the decay time to the asymptotic diffusive behavior depends much more on the extrapolation scheme. Computer simulation results are desirable since they provide an exact test of existing theories and are the basis of comparing experimental data with results from theoretical models for the colloidal interaction [69].

In the following we give some analytical fit formulae [69, 70] for the simulation results that are helpful for a simple comparison with experimental and theoretical data.

Hard Spheres

A hard-sphere system is characterized solely by its packing fraction $\phi = \pi n_p d^3/6$. A simple formula for D_L/D_0 is

$$D_L/D_0 = 1 - 1.94\phi + 1.25\phi^2 - 2.042\phi^3 \tag{8.24}$$

This equation is valid for any ϕ in the fluid phase (i.e., $\phi < 0.494$) and has a relative error of less than 3%. The exact low-density limit $D_L/D_0 = 1 - 2\phi + O(\phi^2)$ is only approximately incorporated into Eq. (24). At the fluid freezing point ($\phi = 0.494$), we get $D_L/D_0 = 0.100$ from Eq. (8.24), which is practically identical to the universal value of 0.098 obtained from the simulations [68, 15].

Yukawa Fluids

We write the Yukawa potential as

$$U(r) = U_0 \ell \, \exp(\kappa^*) \exp(-\kappa r)/r \tag{8.25}$$

where U_0 is an energy and $1/\kappa$ the length scale. The only relevant dimensioness parameters [71] are κ^* and T^*. This means that D_L/D_0 does only depend on the two parameters T^* and κ^*. The fit formula for D_L/D_0 reads as

$$\begin{aligned} D_L/D_0 = {} & 0.6457 \exp(-2.3429 t_r) \\ & + 0.3183 \exp[-(31.70 + 6.053\kappa^*) t_r] + 0.03598 \end{aligned} \tag{8.26}$$

Here t_r is the ratio of the freezing temperature to the actual temperature

$$t_r = T^*_{\text{freez}}(\kappa^*)/T^* \tag{8.27}$$

Meijer and Frenkel have calculated the fluid freezing line in the (T^*, κ^*)-diagram [71]. Their data can be fitted by

$$\begin{aligned} T^*_{\text{freez}}(\kappa^*) = {} & 0.009 + 0.030286\kappa^* - 0.009964\kappa^{*2} \\ & + 0.0033477\kappa^{*3} - 0.0002452\kappa^{*4} \end{aligned} \tag{8.28}$$

such that everything is explicitly known in Eq. (8.26). Equation (8.26) stems from 40 BD simulations for different parameters scanning the whole fluid regime from $\kappa^* = 0$ (unscreened plasma interaction) to 7. It contains a small relative error of 4% for $0 \le \kappa^* \le 7$.

At the freezing transition we have $t_r \equiv 1$, and only the first exponential in Eq. (8.26) gives a relevant contribution such that we see directly the universality of D_L/D_0 with respect to κ^*, where the actual value is $D_L/D_0 = 0.098$.

8.3.4 Dynamical Test of the Effective Interaction between Colloidal Particles

Equation (8.26) was used to analyze experimental data for D_L/D_0 obtained from forced Rayleigh scattering experiments on charged suspensions with varying salt and particle concentrations [69], which then provides a *dynamical test* of the particle interactions. Two different theoretical Yukawa models were tested against the experimental results: the DLVO model defined by Eqs. (8.2)–(8.4), and the Poisson-Boltzmann cell model [8]. The results are shown in Figs. 8.7 and 8.8. All parameters of the experimental sample were known by a careful characterization of the colloidal suspension; in particular, the true bare particle charge Z was determined from other measurements to be $Z = 580$. Once the theoretical Yukawa description of the interaction is fixed, there is no free fitting parameter involved in the comparison between the experimental and simulational

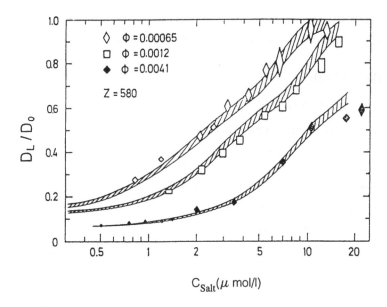

Figure 8.7   Comparison of three series of experimental data for D_L/D_0 with Eq. (26) for the Poisson-Boltzmann-cell model for varying salt concentration c_{salt}. The bare charge is $Z = 580$. The symbols correspond to packing fractions of $\phi = 0.0041, 0.0012, 0.00065$. Hatched areas: prediction using these parameters in Eq.(8.26) with 4% uncertaincy. From Ref. [69].

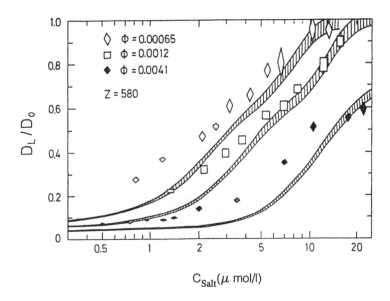

Figure 8.8   Same as Fig. 8.7, but now for the DLVO model. From Ref. [69].

data. The hatched area in Figs. 8.7 and 8.8 corresponds to the predictions of Eq. (8.26) within the given relative error of 4%. The symbols indicate the experimental data points for different particle concentrations n_p (or volume fractions ϕ) and varying salt concentration. One can clearly discriminate between the DLVO and the Poisson-Boltzmann-cell model. As expected, the DLVO model fails since the particles are highly interacting, while the description of the Poisson-Boltzmann-cell model gives remarkably good agreement with the experimental data. Again we emphasize that there is no fit parameter involved. Consequently the nature of the direct effective interparticle forces can be obtained from *dynamical* measurements. The usual route consistes of a comparison with the liquid structure factor $S(Q)$ (see, e.g., [72]. This quantity, however, is drastically affected by charge and size polydispersity [73]. The advantage of the dynamical test of the interaction potentials is that long-time self-diffusion, on the other hand, is rather sensitive to the interparticle forces but rather insensitive to polydispersity effects.

We finally remark that there are further theoretical Yukawa models for the effective interaction between charged colloidal suspensions. Belloni [74] has solved the mean-spherical approximation (MSA) of the primitive model analytically where one also ends up with an effective Yukawa interaction. The result is very close the DLVO potential. Within the experimental error of the data from Ref. [69], one cannot see a big difference between the MSA and the DLVO potential. Thus the MSA model fails as the DLVO model does. Second, there is a modified DLVO approximation that is empirically justified for strongly interacting particles [50]. This Yukawa potential, in turn, is very close to the Poisson-Boltzmann-cell model and is capable of describing the experimental data well.

8.4 Self-Diffusion in Liquid-Crystalline Systems

In this section we study the long-time diffusion of rodlike particles; that is, we are dealing with *anisotropic* liquids. In this case one has translational and rotational long-time diffusion, which are investigated in the framework of a simple model of hard spherocylinders [70]. In order to keep the model simple, we consider Brownian dynamics of hard sphe cylinders neglecting hydrodynamic interactions. The spherocylinders have a total width L_c and diameter d_c, and a rod configuration is characterized by its center-of-mass coordinates $\{\mathbf{R}_i : i = 1, \ldots, N\}$ and orientations $\{\Omega_i : i = 1, \ldots, N\}$, where Ω_i is a unit vector. The potential energy is simple a excluded-volume type: It is zero if the rods do not overlap and infinite otherwise. Due to this simple interaction, the temperature T scales out and the rod number density n_p is the only thermodynamic variable. This density is conveniently expressed in terms of the rod packing fraction $\phi = n_p \pi d_c^2 [d/6 + (L_c - d_c)/4]$. Choosing the rod diameter d_c as the typical length scale, the only additional geometric quantity is the length-to-width ratio

$$p_c = L_c/d_c \tag{8.29}$$

of the rods. Hence all structural and thermodynamic quantities including the bulk phase diagram only depend on ϕ and p_c. For $L_c = d_c$ the isotropic case of hard spheres is recovered.

We adopt Brownian dynamics of the rods and approximate the short-time dynamics by that of one single rod in a solvent characterized by two translational short-time diffusion constants, $D^\perp$ and $D^\|$, perpendicular and parallel to the rod axis, and a rotational short-time diffusion constant D^r. As a function of p_c, these three diffusion constants have been calculated by Broersma [75] and Tirado et al. [76, 77]:

$$D^\perp = \frac{D_0^*}{4\pi}(\ln p_c + 0.839 + 0.185/p_c + 0.233/p_c^{\,2}) \tag{8.30}$$

$$D^\| = \frac{D_0^*}{2\pi}(\ln p_c - 0.207 + 0.980/p_c - 0.133/p_c^{\,2}) \tag{8.31}$$

$$D^r = \frac{3D_0^*}{\pi L_c^2}(\ln p_c - 0.662 + 0.917/p_c - 0.050/p_c^2) \tag{8.32}$$

with

$$D_0^* = k_B T/\eta L_c \tag{8.33}$$

where η the shear viscosity of the solvent.

The corresponding finite difference equations for Brownian dynamics of rods are a bit more complicated than Eq. (8.5) for spheres. At a given time t, the center-of-mass position $\mathbf{R}_i(t)$ of the ith rod can be split into a part $\mathbf{R}_i^\|(t) \equiv [\mathbf{\Omega}_i(t) \cdot \mathbf{R}_i(t)]\mathbf{\Omega}_i(t)$ parallel and another part $\mathbf{R}_i^\perp(t)$ perpendicular to the rod orientation $\mathbf{\Omega}_i(t)$ such that

$$\mathbf{R}_i(t) = \mathbf{R}_i^\|(t) + \mathbf{R}_i^\perp(t) \tag{8.34}$$

The same separation into a parallel and perpendicular part can be done for the total force $\mathbf{F}_i(t)$, acting on the ith rod due to the interaction with the other rods

$$\mathbf{F}_i(t) = \mathbf{F}_i^\|(t) + \mathbf{F}_i^\perp(t) \tag{8.35}$$

Then for a finite time step Δt the temporal evolution of $\mathbf{R}_i(t)$ is given by

$$\mathbf{R}_i^\|(t + \Delta t) = \mathbf{R}_i^\|(t) + \frac{D^\|}{k_B T}\mathbf{F}_i^\|(t)\,\Delta t + (\Delta R^\|)_R\mathbf{\Omega}_i(t) \tag{8.36}$$

where $(\Delta R^\|)_R$ is a random displacement due to solvent kicks, which is a Gaussian-distributed random number with zero mean, $\overline{(\Delta R^\|)_R} = 0$, and variance

$$\overline{(\Delta R^\|)_R^2} = 2D^\|\,\Delta t \tag{8.37}$$

The perpendicular part, on the other hand, diffuses with the perpendicular diffusion constant $D^\perp$:

$$\mathbf{R}_i^\perp(t + \Delta t) = \mathbf{R}_i^\perp(t) + \frac{D^\perp}{k_B T} \mathbf{F}_i^\perp(t) \, \Delta t$$
$$+ (\Delta R_1^\perp)_R \mathbf{e}_{i1}(t) + (\Delta R_2^\perp)_R \mathbf{e}_{i2}(t) \tag{8.38}$$

where again $(\Delta R_1^\perp)_R$ and $(\Delta R_2^\perp)_R$ are Gaussian distributed with zero mean and variance $2D^\perp \Delta t$. Furthermore, $\mathbf{e}_{i1}(t)$ and $\mathbf{e}_{i2}(t)$ are two orthogonal unit vectors perpendicular to $\mathbf{\Omega}_i(t)$.

Finally the orientational degree of freedom diffuses as

$$\mathbf{\Omega}_i(t + \Delta t) = \mathbf{\Omega}_i(t) + \frac{D^r}{k_B T} \mathbf{M}_i(t) \times \mathbf{\Omega}_i(t) \, \Delta t + x_1 \mathbf{e}_{i1}(t) + x_2 \mathbf{e}_{i2}(t) \tag{8.39}$$

where now $\mathbf{M}_i(t)$ is the torque acting on rod i and x_1, x_2 are Gaussian random numbers with zero mean and variance $2D^r \Delta t$.

Using this kind of Brownian motion, extensive BD simulations for the long-time translational and rotational diffusion coefficients were carried out by Löwen [70] over the whole range of the fluid phase in the (ϕ, p_c) phase diagram. The translational long-time self-diffusion coefficient D_L^t is defined as

$$D_L^t = \lim_{t \to \infty} D^t(t) \tag{8.40}$$

with

$$D^t(t) = W(t)/6t \tag{8.41}$$

where

$$W(t) = \frac{1}{N} \sum_{i=1}^{N} < [\mathbf{R}_i(t) - \mathbf{R}_i(0)]^2 > \equiv < [\mathbf{R}_1(t) - \mathbf{R}_1(0)]^2 > \tag{8.42}$$

is the mean square displacement of the center-of-mass coordinate. A natural scale for D_L^t is its short-time limit $D^t = \lim_{t \to 0} D^t(t)$ given by:

$$D^t = \tfrac{1}{3}(2D^\perp + D^{\parallel}) \tag{8.43}$$

The long-time orientational self-diffusion coefficient D_L^r, on the other hand, is defined via the long-time limit of a diffusive process on the unit sphere:

$$D_L^r = \lim_{t \to \infty} D^r(t) \tag{8.44}$$

with

$$D^r(t) = -\frac{1}{2t} \ln[< \mathbf{\Omega}_1(0)\mathbf{\Omega}_1(t) >] \tag{8.45}$$

Of course, the short-time limit $D^r = \lim_{t \to 0} D^r(t)$ provides a natural scale for $D^r(t)$.

It is tempting to check whether there are dynamical freezing rules by looking at the values of D_L^t/D^t and D_L^r/D^r at the freezing line of the liquid. The phase diagram of hard spherocylinders is known by Monte Carlo simulations of Veerman and Frenkel [78] involving stable fluid, crystalline, nematic, and smectic phases. In particular, the location of the fluid-nematic and fluid-crystalline coexistence line is known for $p_c = 1, 2, 4, 6$ [78]. For $p_c > 6$ the theory of Lee [79] gives explicit data for the coexisting fluid and nematic densities including the limit $p_c \to \infty$, which is known exactly from Onsager's theory [80–82]. The Monte Carlo simulations have also revealed that the coexisting phase is crystalline for $p_c \lesssim 4.5$, whereas it is nematic for $p_c \gtrsim 4.5$.

For very high p_c ($p_c > 30$) Onsager's theory yields an asymptotically exact expression for the fluid density ϕ_f at the fluid-nematic transition [81, 82]:

$$\phi_f = 3.2906/p_c \tag{8.46}$$

Furthermore the tube model (see, e.g., Doi and Edwards [83]) yields $D_L^r/D^r = b_c(n_p L_c^3)^{-2}$, where an estimate of b_c can be obtained by fitting the simulation data of Doi et al., resulting in $b_c = 540$ [84]. This yields the asymptotic prediction

$$D_L^r/D^r = 30.8/p_c^2 \tag{8.47}$$

along the fluid-nematic transition line, which becomes valid for $p_c \gtrsim 30$. On the other hand, the translational diffusion tends to $D_L^t/D^t = \frac{1}{3}$ as $p_c \to \infty$ according to the tube model.

In Fig. 8.9, the simulational results of Löwen [70] for D_L^t/D^t and D_L^r/D^r are plotted versus $1/p_c$ for ϕ being on the fluid freezing line. Both simulational data and the asymptotic formula Eq. (8.47) are shown. It turns out that along the freezing line both D_L^t/D^t and D_L^r/D^r are *nonmonotonic* with p_c. The translational diffusion is nonmonotonic in the crystalline region. This is connected to the nonmonotonicity in p_c of the coexisting fluid packing fraction ϕ_f. In the nematic region, D_L^t/D^t increases monotonically to its "tube limit" $1/3$. This limit is practically reached for $p_c \gtrsim 10$.

On the other hand, D_L^r/D^r is strongly decreasing for small increasing p_c in the crystalline region. Then, for $2 \leq p_c \leq 6$, it stays more or less constant ≈ 0.12 irrespective of whether the coexisting phase is crystalline or nematic. If p_c is increased further, D_L^r/D^r increases again and then decreases approaching its asymptotic law given by Eq. (8.47) for $p_c \gtrsim 20$.

Hence it becomes clear that both diffusion ratios D_L^t/D^t and D_L^r/D^r vary significantly along the freezing line, both in the crystalline and nematic region. Consequently a simple universal dynamical freezing criterion that would guarantee a universal value of D_L^t/D^t or D_L^r/D^r for arbitrary p_c is missing. Nevertheless, in the relatively broad region $2 \leq p_c \leq 6$ where the coexisting phase may be crystalline or nematic, D_L^r/D^r is roughly constant. This is the rotational analog of the dynamical freezing rule for spherical systems: It states that *an anisotropic fluid freezes if its long-time rotational self-diffusion coefficient is one order of magnitude smaller than its short-time limit.*

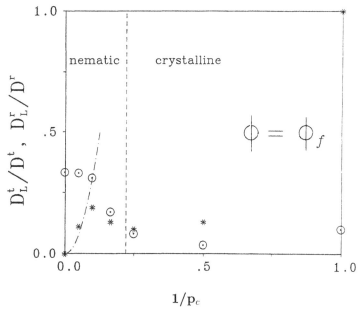

Figure 8.9 Translational and rotational long-time diffusion coefficients, D_L^t (circles) and D_L^r (stars), measured in terms of their short-time limits, D^t and D^r, versus $1/p_c$ at coexistence of the liquid with the crystalline or nematic phase. The dot-dashed line is the asymptotic law Eq. (8.47). The dashed line separates the nematic and crystalline region. From Ref. [70].

8.5 Some Further Open Problems

Let us finally list some fascinating open problems that are relevant to the context of dynamics of colloidal suspensions across the freezing and glass transition. They are mainly connected to systems with reduced geometry.

1. The kinetic glass transitions has not yet been studied in detail for a fluid *confined* to pores or between plates. It is expected that boundaries strongly influence the location of the kinetic glass transition point. If crystallization at the walls can be circumvented, the glass transition should occur at much higher temperatures than in the bulk fluid if the dimensions of the sample (the pore diameter or the distance between the plates) is comparable to the mean interparticle spacing ℓ [85, 86]. Recently, Fehr and Löwen [87] have considered a soft-sphere mixture between two parallel plates. In order to avoid wall crystallization, a lateral disorder was introduced on the walls. The corresponding bulk kinetic glass transition was extensively investigated by Roux, Barrat, and Hansen [88, 89]. The preliminary results indicate a shift of the glass transition temperature to significantly larger values as compared to the bulk glass transition known from Refs. [88] and [89]. Up to now, there

has been no attempt to perform a simple mode-coupling theory in restricted geometry to study the shift of the glass transition theoretically. This is a challenging problem for the future.

2. If the distance between parallel charged plates confining a charged colloidal suspension is enhanced or the surface charge is increased, one finally ends up with a quasi-two-dimensional system. The fundamental question arises whether the kinetic glass transition in two spatial dimensions is fundamentally different from that in three dimensions or not. Recent computer simulational results for a two-dimensional Lennard-Jones liquid by Ranganathan [90] indicate that the qualitative scenario in two dimensions is qualitative similar to that in three dimensions. This is different from the freezing and melting transition, which is definitively different in two dimensions. One should also do quantitative comparison with mode-coupling theory in this case.

3. What about a dynamical freezing rule in two dimensions? Recent investigations of Grier and Murray [5] studying the nonequilibrium formation of quasi-two-dimensional crystal indicate that it is fulfilled in 2D. The fundamental problem is that the nature of the freezing transition is still not completely clear in two dimensions. This concerns mainly the existence of a hexatic phase with long-range orientational order that would intervene between the fluid and solid phase. Strictly speaking [91] there is no translational long-range order in two dimensions, and consequently a crystal phase with long-range order is not thermodynamically stable.

For quasi-two-dimensional colloidal fluids a Yukawa interaction is an approximate description of the interaction, and Brownian dynamics simulations have been performed to get the long-time self-diffusion coefficients by Löwen [92]. In contrast to molecular dynamics in two dimensions, where the long-time self-diffusion coefficient diverges, it is finite for 2D Brownian dynamics.

Another problem is how to locate the fluid coexistence line exactly. Recent simulations with the 2D Yukawa potential by Naidoo and Schnitker [93] are still too inconclusive to decide the exact phase behavior in two dimensions.

4. The glass transition in liquid crystals is more complicated. To date no computer simulation has been reported observing a fluid phase with slow dynamics. Apparently nematic phase formation is fast enough to exclude a glass transition. More detailed work from theory, experiments, and computer simulations is needed to study the slowing down of translational and orientational motion in compressed or supercooled anisotropic fluids.

5. Excellent realizations of charged colloidal rodlike systems are aqueous suspensions of tabacco mosaic viruses. Recently it was shown by Löwen [94] by using an "ab initio" calculation that the Yukawa segment model originally introduced by Klein and co-workers [95, 96] (see also Ref. [97]) is a good description of the interaction. There is a considerable need to explore the complete phase diagram of the Yukawa segment model by computer simulation or density-functional theory and to investigate time-dependent correlation functions by Brownian dynamics simulations. One first step in this direction

was done by Kirchhoff et al., who calculated long-time orientational and translational self-diffusion for the Yukawa segment model [98].

Acknowledgements

I thank J. P. Hansen, P. A. Madden, G. Szamel, T. Kirchhoff, R. Klein, T. Palberg, and J.-L. Barrat for helpful discussions. This work was supported by the Bundesministerium für Forschung und Technologie (BMFT) under Contract No. 03-WA3LMU.

References

1. Russel, W. B., Saville, D. A., and Schowalter, W. R., *Colloidal Dispersions* (Cambridge University Press, Cambridge, 1989).
2. Pusey, P. N., in *Liquids, Freezing and the Glass Transition*, Eds. J.P. Hansen, D. Levesque, and J. Zinn-Justin (North Holland, Amsterdam, (1991), p. 763.
3. Sood, A. K., in *Solid State Physics*, Eds. H. Eherenreich and D. Turnbull (Academic, New York) **45**, 1 (1991).
4. Löwen, H., Physics Reports **237**, 249 (1994).
5. Grier, D. G., and Murray, C. A., in *Structure and Dynamics of Strongly Interacting Colloids and Supramolecular Aggregates in Solution*, Eds. S. H. Chen, J. S. Huang, and P. Tartaglia (Nato ASI, Kluwer, 1992), p. 145.
6. Derjaguin, B. V., and Landau, L. D., *Acta Physicochim. USSR* **14**, 633 (1941).
7. Verwey, E. J. W., and Overbeek, J. Th., *Theory of the Stability of Lyophobic Colloids* (Elsevier, Amsterdam, 1948).
8. Alexander, S., Chaikin, P. M., Grant, P., Moraes, G. J., Pincus, P., and Hone, D., *J. Chem. Phys.* **80**, 5776 (1984).
9. Löwen, H., Maden, P. A., and Hansen, J. P., *Phys. Rev. Lett.* **68**, 1081 (1992).
10. Löwen, H., Hansen, J. P., and Madden, P. A., *J. Chem. Phys.* **98**, 3275 (1993).
11. Löwen, H., and Krampsthuber, G., *Europhys. Lett.* **23**, 673 (1993).
12. Ermak, D. L., *J. Chem. Phys.* **62**, 4189, 4197 (1975).
13. Allen, M. P., and Tildesley, D. J., *Computer Simulation of Liquids* (Clarendon Press, Oxford, 1987).
14. Löwen, H., Roux, J. N., and Hansen, J. P., *J. Phys. Condens. Matter.* **3**, 997 (1991).
15. Löwen, H., Palberg, T., and Simon, R., *Phys. Rev. Letts.* **70**, 1557 (1993).
16. Lindemann, F. A., *Phys. Z.* **11**, 609 (1910).
17. Hansen, J. P., and Verlet, L., *Phys. Rev.* **184**, 151 (1969).
18. Pusey, P. N., and van Megen, W., *Phys. Rev. Lett.* **59**, 2083 (1987).
19. Pusey, P. N., and van Megen, W., *Phys. Rev.* A **43**, 5429 (1991).
20. van Megen, W., Underwood, S. M., and Pusey, P. N., *Phys. Rev. Lett.* **67**, 1586 (1991).
21. van Megen, W., and Underwood, S. M., *Phys. Rev. Lett.* **70**, 2766 (1993).
22. van Megen, W., and Underwood, S. M., *Phys. Rev.* E **47**, 248 (1993).
23. Bartsch, E., Antoinetti, M., Schupp, W., and Sillescu, H., *J. Chem. Phys.* **97**, 3950 (1992).
24. Bartsch, E., Frenz, V., Möller, S., and Sillescu, H., *Physica A* **201**, 363 (1993).
25. Sillescu, H., and Bartsch, E., in *Disorder Effects on Relaxational Processes*, Eds. R. Richert and A. Blumen (Springer, Berlin, 1993), p. 59.
26. Sirota, E. B., Ou-Yang, H. D., Sinha, S. K., Chaikin, P. M., Axe, J. D., and Fuji, Y., *Phys. Rev. Lett.* **62**, 1524 (1989).
27. Meller, A., and Stavans, J., *Phys. Rev. Lett.* **68**, 3646 (1992).
28. Götze, W., in *Liquids, Freezing and the Glass Transition*, Eds. J.P. Hansen, D. Levesque, and J. Zinn-Justin (North Holland, Amsterdam, 1991), p. 287.
29. Götze, W., and Sjögren, L., *Phys. Rev.* A **43**, 5442 (1991).

30. Fuchs, M., Hofacker, I., and Latz, A., *Phys. Rev.* **A 45**, 898 (1992).
31. Szamel, G., and Löwen, H., *Phys. Rev.* **A 44**, 8215 (1991).
32. Fuchs, M., private communication, 1993.
33. Kawasaki, K., *Physica A* **208**, 35 (1994).
34. Löwen, H., Hansen, J. P., and Roux, J. N., *Phys. Rev.* **A 44**, 1169 (1991).
35. Baschnagel, J., Binder, K., and Wittmann, H. -P., *J. Phys. Condens. Matter* **5**, 1597 (1993).
36. Kob, W., and Anderson, H. C., *Phys. Rev.* **E 47**, 3281 (1993).
37. Kob, W., and Anderson, H. C., *Phys. Rev.* **E 48**, 4364 (1993).
38. Migayawa, H., Hiwatari, Y., Bernu, B., and Hansen, J. P., *J. Chem. Phys.* **88**, 3879 (1988).
39. Löwen, H., and Hansen, J. P., in *Recent Developments in the Physics of Liquids*, Eds. W. S. Howells and A. K. Soper (IOP Publishing, Bristol, 1991), p. F283.
40. Walrand, S., Belloni, L., and Drifford, M., *J. Physique* **47**, 1565 (1986).
41. Kops-Werkhoven, M. M., and Fijnaut, H. M., *J. Chem. Phys.* **77**, 2242 (1982).
42. Kops-Werkhoven, M. M., Pathmamanoharan, C., Vrij, A., and Fijnaut, H. M., *J. Chem. Phys.* **77**, 5913 (1982).
43. Ottewill, R. H., and Williams, N. St. J., *Nature (London)* **325**, 232 (1987).
44. van Veluwen, A., and Lekkerkerker, H. N. W., *Phys. Rev.* **A 38**, 3758 (1988).
45. van Megen, W., and Underwood, S. M., *J. Chem. Phys.* **91**, 552 (1989).
46. Dergiorgio, V., Piazza, R., Corti, M., and Stavans, J., *J. Chem. Soc. Faraday Trans.* **87**, 431 (1991).
47. van Megen, W., Underwood, S., Ottewill, R., Williams, N., and Pusey, P. N., *J. Chem. Soc. Faraday Discuss.* **83**, 47 (1987).
48. Härtl, W., Versmold, H., Wittig, U., and Linse, P., *J. Chem. Phys.* **97**, 7797 (1992).
49. Härtl, W., Versmold, H., and Zhang-Heider, X., *Ber. Bunsenges. Phys. Chemie* **95**, 1105 (1991).
50. Dozier, W. D., Lindsay, H. M., and Chaikin, P. M., *J. de Physique, Colloque* **C3**, 165 (1985).
51. Simon, R., Palberg, T., and Leiderer, P., *J. Chem. Phys.* **99**, 3030 (1993).
52. van Blaaderen, A., Peetermans, J., Maret, G., and Dhont, J. K. G., *J. Chem. Phys.* **96**, 4591 (1992).
53. Imhof, A., van Blaaderen, A., Maret, G., Mellema, J., and Dhont, J. K. G., *J. Chem. Phys.* **100**, 2170 (1994).
54. Wiese, H., and Horn, D., *J. Chem. Phys.* **94**, 6429 (1991).
55. Bless, M. H., and Leyte, J. C., *J. Coll. Interfac. Science* **157**, 355 (1993).
56. Zwanzig, R., *Ann. Rev. Phys. Chem.* **16**, 67 (1965).
57. Hess, W., and Klein, R., *Adv. Phys.* **32**, 173 (1983).
58. Klein, R., and Hess, W., in *Ionic Liquids, Molten States and Polyelectrolytes*, Eds. K. H. Bennemann, F. Bouers, and D. Quitmann; Lecture Notes in Physics **172**, 199 (1982).
59. Nägele, G., Medina-Noyola, M., Klein, R., and Arauz-Lara, J. L., *Physica A* **149**, 123 (1988).
60. Medina-Noyola, M., *Phys. Rev. Lett.* **60**, 2705 (1988).
61. Hansen, J.-P., and McDonald, I. R., *Theory of Simple Liquids*, 1st edn. (Academic Press, London, 1976).
62. Cichocki, B., *Physica A* **148**, 165 (1988); **148**, 191 (1988).
63. Löwen, H., and Szamel, G., *J. Phys. Condens. Matter* **5**, 2295 (1993).
64. Leegwater, J. A., and Szamel, G., *Phys. Rev.* **A 46**, 4999 (1992).
65. Szamel, G., and Leegwater, J. A., *Phys. Rev.* **A 46**, 5012 (1992).
66. Indrani, A. V., and Ramaswamy, S., *Phys. Rev. Lett* **73**, 360 (1994).
67. Cichocki, B., and Hinsen, K., *Physica A* **166**, 473 (1990).
68. Cichocki, B., and Hinsen, K., *Physica A* **187**, 133 (1992).
69. Bitzer, F., Palberg, T., Löwen, H., Simon, R., and Leiderer, P., *Phys. Rev.* **E 50**, 2821 (1994).
70. Löwen, H., *Phys. Rev.* **E 50**, 1232 (1994).
71. Meijer, E. J., and Frenkel, D., *J. Chem. Phys.* **94**, 2269 (1991).
72. Härtl, W., and Versmold, H., *J. Chem. Phys.* **88**, 7157 (1988).
73. D'Aguanno, B., and Klein, R., *Phys. Rev.* **A 46**, 7652 (1992).
74. Belloni, L., *J. Chem. Phys.* **85**, 519 (1986).

75. Broersma, S., *J. Chem. Phys.* **32**, 1626 (1960); **32**, 1632 (1960); **74**, 6989 (1981).
76. Tirado, M. M., and de la Torre, J. G., *J. Chem. Phys.* **71**, 2581 (1979); **73**, 1986 (1980).
77. Tirado, M. M., Martinez, C. L., and de la Torre, J. G., *J. Chem. Phys.* **81**, 2047 (1984).
78. Veerman, J. A. C., and Frenkel, D., *Phys. Rev.* A **41**, 3237 (1990).
79. Lee, S.-D., *J. Chem. Phys.* **87**, 4972 (1987).
80. Onsager, L., *Ann. N. Y. Acad. Sci.* **51**, 627 (1949).
81. Lasher, G., *J. Chem. Phys.* **53**, 4141 (1970).
82. Kayser, R. F., and Raveche, H. J., *Phys. Rev.* A **17**, 2067 (1978).
83. Doi, M., and Edwards, S. F., *The Theory of Polymer Dynamics* (Oxford University Press, Oxford, 1986).
84. Doi, M., Yanamoto, I., and Kano, F., *J. Phys. Soc. Japan* **53**, 3000 (1984).
85. Hu, H.-W., Carson, G. A., and Granick, S., *Phys. Rev. Lett.* **66**, 2758 (1991).
86. Thompson, P. A., Grest, G. S., and Robbins, M. O., *Phys. Rev. Lett.* **68**, 3448 (1992).
87. Fehr, T., and Löwen, H., *Phys Rev.* E **52**, 4016 (1995).
88. Roux, J. N., Barrat, J. L., and Hansen, J. P., *J. Phys. Condens. Matter* **1**, 7171 (1989).
89. Barrat, J. L., Roux, J. N., and Hansen, J. P., *Chem. Phys.* **149**, 197 (1990).
90. Ranganathan, S., *J. Phys. Condens. Matter* **6**, 1299 (1994).
91. Fröhlich, J., and Pfister, C., *Commun. Math. Phys.* **81**, 277 (1981).
92. Löwen, H., *J. Phys. Condens. Matter* **4**, 10105 (1992); **5**, 2649 (1993).
93. Naidoo, K. J., and Schnitker, J., *J. Chem. Phys.* **100**, 3114 (1994).
94. Löwen, H., *Phys. Rev. Lett.* **72**, 424 (1994).
95. Schneider, J., Hess, W., and Klein, R., *J. Phys. A: Math. Gen.* **18**, 1221 (1985).
96. Schneider, J., Karrer, D., Dhont, J. K. G., and Klein, R., *J. Chem. Phys.* **87**, 3008 (1987).
97. Deutsch, J. M., and Goldenfeld, N., *J. Physique* **43**, 651 (1982).
98. Kirchhoff, T., Löwen, H., and Klein, R., *Phys. Rev.* E (in press).

CHAPTER
9

Density-Functional Theory of Freezing of Charge-Stabilized Colloidal Suspensions

J. Chakrabarti, H. R. Krishnamurthy, S. Sengupta, and A. K. Sood

This chapter reviews the application of density-functional theory (DFT) to the problems involving the freezing of charge-stabilized colloidal suspensions. The phase diagram of these colloidal suspensions of charged polystyrene spheres show BCC, FCC, and liquid phases. Most of the earlier theoretical work on the phase diagram has been restricted to self-consistent phonon theories, which predict the melting line based on the Lindemann criterion. The calculated phase diagrams from these theories do not agree with the experimental results. In contrast the DFT is able to successfully obtain the full BCC-FCC-liquid phase diagram. The salient features of this theory, like the inclusion of terms involving the three-body direct correlation function of the liquid and a comparison with the experiments are discussed. This chapter also covers the density-functional theory of laser-induced freezing, that is, the freezing of a colloidal liquid into crystalline order in the presence of an external periodic modulating potential V_e. The theory predicts that the initial first-order freezing transition (at small V_e) changes to a continuous one (at large V_e) via a tricritical point provided the modulation wave vectors are suitably chosen. The theory can predict the parameter values of the tricritical points for a realistic colloidal system.

9.1 Introduction

The arrangement and dynamics of colloidal particles, under suitable experimental conditions of particle density and ionic strength of the solvent, can mimic different kinds of phases found in conventional atomic systems such as fluids, crystal, and even glasses [1–5]. The easy tuning of the interparticle interaction makes the aqueous suspension of charged polystyrene spheres an ideal condensed matter system in which to study fluid-to-crystal equilibrium phase transitions.

In this chapter we review the present state of understanding of different equilibrium phases of the polyball suspensions with particular emphasis on how the density-functional theory has been successfully applied to obtain the full body-centered cubic (BCC) - face-centered cubic (FCC) - liquid phase diagram [6, 7]. We shall also discuss the application of DFT to the interesting phenomenon of laser induced freezing [8, 9] wherein the freezing of a colloidal liquid into a crystalline order is assisted by the presence of a periodic modulating external potential induced by a standing-wave pattern of an interfering laser field.

9.2 The Freezing of Charge-Stabilized Colloids: Experiments

9.2.1 The Equilibrium Phase Diagram

The specific polyball system discussed in this review is as follows. Each polyball consists of a large number of styrene polymeric chains entangled in a coil. Each of these chains starts and ends with an acidic group like $-KSO_4$. In a solvent like water with a high dielectric constant ($\epsilon \sim 80$), the end surface groups dissociate and provide a large electrostatic negative charge per particle ($\sim 1000e$ for a particle of diameter of 1000 Å, where e is the fundamental electronic charge). The counterions (cations like K^+ liberated from the polyballs) and the additional ions present in the solvent form a cloud around each polyball and hence screen the Coulomb interaction between them. Thus the interaction between the polyballs over the relevant range of particle separation is predominantly the screened Coulomb interaction and can be written in the form (called the DLVO potential after Derjaguin, Landau, Verwey, and Overbeck) [14]:

$$U(r) = \frac{(Ze)^2}{\epsilon} \left(\frac{\exp(\kappa a)}{1 + \kappa a} \right)^2 \frac{\exp(-\kappa r)}{r} \tag{9.1}$$

where κ is the inverse of the Debye screening length given by:

$$\kappa^2 = \frac{4\pi}{\epsilon k_B T} \left((n_p Z)e^2 + \sum_\alpha n_\alpha (z_\alpha e)^2 \right) \tag{9.2}$$

Here n_p is the number density of the polyballs of radius a, each having surface charge Ze. The additional ions of type α with number density n_α and charge

z_α contribute to the screening in addition to the counterions. Hence the interaction between the polyballs can easily be tuned by changing the polyball density (n_p), or equivalently the volume fraction $\phi = \frac{4}{3}\pi a^3 n_p$ and the impurity (mostly monovalent) concentration (n_α). When κ^{-1} is much smaller than the average interparticle separation $\ell\,(= n_p^{-1/3})$, the interactions between the polyballs are negligible and the particles essentially perform free Brownian motion. As the interaction builds up (i.e., κ^{-1} becomes larger) the particles exihibit spatial correlations over two to three interparticle seperations (liquidlike short-range order). For sufficiently strong interactions, the colloidal particles freeze into a body centered cubic (BCC) or a face-centered cubic (FCC) lattice. When the suspension crystallizes, a beautiful iridescence is observed due to the Bragg diffraction of visible light. Fig. 9.1 shows the experimental phase diagram [2] in (ϕ, n_i) parameter space, determined by using the visible light Kossel diffraction technique for a well-characterized polyball suspension, with particle diameter 1334 ± 10 Åand charge $Z = 1200 \pm 40$.

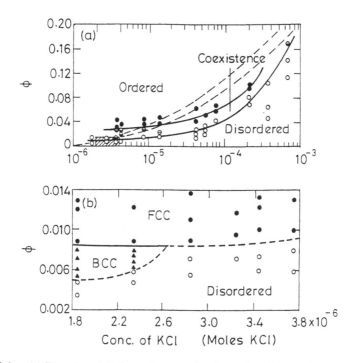

Figure 9.1 (a) Experimental phase diagram showing ordered (crystal) and disordered (fluid) phases. Filled (open) circles correspond to crystals (fluid) occupying the whole volume of the suspension. The solid lines are drawn as guide to the eyes to mark the melting and the freezing boundaries. (b) The hatched portion is shown in a magnified scale. The filled circles correspond to the FCC and the filled triangles correspond to the BCC phase [2]. [Reprinted with permission from Y. Monovoukas, and A.P. Gast, *J. Colloid Interface Sci.* **128**, 533 (1989) Copyright 1989 Academic Press]

The full lines are drawn to guide the eyes to indicate approximately the melting and freezing boundaries. The results of a detailed investigation of the BCC-FCC transition taking place at very low polyball concentration and ionic strength, marked by shaded portion in Fig. 9.1(a), are presented in Fig. 9.1(b). It is seen that as the ionic concentration (n_i) increases, the ordered phase is formed at a higher polyball concentration (n_p). This is easy to understand: As n_i increases, the Debye screening length decreases, and hence to make it comparable to the inteparticle separation, one needs higher n_p.

Figure 9.2 shows another recently determined phase diagram [3] using a small-angle synchrotron x-ray study of charged polystyrene spheres of diameter 910 Åand charge 135e suspended in a solvent consisting of 90% methanol and 10% water (dielectric constant of the solvent is 38). The ordered phases occur at high volume fraction because of small charge on the colloidal particles. It may be noted that in addition to the equilibrium fluid, BCC, and FCC phases, a glassy state is also observed.

Various exotic phases are also observed in two-component colloidal suspensions. For example, even binary mixtures of sterically stabilized colloidal suspensions where the particles interact via a simple hard-core potential show a complex phase diagram. For appropiate diameter ratios one obtains AB_2 and AB_{13} superlattices, as observed experimentally by Bartlett et al. [10]. The latter phase consists of a simple cubic lattice where the basis consists of an A particle

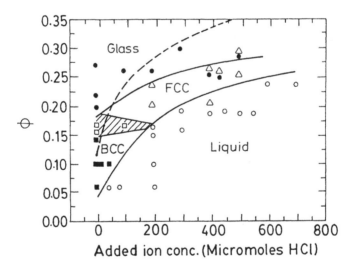

Figure 9.2   (a) Experimental phase diagram of polyball suspension: solid square - BCC; open triangle - FCC; open circles - liquid, and filled circle - glass. The open squares in the hatched area mark the coexistence of BCC and FCC crystals [3]. Reprinted with permission from E. B. Sirota et al., *Phys. Rev. Lett.* **62**, 1524 (1989) Copyright 1989 The American Physical Society]

at $(0, 0, 0)$ and a 13-particle icosahedral B cluster at $(\frac{1}{2}, \frac{1}{2}, \frac{1}{2})$ reminiscent of the $NaZn_{13}$ intermetallic compound. This phase was later confirmed by computer simulations [11] and theoretical density-functional calculations [12]. Equilibrium phase diagrams for binary mixtures of charge-stabilized colloids, however, have not received an equal amount of attention (nevertheless, see, e.g., Kaplan et al. [13]).

9.2.2 The Phenomenon of Laser-Induced Freezing

Several years ago Chowdhury et al. [8] demonstrated the very interesting phenomenon of laser-induced freezing (LIF) in a two-dimensional (2D) suspension of polyballs. They carried out cross-correlation intensity-fluctuation spectroscopy experiments by subjecting a 2D colloidal liquid to crossed laser beams and monitoring the self-scattered field. In their setup two coherent laser beams (at an angle of 8^0 with each other and one of the beams makes an angle of 4^0 with the normal to the sample plain) intersected at the sample. This produces a stationary intensity fringe pattern with peridiocity along the line defined by the intersection of the sample plane with the plane of the incident beams [8, 34] (see Fig. 9.3). The colloidal particles can be thought of as subjected to a 1D periodic modulating potential V_e induced by the standing-wave pattern of the interference pattern. The electric field of the standing wave, given by $E_0 \cos(\mathbf{q}_0 \cdot \mathbf{r}/2)$, induces dipole moments on the polyballs (which are dielectric particles), and these dipole moments interact with the external electric field, leading to an interaction energy given by $-\frac{1}{2}\chi E_0^2 \cos^2(\mathbf{q}_0 \cdot \mathbf{r}/2)$, where χ is the dielectric susceptibility of the polyballs. Clearly, this has a component that is like a periodic potential acting on the particles $V_e \cos(\mathbf{q}_0 \cdot \mathbf{r})$, with $V_e = \chi E_0^2/4$. The particles prefer to sit on the minima of this potential. For low interparticle interaction strength (i.e., large κ, or low values of the first peak of the liquid direct correlation function), one gets a phase called the modulated liquid phase with density modulations with wave vector $\mathbf{q}_0$ and its harmonics. This results in the appearance of maxima in the scattered beam corresponding to $\pm\mathbf{q}_0$ [Fig. 9.3(b)]. Chowdhury et al. [8] showed that, if the wave vector of the modulation $\mathbf{q}_0$ was tuned to be at the first peak in the direct correlation function (DCF) of the colloidal liquid in the absence of V_e, the colloidal system responds to (a sufficiently strong) V_e by forming a 2D (modulated) crystalline phase with predominantly hexagonal order (similar to that of the colloidal crystal obtained in the absence of V_e). The scattered beam intensity now, in addition to the maxima corresponding to $\pm\mathbf{q}_0$, shows secondary maxima corresponding to $\mathbf{k}_2$ and $\mathbf{k}_3$ [Fig. 9.3(b)] so that the pattern has the symmetry of a triangular lattice. It was found that the output intensity of these secondary maxima varies as $\frac{3}{2}$ power of the input intensity, while it has a cubic dependence for the driven modes $(\pm\mathbf{q}_0)$.

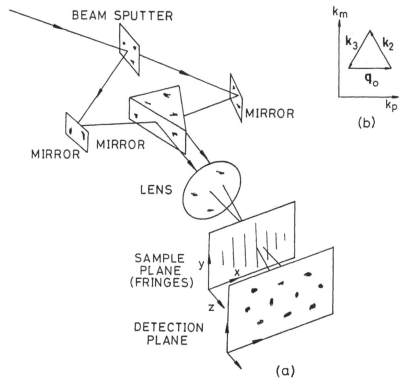

Figure 9.3 (a) Experimental setup for laser-induced freezing [34] [Reprinted with permission from K. Loudiyi and B. J. Ackerson, *Physica A* **184**, 1 (1992) Copyright 1992 Elsevier Science]. (b) The crossed laser beams directly excite the density mode with wave vector $\mathbf{q}_0$ along the $\mathbf{k}_p$ axis [the intersection of the planes of the incident beams and the sample in (a)] and two other modes with wave vectors $\mathbf{k}_2$ and $\mathbf{k}_3$, coupled to $\mathbf{q}_0$ by particle interaction [8]. [Adapted with permission from A. Chowdhury et al. *Phys. Rev. Lett.* **60**, 833 (1985) copyright 1985 The American Physical Society]

9.3 The Freezing of Charge-Stabilized Colloids: Theory

9.3.1 Known Theoretical and Computer Simulation Results

A comprehensive review of various attempts to understand the phase diagram of a polyball suspension has been given in Ref. [1]. Here we shall briefly mention the final results of the analytic theories based on the self-consistent phonon approximation and molecular dynamics computer simulation to motivate the need for the DFT of the phase diagrams.

Rosenberg and Thirumalai [15] have calculated the phase diagram [shown in Fig. 9.4(a)] using the self-consistent phonon theory. The parameters used are $\kappa^* = \kappa \ell$ and $\Gamma^{-1} = (Z^2 e^2 / \ell \epsilon k_B T)^{-1}$. The parameter Γ is related to the relative strength of the Coulomb interaction at the interparticle separation compared

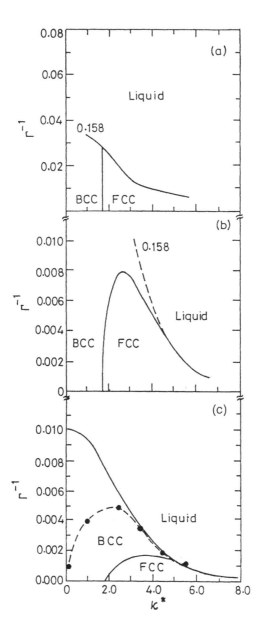

Figure 9.4 (a) Calculated phase diagram based on self-consistent phonon theory. The melting curves are for the Lindemann parameter L_p = 0.1, 0.158 and 0.22 [15]. The dotted line is from the theory of W. Y. Shih et al., *J. Chem. Phys.* **86**, 5127 (1987). (b) Calculated phase diagram. The dotted line is the melting curve calculated from L_p = 0.158 [16] [Reprinted with permission from D. Hone et al., *J. Chem. Phys.* **79**, 1474 (1983) copyright 1983 American Institute of Physics]. (c) The dashed and the solid lines are the MD simulation results as a function of Γ^{-1} and κ^*. The filled dots are the results of density functional theory [30].

to $k_B T$. Figure 9.4(b) shows the calculated phase diagram [16] using a self-consistent harmonic approximation for the BCC -FCC phase boundary and the Lindemann criterion for the melting line. Figure 9.4(c) shows the phase diagram based on a molecular dynamics simulation [17], which agrees well with the experimental data shown in Fig. 9.1(b), provided the effective charge $Z^* e$ on the polyballs are rescaled from $1200e$ to $880e$. The dotted line shown in Fig. 9.4(c) is the BCC-liquid phase boundary calculated by Alexander et al. using one version of the density functional theory (to be discussed later). It is clear from Fig. 9.4 that the theories [shown in Fig. 9.4(a), 9.4(b), and the dotted line in Fig. 9.4(c)] fail completely to match the experimental phase diagram in the low-κ region.

This motivated Sengupta and Sood [6] to reexamine the application of DFT to colloids and their results agree remarkably well with the experiments, as shown in Fig. 9.5. The experimental data are the same as in Fig. 9.1(b).

9.3.2 Density-Functional Theory of Equilibrium Phase Diagrams without External Potential

The most successful theories of the freezing transition are density-functional theories, which attempt to derive an expression for the free energy of the solid as a functional of the time-averaged, nonuniform, local density $\rho(\mathbf{r})$ (which is a constant ρ_0 for a liquid). The freezing transition is treated as a thermodynamic instability of the liquid phase with respect to the formation of static spatially periodic density waves with appropriate crystalline symmetry.

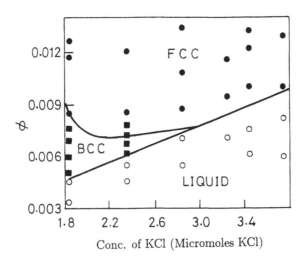

Conc. of KCl (Micromoles KCl)

Figure 9.5   The phase diagram of polyballs. Lines indicate the theoretical phase diagram, while the markers represent experimentally determined stable BCC, FCC, and liquid phases [6].

In the Ramakrishnan and Yussouff theory [18-21] of freezing, the solid is treated as a calculable pertubation on the liquid, and various properties of the solid phase are obtained in terms of the liquid phase parameters. In this theory the grand canonical free energy cost of creating density inhomogeneity in a uniform liquid phase (of density ρ_0) in the presence of an external potential $V(\mathbf{r})$ is approximated by the following functional Taylor expansion in $[\rho(\mathbf{r}) - \rho_0]$:

$$
\begin{aligned}
\mathcal{F}[\rho(\mathbf{r})]/k_B T = {} & \int d\mathbf{r}\, \rho(\mathbf{r})\, \ln[\rho(\mathbf{r})/\rho_0] - \int d\mathbf{r}\, [\rho(\mathbf{r}) - \rho_0] \\
& - \frac{1}{2} \int d\mathbf{r}\, d\mathbf{r}_1 c^{(2)}(\mathbf{r} - \mathbf{r}_1)[\rho(\mathbf{r}) - \rho_0][\rho(\mathbf{r}_1) - \rho_0] \\
& - \frac{1}{6} \int d\mathbf{r}\, d\mathbf{r}_1\, d\mathbf{r}_2\, c^{(3)}(\mathbf{r}, \mathbf{r}_1, \mathbf{r}_2)[\rho(\mathbf{r}) - \rho_0] \\
& \times [\rho(\mathbf{r}_1) - \rho_0][\rho(\mathbf{r}_2) - \rho_0] + \beta \int d\mathbf{r}\, V(\mathbf{r})\rho(\mathbf{r}) \quad (9.3)
\end{aligned}
$$

The freezing problem is reduced to a variational problem of minimizing this free-energy functional. Here $c^{(2)}(\mathbf{r} - \mathbf{r}_1)$ and $c^{(3)}(\mathbf{r}, \mathbf{r}_2, \mathbf{r}_3)$ are the second- and third-order direct correlation functions of the liquid. The whole point is that, under suitable conditions, $\mathcal{F}$ develops minimum corresponding to a periodic $\rho(\mathbf{r})$, which has the symmetry of a crystalline lattice. When the value of $\mathcal{F}$ at the crystalline minimum equals that at the liquid minimum $[\rho(\mathbf{r}) = \rho_0]$, the liquid freezes to a solid. The process of freezing of a liquid to a solid is thus attributed to the spontaneous generation of density waves with crystalline symmetry in a liquid of uniform density.

Let us define a molecular field $\xi(\mathbf{r}) = \ln[\rho(\mathbf{r})/\rho_0]$. The self-consistency condition obtained by the minimization of Eq. (9.3) with respect to $\rho(\mathbf{r})$ can be written in terms of the Fourier components $\mu_{\mathbf{G}}, \xi_{\mathbf{G}},$ and $V_{\mathbf{G}},$ of $\rho(\mathbf{r})/\rho_0, \xi(\mathbf{r}),$ and $V(\mathbf{r})$, respectively, where $\mathbf{G}$ is a reciprocal lattice vector (RLV) of the crystal, as:

$$
\xi_{\mathbf{G}} = c_{\mathbf{G}}^{(2)} \mu_{\mathbf{G}} + \frac{1}{2} \sum_{\mathbf{G}_1} c_{\mathbf{G}, \mathbf{G}_1, \mathbf{G} - \mathbf{G}_1}^{(3)} \mu_{\mathbf{G} - \mathbf{G}_1} \mu_{\mathbf{G}_1} - \beta V_{\mathbf{G}} \qquad (9.4)
$$

While this is formally an infinite set of coupled nonlinear equations, the number of relevant equations is actually finite since the liquid DCF falls to zero for sufficiently large G. However, this number is still rather large as the decay of $c_G^{(2)}$ is slow. The handling of a large number of coupled nonliear equations in Eq. (9.4) can be avoided by variational ansatzes. The simplest and the most widely used of these is to assume that the single-particle density in the crystalline phase to be a sum of nonoverlapping Gaussians centered on the given lattice sites $\mathbf{R}$, viz., $\rho(\mathbf{r}) = (\alpha/\pi)^{3/2} \sum_{\mathbf{R}} \exp[-\alpha(\mathbf{r} - \mathbf{R})^2]$, where α is the width of the Gaussian and is the only variational parameter to minimize the free energy in Eq. (9.3) [22]. However, this approach is fraught with the trouble that the self-consistency equations in Eq. (9.4) is not guaranteed to be satisfied [23].

In the work of Sengupta and Sood [6], the effect of the higher-order correlations and the deviations from the Gaussian density profile ansatz have been taken care of by a new method [23], which can be motivated as follows. For simplicity, let us exclude terms containing $c^{(3)}$, $c^{(4)}$, etc. The molecular field is sharply peaked at the lattice site in each unit cell (which can be chosen as the origin). We note that the Fourier component of $\rho(\mathbf{r})/\rho_0$ can be written as $\mu_{\mathbf{G}} = \int \exp(i\mathbf{G} \cdot \mathbf{r}) \exp[\xi(\mathbf{r})]$. By Taylor expansion of $\xi(\mathbf{r})$ about $r = 0$ and performing the integral, one can easily see that $\mu_{\mathbf{G}}$ approaches asymptotically to a Gaussian form $A \exp(-G^2 a^2/4\alpha)$, where a is the lattice parameter, determined from the smallest RLV at the first peak of the liquid DCF. Here A and α describe the asymptotic Gaussian. For small G, however, there will be corrections to the Gaussian form. This immediately leads to the following variational ansatz:

$$\xi_{\mathbf{G}} = c_{\mathbf{G}}^{(2)} A \exp(-G^2 a^2/4\alpha), \qquad G > G_{\max}$$
$$\xi_{\mathbf{G}} = c_{\mathbf{G}}^{(2)} A \exp(-G^2 a^2/4\alpha) + \xi_{\mathbf{G}}^c, \qquad G < G_{\max} \tag{9.5}$$

for the Fourier components of $\xi(\mathbf{r})$. Here $\xi_{\mathbf{G}}^c$ is the correction to the Gaussian part. The equations that the variational parameters have to satisfy can be easily shown to be [23]:

$$\alpha = a^2/2\xi''$$
$$A = v^{-1} \exp[\xi(0)](\pi a^2/\alpha)^{3/2}$$
$$\xi_{\mathbf{G}}^c = \left(v^{-1} \int d\mathbf{r} \exp(i\mathbf{G} \cdot \mathbf{r}) \exp\left(\sum_{\mathbf{G}'} \xi_{\mathbf{G}'} \exp(i\mathbf{G}' \cdot \mathbf{r}) \right) \right.$$
$$\left. -A \exp(-G^2 a^2/4\alpha) \right) c_{\mathbf{G}}^{(2)} \tag{9.6}$$

Note that in the last relation of Eq. (9.6), ξ_G is given by Eq. (9.4). Here ξ'' is the curvature of the molecular field at the origin and v is the volume of the unit cell. The set of Eq. (9.6) is now solved iteratively. Note that the above ansatz is made in the context of ξ_G and not μ_G, and no further assumptions such as the nonoverlap of the different Gaussians in the decomposition of $\rho(\mathbf{r})$ are made. This method is essentially an exact procedure if one keeps the correction fields up to sufficiently large G.

Inputs to the DFT

It has been shown [24, 25] that the measured two-point DCF [which is related to the structure factor $S_{\mathbf{G}} = 1/(1 - c_{\mathbf{G}}^{(2)})$] of polyballs can be understood in terms of the liquid state theory using a rescaled mean spherical approximation (RMSA). The mean spherical approximation (MSA) uses the following closure condition:

$$c^{(2)}(r) = -\beta U(r), \qquad \text{for } r > 2a$$
$$h(r) = -1, \text{ for } r < 2a \tag{9.7}$$

where $h(r)$ is the pair correlation function. The MSA gives reliable DCF only for sufficiently large concentration ($\phi \geq 0.2$). For dilute systems, the MSA yields unphysical negative contact values for the radial distribution function. This difficulty is circumvented in the RMSA by rescaling the particle radius a to a' while maintaining the Coulomb coupling [$\Gamma_c = 2\beta U(r), r = 2a'$]. The physical argument rests on the observation that for large Γ_c, the contact configurations are extremely rare, because the value of the potential at the contact is much larger than the thermal energy. This implies that the Coulomb interaction alone makes $g(r) = 0$ because $r \leq a'$ without the hard core playing any role.

It is observed that from the two-particle correlation alone it is impossible to get a stable or even a metastable BCC solid [26, 27] because of the large positive contribution to the free energy from the (2,0,0) reciprocal lattice vector (RLV) of the BCC lattice (see Fig. 9.6). The simplest way that (2,0,0) density wave can form is by a nonlinear effect involving three-body correlations, $c^{(3)}$. The three-body term corresponding to the isosceles triangle $(1, 1, 0) + (1, \bar{1}, 0) = (2, 0, 0)$ is a key component in promoting this effect since it involves RLVs of the smallest possible size, and the corresponding $c^{(3)}$ is positive as seen from different computer simulations [29]. Nothing, however, is

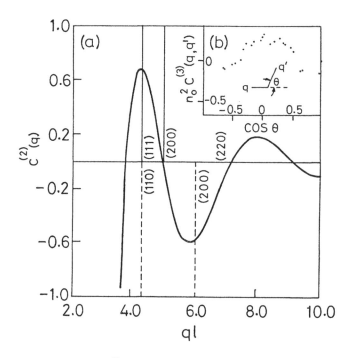

Figure 9.6   Typical RMSA $c^{(2)}(q)$ versus $q\ell$ plot for polyballs near freezing. The solid lines mark the positions of the first few RLVs for FCC lattice, and the dashed lines do the same for BCC lattice [6].

known about this quantity for colloids. Still it is generally held that the higher-order correlation functions are relatively insensitive [28] to the exact form and amount of couplings. Accordingly, in the thory of Sengupta and Sood [6], the $c^{(3)}$ terms corresponding to $(1, 1, 0) + (1, \bar{1}, 0) = (2, 0, 0)$ for the BCC case and those for $(1, 1, 1) + (1, \bar{1}, \bar{1}) = (2, 0, 0)$ and $(1, 1, 1) + (1, 1, \bar{1}) = (2, 2, 0)$ for the FCC case are retained. Computer simulations indicate that the three-body terms are roughly equal for the two triangles for the FCC case. This reduces the three-body contribution to two numbers c_b and c_f corresponding to the BCC and the FCC case, respectively.

Comparision with Experimental Results

The parameters c_b and c_f were so chosen that the liquid-BCC and BCC-FCC transitions are reproduced for a particular value of n_i $(= 1.848 \times 10^{-9}$ moles/cm$^3)$ for the experimental system of Monovoukas and Gast shown in Fig. 9.1(b). The values were $c_b = 0.23$ and $c_f = 0.116$. The entire phase diagram was calculated by keeping c_b and c_f fixed at these values and only putting calculated $c_G^{(2)}$ using RMSA. The results are shown in Fig. 9.4. The Lindemann parameter L_p was also calculated using the formula: $L_p = [v^{-1} \int^{\text{cell}} d\mathbf{r} \, r^2 \exp \xi(\mathbf{r})]^{1/2}/d_{nn}$, where d_{nn} is the nearest-neighbor distance. Typical freezing parameters from their calculation at the triple point were $L_p = 0.197$ for BCC and $L_p = 0.131$ for FCC. The fractional change in density is around 1.6% in the BCC transition and around 2.6% for FCC transitions. This calculation [6] shows that while three-body effects are required to obtain a stable BCC phase, the BCC-FCC transition can be understood in terms of the shape of $c^{(2)}$ alone. As ϕ increases (i.e., the steepness of the potential increases), the first minimum becomes too negative and the system begins to prefer the FCC structure that avoids this minimum. The calculation also illustrates that a very simple parametrization of the three-body term can reproduce the experimentally observed phase diagram.

Other Results of DFT

At this stage we shall comment on the earlier DFT work of Alexander et al. [30], whose results as shown by the dotted line in Fig. 9.4(c), do not agree with the experiments. Their theory is equivalent to truncating the local, nonpolynomial part of the density-functional free energy in Eq. (9.3) at the second order. Two sets of reciprocal lattice vectors corresponding to the first and the second peak region of the liquid DCF are retained. The liquid-BCC freezing line shown by the dotted line in Fig. 9.4(c) deviates sharply from the experimental as well as computer simulation results for small values of κ.

Salgi and Rajagopalan [7] have used another formalism of the DFT, called the modified weighted density-functional (MWDA) theory. In weighted density-functional theories [31], the free-energy density $\mathcal{F}$ of a solid with an inhomogeneous density is approximated by setting it equal to that of a liquid with an inhomogeneous weighted density $\tilde{\rho}$ that is a functional of ρ given by:

$$\tilde{\rho} = \int d\mathbf{r}_1 \, w(|\mathbf{r} - \mathbf{r}_1|; \, \tilde{\rho})\rho(\mathbf{r}_1) \tag{9.8}$$

The total free energy of the solid is then given by:

$$\mathcal{F} = \int d\mathbf{r} f_0(\tilde{\rho})\rho \tag{9.9}$$

where $f_0(\tilde{\rho})$ is the excess free energy per particle in the homogeneous liquid phase of density ρ. The normalized weighting function w is so chosen that the DCF is reproduced in the homogeneous limit. This requirement yields a self-consistent equation for ρ that has to be satisfied. This computation turns out to be time consuming. In a modified version (MWDA) [32], which is computationally easier, the excess free energy of the solid is given by the free energy of a homogeneous system calculated at a homogeneous modified weighted density $\tilde{\rho}$:

$$\mathcal{F} = N f_0(\tilde{\rho}) \tag{9.10}$$

where now

$$\tilde{\rho} = \int d\mathbf{r}_1 \, d\mathbf{r} \, w(|\mathbf{r} - \mathbf{r}_1|; \, \tilde{\rho})\rho(\mathbf{r}_1)\tilde{\rho})\rho(\mathbf{r}) \tag{9.11}$$

The weighting function w is determined from the same requirement as WDA. The weighted density-functional theories are nonperturbative in approach, and a subset of higher order liquid DCF are retained here.

Salgi and Raj Rajagopalan considered only up to $c^{(2)}(\mathbf{G})$ terms and used the Gaussian density profile ansatz for $\rho(\mathbf{r})$. Their phase diagram is qualitatively similar to the experimental phase diagram of Sirota et al. [3] shown in Fig. 9.2, provided the charge of the colloidal particles is scaled to a higher value. However, the authors have noted that in quantitative detail there is a difference between their theoretical results and the experiment in the low-ϕ region. The deviation in the BCC-FCC coexistence line can probably be due to deviation from the Gaussian density profile ansatz. The stable BCC phase was obtained in this theory because MWDA includes some of the higher correlation functions.

9.3.3 Density-Functional Theory for Laser-Induced Freezing Early Results

In the original paper where the experimental results were reported, Chowdhury et al. [8] also theoretically analyzed the phenomenon of laser-induced freezing in terms of a simple Landau-Alexander-McTague [33] theory. In their theory, they took the density modes ξ_1, ξ_2, and ξ_3 corresponding to three smallest RLVs of the triangular lattice as the order parameters [Fig. 9.3(b)]. They considered a Landau free energy of the following form:

$$F = 2A\xi_1 + B\sum_i \xi_i^2 + 2C\xi_1\xi_2\xi_3 + D\left(\sum_i \xi_i^2\right)^2 + E\left(\sum_i \xi_i^4\right). \quad (9.12)$$

The first term represents the coupling to the external potential (strength being given by A). The second term stabilizes the liquid phase ($\xi_i = 0$) for large B. The third term gives the first-order transition in the absence of the external potential. The remaining two terms tend to stabilize the free energy. They considered the fact that in the presence of the external potential ξ_1 would differ from the degenerate modes ξ_2 and ξ_3. Using this free energy, they concluded that the transition from the 1D modulated liquid phase (i.e., the liquid phase with weak modulation induced by V_e) to the 2D (modulated) crystalline phase can be made continuous for sufficiently large laser fields. This, if true, would make for a fascinating critical phenomenon. However, while later studies using direct microscopic observations [34] and Monte Carlo simulations [35] have confirmed the existence of LIF, their conclusions regarding the nature of the transition between the modulated liquid and the crystal have not been definitive.

It is of obvious interest to ask what the DFT has to say about Laser Induced Freezing. The first work that we know of in this direction is that of Xu and Baus [36]. They studied the DFT of a 3D hard sphere system with its DCF being modelled by the Percus-Yevick theory, with the Gaussian density profile ansatz for $\rho(\mathbf{r})$. They considered the effect of a 3D modulating potential V_e commensurate with the FCC lattice to which the system freezes in the absence of V_e. Their results seemed to indicate that the fluid to crystal transition remains first-order no matter how large V_e is, in clear contradiction to the results of ref. [8].

Recently Chakrabarti et al. have reexamined the DFT for laser-induced freezing [9]. This work shows definitively that, with increasing V_e, the transition from the modulated liquid to the modulated crystalline phase changes from first order to continuous via a tricritical point (TCP) provided certain conditions, stated below, are satisfied by the wave vectors of the modulating potential.[1] This theory reestablishes the conclusion of Chowdhury et al, but now liberated from the bounds of the Landau theory [8, 33] (which, as has been shown, has no stable crystalline phase for large V_e). Furthermore, the theory has been applied to real colloidal systems to predict the parameter ranges for which such continuous $liquid \rightarrow crystal$ transitions can be found. This theory is reviewed in some detail in the following four subsections.

---

[1] Thus we disagree with the conclusions of [36] that the transition is always first order except in the case where V_e has the full symmetry of the crystal (for which, trivially, there is no phase transition). We believe that the discrepancy arises because various approximations used in [36] are inapplicable to the problem at hand.

Simple DFT for Laser-Induced Freezing for Incompressible Liquids

In order to understand the qualitative features of DFT theory of real colloids, it is useful to consider first the simplest version of DFT, for an incompressible 2D liquid, which when $V_e = 0$ freezes into a triangular lattice [37, 8] with its 6 smallest RLV, $\{\mathbf{g}_{oi}\} = (0, \pm 1)q_0, (\pm\sqrt{3}/2, \pm 1/2)q_0$. Here q_0 corresponds to the first peak of the liquid DCF $c^{(2)}(q)$. We take V_e to be 1D modulated, with wave vectors $\{\mathbf{g}_{fi}\} = (0, \pm 1)q_0$ from this set $\{\mathbf{g}_{oi}\}$. The order parameters in the simplest DFT are the Fourier components, with wave vectors $\{\mathbf{g}_{oi}\}$, of the molecular field $\xi(r) \equiv \ln[\rho(r)/\rho_0]$, where $\rho(r)$ is the local density in the modulated liquid or crystal [18, 21, 23]. From symmetry considerations it follows that for the crystal with V_e, the order parameters for the RLV set $\{\mathbf{g}_{fi}\}$ will have one value, say, ξ_f, and those for the rest of the RLVs in the smallest set $\{\mathbf{g}_{di}\}$ [$\equiv \{\mathbf{g}_{oi}\} - \{\mathbf{g}_{fi}\}$], another value, say, ξ_d. Let $c_1^{(2)} \equiv \rho_0 c^{(2)}(q_0)$. From Eq. (9.4), one can easily see that the self-consistency equations for ξ_f and ξ_d in the incompressible limit (and retaining terms upto $c^{(2)}$) are:

$$\xi_f = \frac{c_1^{(2)}}{2n_f} \frac{\int^{\text{cell}} d^d\mathbf{r}\, \phi_f \exp[\xi_f \phi_f(\mathbf{r}) + \xi_d \phi_d(\mathbf{r})]}{\int^{\text{cell}} d^d\mathbf{r}\, \exp[\xi_f \phi_f(\mathbf{r}) + \xi_d \phi_d(\mathbf{r})]} - \beta V_e$$

$$\xi_d = \frac{c_1^{(2)}}{2n_d} \frac{\int^{\text{cell}} d^d\mathbf{r}\, \phi_d \exp[\xi_f \phi_f(\mathbf{r}) + \xi_d \phi_d(\mathbf{r})]}{\int^{\text{cell}} d^d\mathbf{r}\, \exp[\xi_f \phi_f(\mathbf{r}) + \xi_d \phi_d(\mathbf{r})]} \qquad (9.13)$$

These self-consistency equations can be thought of as arising from the minimization of the following free energy functional [23] :

$$\beta F = -\ln \Phi + \frac{(n_f \xi_f^2 + n_d \xi_d^2)}{2c_1^{(2)}} - \frac{n_f \xi_f}{c_1^{(2)}} \beta V_e \qquad (9.14)$$

in the space of ξ_f and ξ_d (by setting $\partial F/\partial \xi_f = \partial F/\partial \xi_d = 0$), where $\phi_f(\mathbf{r}) \equiv \sum_j \exp(i\mathbf{g}_{fj} \cdot \mathbf{r})$ and $\phi_d(\mathbf{r}) \equiv \sum_j \exp(i\mathbf{g}_{dj} \cdot \mathbf{r})$, and n_f and n_d are the number of wave vectors in the set $\{\mathbf{g}_{fi}\}$ and $\{\mathbf{g}_{di}\}$, respectively. Here $\Phi \equiv v^{-1} \int^{\text{cell}} d^d\mathbf{r}\, \exp[\xi_f \phi_f(\mathbf{r}) + \xi_d \phi_d(\mathbf{r})]$. This free energy [2] is to be minimized in the space of the order parameters for a given $-\beta V_e$ and $c_1^{(2)}$ to find the phase diagram. This is done by solving numerically the two self-consistency equations, Eq. (9.13) and the phase diagram is obtained as shown in Fig. 9.7. This has two phases: (1) The modulated liquid with $\xi_f \neq 0$ but $\xi_d = 0$; this goes continuously to the ordinary liquid phase with $\xi_f = \xi_d = 0$ as $V_e \to 0$; and (2) the (modulated)

---

[2] We note that in contrast to the Landau-type [33] free energy functions, the free energy (9.14) is a nonpolynomial function, with arbitrary powers in a Landau expansion being present. Furthermore, the coefficients of the expansion are uniquely fixed, given $c_1^{(2)}$ and $-\beta V_e$.

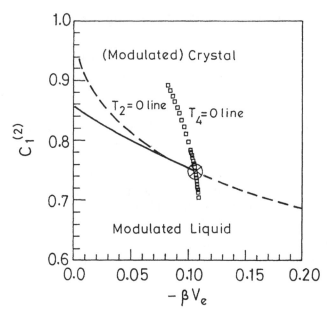

Figure 9.7   DFT phase diagram of a (incompressible) 2D system subjected to 1D modulation, showing the first-order (solid line) and continuous (dashed line) modulated liquid→crystal transitions. These are separated by the tricritical point (⊗), which is the intersection of the $T_2 = 0$ line and $T_4 = 0$ line (see text) [9].

crystalline phase with $\xi_f \neq \xi_d \neq 0$; as $V_e \to 0$, this goes into the ordinary crystalline phase with $\xi_f = \xi_d \neq 0$. As is known [37], at $V_e = 0$, the liquid→crystal transition is first order and takes place at $c_1^{(2)} = 0.857$ (which corresponds to the first peak height of the liquid structure factor $S_{max} = 7.14$.) When V_e is turned on, the transition remains first order, but moves to smaller values of $c_1^{(2)}$, as indicated by the solid line in Fig. 9.7. Thus V_e facilitates the liquid→crystal transition, which is the phenomenon of laser-induced freezing. This transition is characterized by a discontinuous change in ξ_f as well as by a discontinuous development of ξ_d. *However, the jumps in ξ_d and ξ_f decrease with increasing V_e and finally vanish at the TCP given by: $-\beta V_e = 0.106$ and $c_1^{(2)} = 0.748$. Thereafter, one has a continuous transition from the modulated liquid to the (modulated) crystalline phase across the dashed line in Fig. 9.7.* Illustrative behavior of ξ_f and ξ_d as $c_1^{(2)}$ increases for two fixed values of $-\beta V_e$ (0.015 and 0.15) characterizing the two types of transition is shown in Fig. 9.8(a) and (b).

   Qualitatively similar results arise in a similarly simplified DFT in three dimensions (3d), if V_e is taken to be two dimensional. Here, when $V_e = 0$, the liquid freezes [18, 21] at $c_1^{(2)} = 0.7$ ($S_{max} = 3.3$) into a BCC crystal with the 12 RLVs $\{\mathbf{g}_{oi}\} = (\pm 1, \pm 1, 0)q_0/\sqrt{2}$, $(0, \pm 1, \pm 1)q_0/\sqrt{2}$, $(\pm 1, 0, \pm 1)q_0/\sqrt{2}$. When a 2D modulation V_e characterized by the four wave vectors $\{\mathbf{g}_{fi}\} = (\pm 1, \pm 1, 0)q_0/\sqrt{2}$

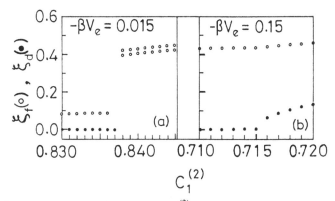

Figure 9.8 ξ_f and ξ_d as functions of $c_1^{(2)}$ showing (a) a first-order modulated liquid→crystal transition for $-\beta V_e = 0.015$ and (b) a continuous transition for $-\beta V_e = 0.15$ [9].

is turned on, once again the first order transition from the modulated liquid to the (modulated) crystal changes into a continuous transition in a way similar to that shown in Fig. 9.7, at a TCP given by $-\beta V_e = 0.22$ and $c_1^{(2)} = 0.55$.

It is not hard to establish the general criterion the wave-vector set $\{\mathbf{g}_{fi}\}$ characterizing V_e and its complementary set $\{\mathbf{g}_{di}\}$ must satisfy, in order that the (modulated) liquid→crystal transition changes from first order to a continuous one as V_e increases. The criterion is that

An odd combination of vectors of $\{\mathbf{g}_{di}\} \neq$ any integer combination of the vectors of $\{\mathbf{g}_{fi}\}$.   (A)

It is easy to verify that this condition is satisfied in the context of the modulations discussed. Note in particular that in the 3D BCC case, a 1D set or any arbitrary 2D set for $\{\mathbf{g}_{fi}\}$ picked out from $\{\mathbf{g}_{oi}\}$ would not satisfy (A). *The condition (A) is all-important, for, as we show, if it is satisfied, the existence of the TCP is robust with respect to the inclusion of the effect of finite compressibility and of the larger RLV-shell order parameters in the DFT (and hence in the context of an exact DFT).*

In order to see how condition (A) arises, consider generating from Eq. (9.14) the Landau expansion for βF in powers of ξ_d, given that $\xi_f \neq 0$ (with ξ_f treated nonperturbatively) *The key point is that if condition (A) is satisfied, this expansion has only even powers of ξ_d, the coefficients (which we can compute numerically), being functions of $-\beta V_e$ and $c_1^{(2)}$.* It turns out that as $c_1^{(2)}$ and $-\beta V_e$ increase, the coefficient of the second-order term in this Landau expansion, T_2, changes sign and becomes negative, leading to an instability with respect to the formation of ξ_d. In the region of the parameter space where the fourth-order coefficient, T_4, is positive, this leads to a continuous transition. However, if there is also a region of parameter space where T_4 is negative (but T_6 is positive), the continuous transition is preempted by a first-order transition. The TCP thus arises when both T_2 and T_4 become zero. In the 2D case, the lines along which $T_2 = 0$ (marked by dashes) and $T_4 = 0$ (marked by small squares) are shown in Fig. 9.7. To the left of the $T_4 = 0$ line, where T_4 is negative, the first-order transition (solid line) preempts the continuous transition. The $T_4 = 0$ line meets the $T_2 = 0$ line at the tricritical point. In this way one obtains the precise location of the tricritical points quoted earlier.

It is interesting at this point to compare the above phase diagram with that obtainable in the Landau theory of Chowdhury et al. [8]. For this purpose consider expanding the conventional Ramakrishnan-Yussouff free energy functional in powers of the Fourier components of $\rho(\mathbf{r})/\rho_0$ for the wave vectors $\{\mathbf{g}_{fi}\}$ and $\{\mathbf{g}_{di}\}$ in the 2D context. The phase diagrams obtained by minimizing the free energy that results by truncating the power series at different powers, up to the twelfth order, are shown in Fig. 9.9. In each of these cases one obtains a TCP as marked; however, the numbers are very different from, and converge very slowly to, those of the full density-functional theory. *More importantly, the continuous transition line eventually bends upward for large enough V_e, indicating that there is no stable crystal phase for large V_e in these truncated Landau theories, in contrast to the DFT result*, where the critical line asymptotes to $c_1^{(2)} = 0.5$ for large $-\beta V_e$.

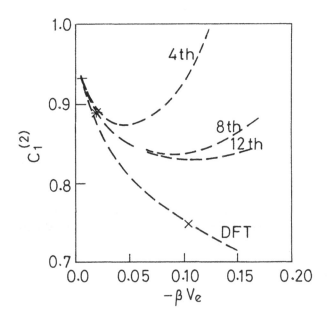

Figure 9.9 $T_2 = 0$ lines of the truncated Landau theories (see text) and that of the DFT. The TCPs are marked by + (fourth-order truncation), *(eighth and twelfth order truncations) and × (full DFT).

Effect of More Order Parameters

Next consider the robustness of the above results with respect to the inclusion of larger RLV-shell order parameters relevant in the context of real colloids. In this case one partitions the RLV set of the crystal into $\{g_{fi}\}$, $\{g_{di}\}$, the set $\{g_\alpha\}$ obtained by (new) integer combinations of $\{g_{fi}\}$, and the rest $\{G_l\}$; with order parameters ξ_f, ξ_d, ξ_α, and ξ_l, respectively. The DFT free energy to be minimized is:

$$\beta F = -\ln \Phi + \frac{1}{2c_1^{(2)}}(n_f \xi_f^2 + n_d \xi_d^2)$$

$$+ \sum_\alpha \frac{1}{2c_\alpha^{(2)}} \xi_\alpha^2 + \sum_l \frac{1}{2c_l^{(2)}} \xi_l^2 + \frac{n_f \xi_f}{c_1^{(2)}} \beta V_e \qquad (9.15)$$

where now $\Phi \equiv v^{-1} \int_\mathbf{r}^{\text{cell}} d^d \mathbf{r} \exp[\xi_f \phi_f(\mathbf{r}) + \xi_d \phi_d(\mathbf{r}) + \sum_\alpha \xi_\alpha \exp(i\mathbf{g}_\alpha \cdot \mathbf{r}) + \sum_l \xi_l \exp(i\mathbf{G}_l \cdot \mathbf{r})]$. In the modulated liquid phase $\xi_d = \xi_l = 0$. With respect to this phase, consider the Landau expansion for the free energy in terms of ξ_d and ξ_l to quadratic order. This will now have cross terms, determined by $\int^{\text{cell}} \phi_d \exp(i\mathbf{G}_l \cdot \mathbf{r}) \exp[\xi_f \phi_f + \sum_\alpha \xi_\alpha \exp(i\mathbf{g}_\alpha.\mathbf{r})]$. Now only those ξ_l for which such integrals are nonzero can affect the quadratic instability of ξ_d. This happens only if $\mathbf{G}_l$ is a linear combination of an arbitrary number of vectors from set $\{g_{fi}\}$ and *one* from the set $\{g_{di}\}$. The resulting quadratic form can be written in the form $T_{dd}^{(2)} \xi_d^2 + 2T_{dl}^{(2)} \xi_d \xi_l + T_{lm}^{(2)} \xi_l \xi_m$. Now the dominant instability of the modulated liquid phase will be determined by the order parameter mode $\tilde{\xi}_d$, which is that particular linear combination of ξ_d and ξ_l (typically, mostly ξ_d plus a little bit of other ξ_l), which corresponds to the lowest eigenvalue of this (Hessian) matrix $\mathbf{T}^{(2)}$. The instability sets in when this lowest eigenvalue of $\mathbf{T}^{(2)}$ crosses zero and goes negative. *The key point is that if condition (A) is satisfied, then the Landau expansion does not contain any odd powers of $\tilde{\xi}_d$.* For the coefficient of $(\tilde{\xi}_d)^n$ will be determined by sums over integrals of the form $\int^{\text{cell}} \phi_d^{n-p}[\exp(i\mathbf{G}_l \cdot \mathbf{r})]^p \exp[\xi_f \phi_f + \sum_\alpha \xi_\alpha \exp(i\mathbf{g}_\alpha \cdot \mathbf{r})]$. By virtue of condition (A), this integral will vanish for any odd n. So the TCP in the phase diagram is retained, but relocated, being now given by the condition of simultaneous vanishing of the coefficients of $(\tilde{\xi}_d)^2$ and $(\tilde{\xi}_d)^4$ in the Landau expansion.

Effect of Finite Compressibility

The effect of finite compressibility of liquid on the tricritical point is also accounted for. For a liquid with finite compressibility, typically $c_0^{(2)}$ is large negative, and one can consider its effect perturbatively. Up to $O(1/c_0^{(2)})$, the free energy defined in Eq. (9.15) will have a correction term $(\ln \Phi)^2/c_0^{(2)}$. All the above arguments hold good even for this case. From the symmetry of the problem, it is obvious that the Landau expansion of the free energy again will have only even powers of ξ_d. As a result of the correction term in the free energy, the coefficients

T_2 and T_4 will have corrections, and consequently the TCP will be shifted as a result of finite $c_0^{(2)}$.

These conclusions have been explicitly verified by numerical calculations including the effects of the order parameters corresponding to wave vectors $\{g_{2i}\}$ near the second peak in $c^{(2)}(q)$ and finite compressibility. The resulting relocated TCPs are listed in Table 9.1. When $c_0^{(2)} \neq -\infty$, there is a change in density at the first-order liquid$\rightarrow$solid transition, but this decreases with increasing V_e and vanishes at the TCP (see footnote 1 and Fig. 9.10).

Calculation for Realistic Polyball Systems

Chakrabarti et al. have also reported calculations for a model that describes the experimental system of Monovoukas and Gast [2] for which the DFT phase diagram is known in the absence of V_e [6] (see Fig. 9.5). As before, the liquid DCF is obtained from the rescaled mean spherical approximation of Hansen and Hayter [24, 1, 6] using the model DLVO potential. They focus their attention on the portion of the phase diagram where there is a first-order transition from the liquid to a BCC phase in the absence of V_e. Taking the modulation wave vectors $\{g_{fi}\}$ of the external potential to be along $(\pm1, \pm1, 0)q_0/\sqrt{2}$ as before, calculations are done retaining order parameters corresponding to 10, 20, and 50 shells of the RLVs of the BCC lattice. The three-body terms have to be retained in the DFT [18, 21, 6] in the same spirit as discussed before in Section 9.3 with $c^{(3)} = 0.23$. In the three-dimensional parameter space of impurity concentration n_i, volume fraction ϕ, and $-\beta V_e$, the first-order and continuous transitions between the modulated liquid and the crystal now take place across *surfaces*, which meet in a *line of TCP*. They find that in the ranges of n_i from $(1.8\text{-}2.6)\times10^{-9}$ mole/cm³, $-\beta V_e$ is almost a constant $\simeq 0.198$ along the tricritical line. The corresponding electric field strength in the laser beam is $E_c \simeq 0.168$ (J/m³)$^{1/2}$.[3] So in the above

Table 9.1 TCP coordinates for various versions of DFT

Crystal lattice when $V_e = 0$		2D triangular $(\pm1, \pm\sqrt{3})q_0$			3D BCC $(2, 1, 1)q_0$		
	$\{g_{2i}\}$						
	$c_2^{(2)} \equiv c^{(2)}(g_{2i})$	0	0	0.16	0	0	0.07
	$c_0^{(2)}$	$-\infty$	-51.5	$-\infty$	$-\infty$	-40	$-\infty$
TCP	$-\beta V_e$	0.106	0.114	0.117	0.219	0.237	0.228
	$c_1^{(2)}$	0.748	0.741	0.731	0.520	0.507	0.508

---

[3] Our estimate of the laser electric field E has been made using $V_e = \frac{1}{2}\chi E^2$, where χ is the dielectric susceptibility of the colloidal particles. χ has been estimated as $\frac{1}{2}\{(n_1^2 - n_2^2)/[(n_1/n_2)^2 + 2]a^3$, [see M. M. Burns, J. M. Fournier, and J.A. Golovchenko, *Science* **249**, 749 (1990)], where n_1 and

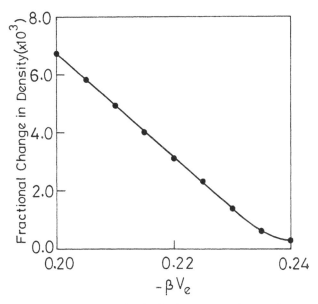

Figure 9.10   Fractional change in density in the context of one order parameter theory of LIF in the 3D BCC case with finite compressibility.

range of ϕ and n_i values, this system should show first-order and continuous liquid→solid transitions for $E < E_c$ and for $E > E_c$.

9.4 Conclusions

We have discussed two important applications of DFT to understand the interesting phenomena observed in polyball suspensions. It explains experimentally observed bulk phases of polyballs in three dimensions without any adjustable parameters. One can also apply DFT to different types of ordering in two dimensions as a function of number of layers [38]. The power of DFT is also apparent in its application to laser-induced freezing, which shows the freezing transition to be a continuous one and predicts the region of the parameter space where one may hope to get the tricritical points and critical lines in the transition from the modulated liquid to the (modulated) crystal. This raises many other fascinating questions that can be asked about these critical points, namely, their robustness with respect to the inclusion of fluctuation effects (especially in the 2D context), their light scattering characteristics, their universality class, etc. This also calls for fresh experimental studies of colloidal systems with a view to investigate these

---

n_2 are, respectively, the refractive indices of the colloidal particles and the solvent, and a is the radius of the particles.

issues. Also, it will be very interesting to study the situation where the wave vector of the applied modulation is incommensurate with the ordering wave vector of the liquid.

Acknowledgements

JC and AKS thank the Indo-French Centre for the Promotion of Advanced Research for financial support (Project No. 607-1).

Note added in Proof

Recently Chakrabarti, Krishnamurthy, Sood, and Sengupta (Phys. Rev. Lett. **75**, 2232 (1995)) have reported results from a Monte Carlo simulation study of laser modulated colloids. Their results corroborate the conclusions of DFT for the laser induced freezing. However, in contrast to the DFT results, they find in the simulation a novel reentrant laser induced melting transition from the crystal to the modulated liquid at large βV_e.

Choudhury and Ghosh (Phys. Rev. **E 51**, 4503 (1995)) have carried out a DFT study of the freezing of polyballs using the MWDA and RY scheme. Their results also agree with the experiments shown in Figs. 9.1 and 9.2.

References

1. Sood, A. K., in *Solid State Physics*, Eds. H. Ehrenreich and D. Turnbull (Academic Press, NY, 1991), Vol. **45**, p. 1.
2. Monovoukas, Y., and Gast, A. P., *J. Colloid Interface Sci.* **128**, 533 (1989).
3. Sirota, E. B., Ou-Yang, H. D., Sinha, S. K., Chaikin, P. M., Axe, J. D., and Fuji, Y., *Phys. Rev. Lett.* **62**, 1524 (1989).
4. Kesavamurthy, R., Sood, A. K., Tata, B. V. R., and Arora, A. K., *J. Phys. C* **21**, 4737 (1982).
5. Meller, A., and Stavans, J., *Phys. Rev. Lett.* **68**, 3646 (1992).
6. Sengupta, S., and Sood, A. K., *Phys. Rev.* **A 44**, 1233 (1991).
7. Salgi, P., and Rajagopalan, R., *Langmuir* **7**, 383 (1991).
8. Chowdhury, A., Ackerson, B. J., and Clark, N. A., *Phys. Rev. Lett.* **60**, 833 (1985); also Ackerson, B. J., and Chowdhury, A. H., *Faraday Discuss. Chem. Soc.* **83**, 309 (1987).
9. Chakrabarti, J., Krishnamurthy, H. R., and Sood, A. K. *Phys. Rev. Lett.* **73**, 2923 (1994).
10. Bartlett, P., Ottewill, R. H., and Pusey, P. N., *Phys. Rev. Lett.* **68**, 3801 (1992).
11. Eldridge, M. D., Madden, P. A., and Frenkel, D., *Molec. Phys.* **79**, 105 (1993).
12. Xu, H., and Baus, M., *J. Phys. Condens. Matter* **4**, L663 (1992).
13. Kaplan, P. D., Rouke, J. L., Yodh, A. G., and Pine, D. J., *Phys. Rev. Lett.* **72**, 582 (1994).
14. Alexander, S., Chaikin, P. M., Grant, P., Morales, G. J., Pincus, P., and Hone, D., *J. Chem. Phys.* **80**, 5776 (1984).
15. Rosenberg, R. O., and Thirumalai, D., *Phys. Rev.* **A 36**, 5690 (1987).
16. Hone, D., Alexander, S., Chaikin, P. M., and Pincus, P., *J. Chem. Phys.* **79**, 1474 (1983).
17. Robbins, M. O., Kremer, K., and Grest, G. S., *J. Chem. Phys.* **88**, 3286 (1988).
18. Ramakrishnan, T. V., and Yusouff, M., *Phys. Rev.* **B 19**, 2775 (1979); Ramakrishnan,T. V., *Pramana - J. Phys.* **22**, 365 (1984).
19. Haymet, A. D. J., and Oxtoby, D. W., *J. Chem. Phys.* **74**, 2559 (1981).
20. Ebner, C., Krishnamurthy, H. R., and Pandit, R., *Phys. Rev.* **A 43**, 4355 (1990).

21. Singh, Y., *Phys. Reports* **207**, 351 (1991).
22. Jacobs, R. L., *J. Phys. C* **16**, 273 (1983).
23. Krishnamurthy, H. R., (unpublished); Sengupta, S., Ph.D Thesis, Indian Institute of Science, India (1991) (unpublished).
24. Hansen, J. P., and Hayter, J. B., *Molec. Phys.* **25**, 651 (1982).
25. Tata, B. V. R., Sood, A. K., and Kesavamoorthy, R., *Pramana- J. Phys.* **34**, 23 (1990).
26. Rovere, M., and Tosi, M. P., *J. Phys. C* **18**, 3445 (1985).
27. Barrat, J. L., *Europhys. Lett.* **3**, 523 (1987).
28. Barrat, J. L., Hansen, J. P., and Pastore, G., *Molec. Phys.* **63**, 747 (1988).
29. Barrat, J. L., Hansen, J. P., and Pastore, G., *Molec. Phys.* **63**, 747 (1988).
30. Alexander, S., Chaikin, P. M., Grant, P., Morales, G. J., Pincus, P., and Hone, D., *J. Chem. Phys.* **80**, 5776 (1984).
31. Tarazona, P., *Molec. Phys.* **52**, 81 (1984); Curtin, W. A., and Ashcroft, N. W., *Phys. Rev.* **A 32**, 2909 (1985).
32. Denton, A., and Ashcroft, N. W., *Phys. Rev.* **A 39**, 4709 (1989).
33. Landau, L., and Lifshitz, E. M., *Statistical Physics - Course on Theoretical Physics, Vol. 5*, Part 1, 3rd Edn. (Pergamon, NY, 1986); Alexander, S., and McTague, J., *Phys. Rev. Lett.* **41**, 702 (1978).
34. Loudiyi, K., and Ackerson, B. J., *Physica A* **184**, 1 (1992).
35. Loudiyi, K., and Ackerson, B. J., *Physica A* **184**, 26 (1992).
36. Xu, H., and Baus, M., *Phys. Lett. A* **117**, 127 (1986); Barrat, J. L., and Xu, H., *J. Phys.: Condens. Matter* **2**, 9445 (1990).
37. Ramakrishnan, T. V., *Phys. Rev. Lett.* **8**, 541 (1982).
38. Pansu, P., Pieransky, P., and Strzelecki, L., *J. Physique* **44**, 531 (1983).

Photothermal Stability of Colloidal Crystals

R. Kesavamoorthy

Electrostatically stabilized colloidal crystals can be compressed locally when illuminated by intense laser light. This is because of local heating of the aqueous suspension due to absorption of the laser radiation by the particles. If the absorption is sufficiently strong, colloidal crystals can even melt. This chapter reviews the recent experimental and theoretical investigations of the photothermal compression of colloidal crystals formed by polystyrene particles that also incorporate dye molecules to facilitate absorption of laser radiation. The particles in the illuminated volume absorb the laser light and are heated. These hot particles heat the medium in the illuminated volume, change the dielectric constant, and hence alter the electrostatic interaction between the particles. This changes the lattice parameter of the colloidal crystals. Upon removing the intense laser beam (heat beam), the deformed crystal relaxes back. The time dependence of the lattice spacing is studied by monitoring the transmission profile of a weak laser beam (probe beam) incident on the crystal at an angle close to the Bragg angle. It is found that the changes in the lattice spacings are proportional to the local heating at low intensities; however, at large intensities colloidal crystals exhibit local melting. The phenomenon is theoretically analyzed by considering the thermal profile and the temperature dependence of the screened Coulomb interaction among the particles. Instead of perturbing the colloidal crystals by the heat beam only at one spot, it is also possible to create a one-dimensional periodic particle concentration profile by illuminating the colloidal crystals by two interfering heat beams. This results in a spatially modulated heat beam intensity (line grating) in the sample. The measurement of collective diffusion coefficient of the particles obtained from the relaxation of concentration profiles is discussed.

10.1 Introduction

Aqueous suspensions of charged monodisperse polymer spheres are often used as model systems to study interparticle interactions [1-5]. Depending upon the sphere charge, sphere number concentration, and the solution ionic strength, these suspensions can show gas, liquid, glass, and crystalline order with macroscopic interparticle seperation distances of the order of submicrometers. The dominant interparticle interaction derives from a screened Coulomb repulsion between the spheres due to charged groups on their surfaces. These groups result from ionization of surface functional groups. The repulsive interaction is attenuated by screening of the charges by the diffused counterion cloud in the medium surrounding the spheres. The repulsive potential decreases monotonically with distance from the sphere surface, and the structural order of the system is determined by parameters such as the sphere charge, sphere diameter, the particle concentration, the suspension ionic strength, dielectric constant, and the temperature. At low ionic strengths ($<10^{-5}$ M) and at sufficiently high particle concentrations ($>10^{12}$ particles cm^{-3}), aqueous dispersions of charged polymer spheres adopt a regular crystal structure in order to minimize the interparticle repulsive interactions. The interactions between colloidal particles having a small surface charge is explained by the screened Coulomb pair potential (DLVO theory) [6,7]. When the surface charge on the particle is high, the interaction potential falls off much more rapidly than would be predicted by this model. It has been suggested that it is possible to use this model for highly charged particles if the surface charge is renormalized to a smaller effective charge [5, 8].

10.1.1 Thermal Stability of Colloidal Crystals

Colloidal crystals can be prepared with lattice parameters of the order of hundreds of nanometers. These crystals can Bragg diffract radiation in the ultraviolet, visible, and near-infrared spectral regions [9, 10]. This unique property of colloidal crystals has been used to make practical devices such as narrow-band optical rejection filter [11]. Bragg diffraction from the crystals is a sensitive function of the lattice parameter, the sphere diameter, the refractive index mismatch between the sphere and the medium, the crystal Debye-Waller factor, and crystal defects [12, 13]. Thus optical diffraction measurements have been extensively utilized to study interparticle interactions in colloidal crystals under shear [14], gravitational effects [15-17], on deionization [18, 19], in the presence of electric fields [20] and as a function of temperature T [21-24].

The temperature dependence of the interparticle interaction (and hence the structural order and stability of the colloidal crystal) derives from the temperature dependence of the dielectric constant of water, the double-layer thickness, and the magnitude of the interaction energy relative to the thermal energy $k_B T$. However, until now temperature has not been considered to be an useful variable in the study of interparticle interactions in colloidal crystals because of the limited range of experimentally accessible values. The effects of temperature on colloidal crystals can be strikingly opposite to that observed in atomic and molecular crystals. This difference in behavior occurs because the interparticle interactions depend intimately upon temperature. For example, the repulsive interaction scales with the dielectric constant, and the dielectric constant of water decreases faster than T^{-1}. In fact, it has been predicted that colloidal crystals will melt upon cooling [25].

10.1.2 Diffusion in Colloidal Crystals

Probing the structure and dynamics of colloidal systems, including colloidal crystals, by laser light scattering is now routinely done [1, 26]. Both collective (mutual) and self-(tracer) diffusion is sensitive to interparticle and particle-solvent interactions, which can be altered easily by changing the suspension composition or temperature. Collective diffusion describes how a concentration fluctuation relaxes, and hence is a measure of bulk particle motion over a distance large compared to the interparticle separation [26]. Consequently, the collective diffusion coefficient is related to the bulk modulus. Self-diffusion, on the other hand, refers to the single-particle motion and is determined by interparticle interactions [26]. In the dilute limit, both the collective and self-diffusion coefficients approach the free-particle diffusion. At higher particle concentrations, such as particle concentrations needed for forming a crystalline phase, collective and self-diffusion are distinctly different processes [1, 27–30]. Therefore, in order to characterize colloidal crystals fully, both the collective and self-diffusion coefficients are required. Quasielastic light scattering is often used to monitor collective and self-diffusion at low particle concentrations. At higher particle concentrations, however, multiple scattering complicates the analysis. Techniques like forced Rayleigh scattering have been developed to measure the self-diffusion coefficient [28, 31, 32]. However, while techniques have been developed to create macroscopic particle concentration gradients, these methods have not been used to measure the collective diffusion coefficient. Radiation pressure, for instance, has been used to induce a particle concentration grating in disordered colloidal suspensions, which results in local ordering [33]. In addition, concentration gradients have been induced by an application of electric [34–36] and gravitational fields [15–17], centrifugation [37], and shearing [14, 38–40].

In this chapter, the photothermal stability of colloidal crystals is reviewed. Colloidal spheres containing light-absorbing dyes are used, and the crystals are heated by illumination with incident laser light at a wavelength absorbed by the dye. The local heating of the colloidal crystal using a laser beam results in a localized compression of the crystal within the heated region. Bragg diffraction is used dynamically to monitor the change in crystal ordering as a function of incident intensity and examine the dynamics of the lattice compression. Localized compression of the dyed colloidal crystals under the influence of a temperature field created by a laser beam is a direct consequence of the negative temperature dependence of the screened Coulomb pair potential; the temperature increase perturbs the equilibrium lattice of the colloidal crystal, which is defined completely by the temperature-dependent interaction potential. For the crystals studied here the perturbation (or the deformation) results in a compression of the lattice (without any change in the crystal structure). A linear response model is used to describe the dynamics of the colloidal response to this temperature perturbation. The only temperature-dependent parameter in the model is the electrostatic interaction between the particles. Using this model the linear deformation and temperature rise in the heated region are obtained. The necessity for renormalizing the charge on the particles and a procedure to estimate it are described. Further, a novel technique to determine the collective diffusion coefficient in colloidal crystals is presented that employs the phenomenon of local compression induced by the laser beam local heating. The incident laser beam intensity is modulated as one-dimensional periodic dark and bright fringes. This intensity profile induces a one-dimensional periodic particle concentration profile in colloidal crystals. The induced particle concentration gradient relaxes after the incident beam is turned off. The relaxation time constant is measured, and the collective diffusion coefficient is estimated.

10.2 Experimental

10.2.1 Synthesis of Colloidal Crystals

The sulfonated 83 nm ($\pm 5\%$) diameter polystyrene spheres used in this study were synthesized by emulsion polymerization [12]. The polystyrene polymers were cross-linked with divinyl benzene in order to obtain rigid spheres that will not deform on heating. The surface charge on the sphere was estimated by measuring the conductivity of the purified suspension having various particle concentrations. Each sphere contained $\sim 2340 \pm 80$ surface charge groups. The polymer spheres were cleaned by dialysis against deionized water containing Bio-Rad Ag 501-X8 mixed-bed ion-exchange resins. The dialysis was carried out for 2-4 weeks using 50000 MWCO Spectra/Por 6 dialysis tubing (Fisher). The dialysis water was changed every 24 h. The cleaned suspensions were stored in Nalgene bottles to minimize contamination from ionic impurities. The ion-exchange resin was cleaned using the procedures of Van den Hul and Vanderhoff [41].

The particle diameter was determined by quasielastic light scattering (QELS) using a Langley-Ford 1096 correlator and confirmed by transmission electron microscopy (TEM). The number of surface charge groups per sphere was determined by measuring the conductivity of the deionized suspension as a function of particle volume fraction [42] using a Radiometer-Copenhagen CDM83 conductivity meter.

The sulfonated polystyrene spheres were dyed with Oil Red O (Aldrich). The inset of Fig. 10.1 shows the absorption spectrum of the dye in ethyl benzene. The dye was dissolved in acetone, and aliquots of the dye solution were mixed with the cleaned polystyrene suspension. The mixture was agitated for 30-60 min and then dialyzed against deionized water for 2-3 d. Dialysis quantitatively removes the acetone from the suspension mixture. The absorption of the dyed particle suspension was determined to be 30 cm^{-1} at 514.5 nm (for a particle volume fraction of 2%). The particle size and charge did not change upon dye incorporation.

Large body-centered cubic (BCC) crytals of the dyed and undyed 83 nm diameter polystyrene particles were prepared according a procedure described elsewhere [11]. The crystal dimensions were $2.5 \times 2.5 \times 0.005$ cm^3. The sphere volume fraction was maintained at 2% to remain below the BCC/FCC (face-centered cubic) coexistence region. This was confirmed by the single sharp peak

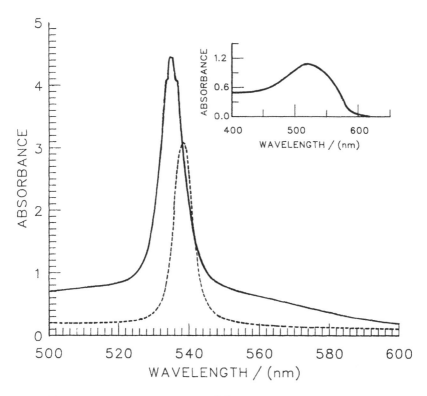

Figure 10.1 Visible absorption spectra of the large BCC colloidal crystals prepared from dyed (solid line) and undyed (dashed line) polystyrene spheres. The inset shows the visible absorption spectrum of Oil Red O in ethyl benzene (1.1×10^{-4}M).

in the diffraction profile of the crystals and the twinned BCC structure Kossel ring pattern [12]. For some of the measurements the crystals were prepared in 0.05 cm quartz cells (NSG Precision Cells). The refractive indices of the suspensions were determined with a temperature-controlled Abbe refractometer (Bausch and Lomb). A Perkin Elmer Lambda 9 UV-Visible-Near IR spectrophotometer was used to obtain transmission spectra of the crystals.

10.2.2 Setup to Investigate Photothermal Stability

Figure 10.2 illustrates the schematic of an experimental apparatus used to investigate the photothermal stability of the colloidal crystals. A Spectra Physics Model 164 argon ion laser with a broadband rear reflector was used as the source for the time-dependent intensity measurements. The light was dispersed outside the laser cavity by a diffraction grating, and the 514.5 nm beam was reflected by a mirror through a Newport Model 846 HP electronic shutter controller. The shutter, which has a rise time of 3 ms, determined the minimum temporal pulse width of the pump beam. A 30 cm focal length lens focused the light to a 0.2 mm spot size in the crystal. The 488 nm probe beam was attenuated by neutral-density filters and was focused by a 50 cm focal length lens to overlap the pump beam in the crystal.

The transmitted probe intensity was detected with a Hamamatsu S1226-18BQ photodiode, and the data were collected using a Hewlett-Packard Model 54201A digitizing oscilloscope operating in single-shot mode. A narrow-bandpass interference filter was placed in front of the photodiode so that only the probe intensity was detected. The eletronic shutter triggered the oscilloscope to begin collection.

The samples were oriented such that the transmitted probe beam fell on the inner or outer edge of the projected Kossel ring at the detector, resulting in an angle of incidence that was slightly greater than or slightly less than the Bragg angle, respectively. The crystal was allowed to relax for several min between measurements at different powers. The Kossel ring oriented about the axis normal to the sample cell derives from BCC (110) planes, and the Kossel ring diameter depends essentially only upon the lattice parameter. If the probe beam is incident close to the Bragg angle, even a small lattice parameter change results in a large change in the transmitted and diffracted intensities. The ambient temperature for these experiments was 18.5°C.

10.2.3 Setup to Determine the Collective Diffusion Coefficient Using the Photothermal Effect

A schematic of the experimental apparatus is illustrated in Fig. 10.3. A Spectra Physics Model 164 argon ion laser with a broadband rear reflector was used as the source. The laser output was expanded using two short focal length lenses and dispersed with a diffraction grating. The vertically polarized 514.5 nm pump beam was reflected by a mirror and focused partially with a spherical lens through a

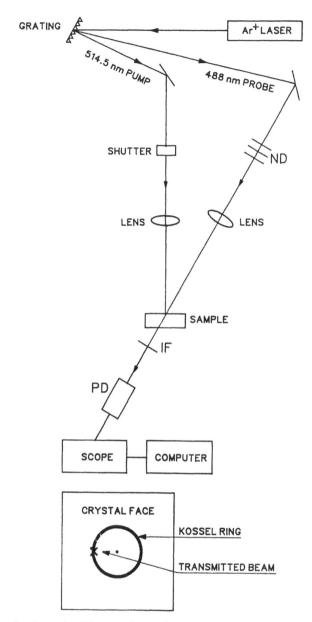

Figure 10.2 A schematic of the experimental apparatus to demonstrate the photothermal compression.

Ronchi ruling (100 lines/inch, Edmond Scientific). The resultant one-dimensional spatially periodic array of bright spots was imaged with a 44 mm cylindrical lens to create a one-dimensional periodic array of bright fringes. A Newport Model

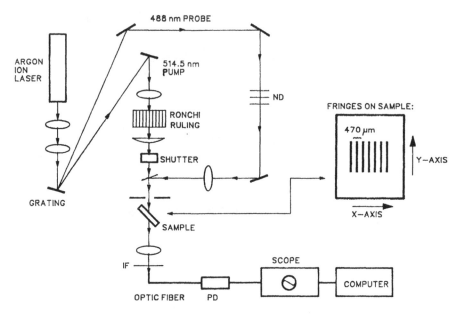

Figure 10.3 A schematic of the experimental setup to determine the collective diffusion coefficient. The inset schematically shows the fringe pattern on the sample.

846 HP electronic shutter, controlled by a Newport Model 845 digital shutter controller, was placed before the sample to turn the pump beam on and off and to trigger detection. An aperture after the shutter limited the number of fringes on the sample to be around 20. The bright fringe periodicity (470 μm $\pm 4\%$) was determined by translating a razor blade, mounted at the sample position, through the intensity grating. The height of the bright fringes at the sample was determined to be 2 mm. The incident power density per fringe was determined using a Scientech model 361 laser power meter.

Neutral-density filters were used to decrease the intensity of the 488 nm probe beam. A 300 mm focal length lens focused the probe beam on the sample. The spot size at the sample was determined to be less than 50 μm in diameter by translating a razor blade through a spot at the focal point. A beam splitter was oriented between the shutter and the aperture such that the probe beam was collinear with the pump beam and the probe spot fell on a bright fringe in the sample when the crystal was rotated close to the Bragg angle for the 488 nm light. For the data reported here, the angle of incidence is one degree greater than the Bragg angle. Under these conditions, the probe beam spot fell on the inner edge of the projected Kossel ring deriving from the BCC (110) planes, and a thermally induced local compression decreases the transmitted probe beam intensity, while relaxation results in the increase of transmitted intensity.

The transmitted probe light was collected with a condensing lens and focused onto a 1 mm diameter optic fiber coupled to a Hamamatsu S1226-18BQ photodiode. An interference filter was placed prior to the optic fiber to limit the detected intensity to the 488 nm light. The measured time-dependent transmitted intensity was stored in a Hewlett-Packard model 54201A digitizing oscilloscope, which was operated in single-shot mode and triggered by the electronic shutter. Opening the shutter allowed the pump beam fringe pattern to fall on the sample triggered the oscilloscope, but collection was delayed to begin when the shutter closed. A background was collected prior to each data set under identical conditions, except that the pump beam was physically blocked. The digitized data were downloaded to a computer for analysis. The pump intensities used are sufficiently low such that we can neglect particle motion due to photophoresis, thermophoresis, and radiation pressure [43]. No change in the transmitted intensity was observed for the crystal prepared with undyed polystyrene spheres [44]. These experiments were carried out at the ambient temperature of 23.5°C.

10.3 Results and Discussion

10.3.1 Visible Transmission Spectra of Colloidal Crystals

Figure 10.1 shows the visible transmission spectra (plotted on an absorbance scale) of a large colloidal crystal prepared with dyed and undyed polystyrene spheres. The inset in Fig. 10.1 shows the absorption of the dye dissolved in ethyl benzene; this should be essentially identical to the absorption spectrum of the dye within the polystyrene spheres. The transmission of the dyed crystal differs from that of the undyed crystal mainly due to the absorption by the dye. The sharp peaks in Fig. 10.1 correspond to the transmission minima for those wavelengths satisfying the Bragg diffraction condition

$$\lambda = 2\mu_s d_{hkl} \sin(\theta_B^{xtl}) \tag{10.1}$$

where λ is the wavelength of the incident light in air, μ_s is the suspension refractive index, and d_{hkl} is the interplanar spacing of the BCC (110) planes. For normal incidence the Bragg angle in the crystal, θ_B^{xtl}, is 90°. The spectra were measured by placing the sample cell normal to the light beam of the spectrometer. The beam is slightly divergent; so the absorption bandwidth partially derives from the angular spread around the 90° normal incident angle, and the actual attenuation for plane-wave incident light ($\sim 10^{-9}$) is much greater than the value measured in the spectrophotometer [12]. The suspension refractive index μ_s was measured at 589.3 nm using a temperature-controlled Abbe refractometer and was related to the refractive indices of pure water n_w and polystyrene μ_{ps} by

$$\mu_s = \mu_w(1 - \phi) + \mu_{ps}\phi \tag{10.2}$$

where ϕ is the volume fraction of polystyrene in the suspension. The Cauchy relations [45, 46], which are given by

$$\mu_w = 1.324 + 3046/\lambda^2 \tag{10.3}$$

$$\mu_{ps} = 1.5683 + 10087/\lambda^2 \tag{10.4}$$

have been used to calculate the suspension refractive index at the wavelength of incidence. Because the diffraction angle and incident wavelength are related by Eq. (10.1), one can quantitatively measure the change in the lattice parameter by monitoring the change in the wavelength of the absorption maximum at fixed incident angle using the spectrophotometer. Figure 10.4(a) shows the absorption spectrum of an absorbing colloidal crystal in a sealed 0.05 cm path length quartz cell. We illuminate the absorbing colloidal crystal with 500 mW of a 514.5 nm pump beam for 15 min. The incident beam diameter was 4 mm, and a 5 mm diameter circular aperture was placed over the cell to limit the volume sampled in the spectrometer to the laser-heated region only. Figure 10.4(b) shows the absorption spectrum taken 2 min after heating. Figure 10.4(c) gives the difference spectrum. The peak position blueshifts by 3 nm immediately after heating. In addition, the peak maximum decreases, and the bandwidth increases. Since the wavelength corresponding to the peak position decreases after heating, Eq. (10.1) indicates that the product $\mu_s d_{hkl}$ decreases. The undyed colloidal crystal did not show any difference in the absorption spectrum after laser illumination.

10.3.2 Probe Beam Transmission during and after Locally Heating the Colloidal Crystal

For a given wavelength of light, the incident angles that satisfy the Bragg diffraction condition map as a cone of minimum transmitted intensity, called a Kossel ring, for each set of reflecting planes. The Kossel rings appear as dark circles or ellipses in a diffuse light background when projected on a screen. The diffuse background derives from light scattered by the lattice defects and phonons [12]. The pattern of rings on the screen is diagnostic of crystal structure, and the diameter of any ring is related to the lattice parameter for the set of reflecting planes [47]. For a crystal of undyed spheres the Kossel ring diameter is independent of the incident intensity. At low incident intensities (<25 W/cm^2) the diffracted intensities and bandwidths of the crystal of dyed polystyrene spheres are similar to those of the nonabsorbing crystal, while the transmitted intensity is attenuated somewhat by absorption by the dye [48]. On the other hand, when the incident laser intensity (at 514.5 nm) is increased to above 25 W/cm^2, the diffracted and transmitted intensities for the absorbing crystal change appreciably as a function of time. Upon high-intensity illumination the Kossel ring decreases in diameter as a function of time, and finally disappears. The rate of contraction of the Kossel ring increases as the incident intensity increases. After the laser beam is turned off, the rings reappear and return to the original diameter. The recovery time depends on the length of time the crystal is exposed to the pump beam.

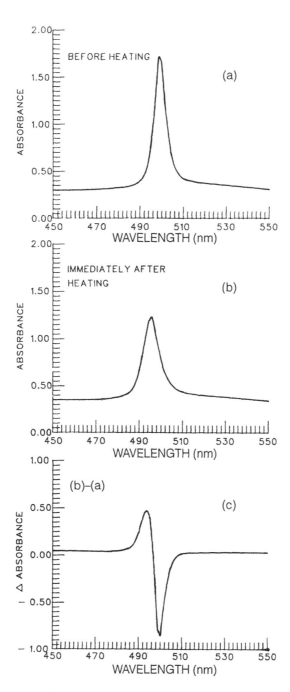

Figure 10.4 Visible absorption spectra of an absorbing colloidal crystal in a 0.05 cm path length flow cell: (a) before heating, (b) immediately following heating, and (c) the difference spectrum (b)−(a).

Photographs and video tapes of the central Kossel ring deriving from the BCC (110) set of planes show a 20% change in Kossel ring diameter upon laser heating. This must result from $\sim 6^o$ change in the Bragg angle. In addition, the width of the Kossel ring, measured from the inner edge to the outer edge of the dark ring, increases with time. The transmission of the 488 nm probe beam through the large BCC crystals during and heating with the 514.5 nm pump beam was also measured. Figure 10.5 shows the transmission as a function of time where the angle of incidence of the probe beam, θ_i^{air}, is typically 1.5° greater than the Bragg angle, θ_B^{air} (66°), for the 488 nm probe beam. The inset in Fig. 10.5 schematically illustrates the geometry of the Bragg and incidence angles relative to the lattice planes. For curve A the crystal was heated for 1 s with 159 W/cm² of 514.5 nm light focused down to 200 μm in diameter. When the pump beam is turned on, the transmitted probe intensity increases initially for up to almost 100 ms and then monotonically decreases. Even after the pump beam is turned off, the transmitted probe intensity continues to decrease further up to 200 ms and then increases.

The decrease in transmitted intensity is due to the shift of the Bragg condition toward the incident angle of the probe; this is obvious from the Kossel ring contraction upon pump heating. Thus the light is diffracted away rather than transmitted through the crystal. Upon removal of the pump beam the crystal relaxes and the Bragg angle shifts away from the angle of incidence of the probe beam, and the probe beam transmittance increases. The extent of the transmittance change and the lattice parameter contraction is a function of the extent of crystal

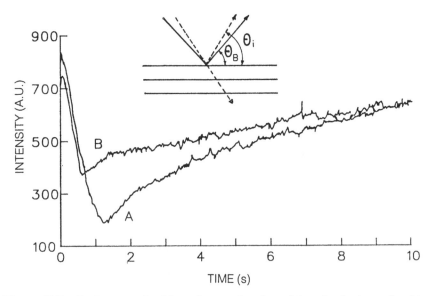

Figure 10.5   Probe transmitted intensity as a function of time for the large absorbing crystal with $\theta_i > \theta_B$. A, The crystal was heated with 159 W/cm² for 1 s, and B, the crystal was heated with 159 W/cm² for 500 ms, The inset shows the experimental geometry.

heating, as shown by curve B, in which the pump beam only heats the crystal for 0.5 s compared to the 1 s for curve A.

Figure 10.6 shows a time-dependent pump/probe measurement where the angle of incidence is slightly less ($\sim 1.5°$) than the Bragg angle. The pump power flux for this measurement was 382 W/cm^2. The transmitted probe intensity monotonically increases until the pump beam is shut off, then decreases as the crystal relaxes. Here the transmitted intensity increases because the Bragg angle moves away from the probe beam angle of incidence upon heating. When the pump beam is turned off, the crystal relaxes, resulting in a decrease in the mismatch between the Bragg angle and the angle of incidence, and the transmitted intensity decreases. An increase in the transmitted intensity is observed over a 100 ms time scale when the Bragg angle is less than the angle of incidence. This increase derives from a short-time lattice parameter increase, which will be discussed in the next section. Undyed BCC colloidal crystals showed no changes in their diffraction properties upon pump beam illumination.

Absorption of high incident intensities (> 25 W/cm^2) of light by a colloidal crystal causes the Kossel rings to contract, which indicates an increase in the Bragg angle. This Bragg angle increase is almost entirely due to a decrease in the lattice parameter. The decrease in lattice parameter, and the subsequent relaxation after the pump beam is turned off, depends on the total energy deposited into the crystal. Because the wavelength of the incident light in air is determined by the

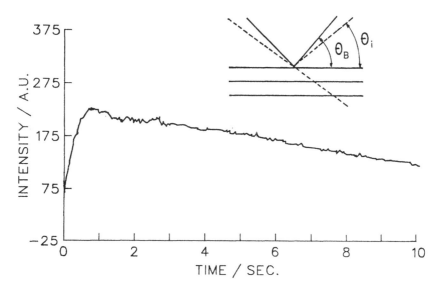

Figure 10.6 Probe transmitted intensity as a function of time for the large absorbing crystal with $\theta_i < \theta_B$. The crystal was heated with 382 W/cm^2 for 1 s. The inset shows the experimental geometry.

source and is therefore fixed, the Bragg angle increase derives from a decrease in the product $\mu_s d_{hkl}$.

An upper limit of about a 0.8% decrease in the product $\mu_s d_{hkl}$ is estimated for the pump/probe beam data presented here. This value is determined by comparing the width of the Kossel ring (measured from the inner to the outer edge of the dark ring) to the Kossel ring diameter. For the measurements where the angle of incidence is slightly greater than the Bragg angle, an initial increase in the transmitted probe intensity is observed for 200 ms, and a subsequent monotonic decrease is noted until the pump beam is turned off. If the Kossel ring diameter exceeds half of the Kossel ring width, then one must observe a decrease in transmitted intensity to a minimum when the Bragg angle equals the incidence angle. This would then be followed by an increase in transmission as the Bragg angle exceeds the angle of incidence. However, because the observed transmittance decrease is monotonic, the Bragg angle cannot exceed the probe beam incident angle during the pump beam duration. Therefore, the change in the product $\mu_s d_{hkl}$ is estimated, using Eq. (10.1), to be less than half of the Kossel ring width, or ~0.8%.

As the sample consists of only 2% by volume polystyrene, the temperature dependence of the suspension refractive index is essentially identical to that of the water. Because of the similarity to the formation of a thermal lens in an absorbing sample, the equations of Dovichi [49] are used to calculate the thermal equilibrium time after the pump beam is turned off for the absorbing colloidal crystals. For an incident beam radius $W_b = 0.01$ cm and a thermal diffusivity of water (at 20°C) of $D_T = 1.42 \times 10^{-3}$ cm^2 s^{-1}, the thermal relaxation time is calculated to be 18 ms using the expression $\tau = W_b^2/4D_T$. The 18 ms time calculated is much faster than the observed relaxation time (of the order of seconds). Thus the dominant phenomenon giving rise to the change in the Bragg angle is the thermally induced change in the interplanar spacing.

10.3.3 Linear Response Theory of Photothermal Effects

Consider a colloidal crystal illuminated by a pulsed Gaussian laser beam. The absorption of this laser beam by the colloidal crystal leads to an approximately Gaussian temperature profile in the crystal [49]. For simplicity, we approximate this temperature profile by a triangular function, as shown in Fig. 10.7. The maximum temperature rise, ΔT_{max}, in the sample is at the center of the beam and falls off linearly to room temperature (291.5 K) at the edge of the beam (0.1 mm from the center). The temperature of different regions in the illuminated volume lie between 291.5 and $(291.5 + \Delta T_{max})$ K. This gradient in temperature also gives rise to a gradient in the interparticle interaction potential. The differences in the interaction potential between adjacent regions result in a driving force that evolves the colloidal crystal toward a new equilibrium configuration. We assume a linear response of the colloidal crystal to the temperature increase. Hence the deformation of the crystal is linear since the temperature profile is linear.

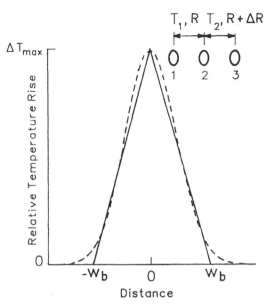

Figure 10.7   The modeled triangular spatial temperature profile in the colloidal crystal induced by the pump laser beam. The inset is a schematic representation of a linear array of particles within the heated region.

The temperature-dependent screened Coulomb pair potential is given by [2]

$$U(r, T) = \frac{Z^2 e^2}{\epsilon} \left(\frac{e^{\kappa a}}{1 + \kappa a} \right)^2 \frac{e^{-kr}}{r} \tag{10.5}$$

where Ze and a are the charge and the radius of the particle, r is the interparticle separation, and the inverse Debye screening length κ is given by

$$\kappa^2 = \frac{4\pi e^2}{\epsilon k_B T} (n_p Z + n_i) \tag{10.6}$$

In this equation n_p is the particle concentration, n_i ionic impurity concentration, and k_B the Boltzmann constant. Because the surface charge is provided by a strong acid group (sulfonic acid), we assume that the extent of dissociation is not temperature dependent; the dielectric constant ϵ is the only parameter that is directly sensitive to the temperature. The temperature dependence of ϵ for water is given by

$$\epsilon_{fit} = \frac{A_{fit}}{T^{1.5}} + \frac{B_{fit}}{T} + \frac{C_{fit}}{T^{0.5}} + D_{fit}. \tag{10.7}$$

The best fit to Eq. (10.7) is obtained for $A_{fit} = -439481.41$, $B_{fit} = 72528.68$, $C_{fit} = -314.08$, and $D_{fit} = -61.45$ in the temperature range 290 to 360 K.

Figure 10.8 shows the temperature dependence of the interaction potential energy between two 83 nm diameter spheres for an interparticle separation of 2.35×10^{-5} cm. A renormalized surface charge $Z = 1150$ per sphere [8] was used in this calculation, and the ionic impurity concentration was assumed to be Zn_p. The temperature dependence of the interaction potential energy remains similar if the measured surface charge is used in the calculation. In addition, increasing or decreasing ionic concentration, which changes the magnitude of the interaction potential energy, will not affect the trend. It is clear that the interparticle repulsive potential decreases with temperature for temperatures above 290 K. The temperature dependence of the derivative of the interaction potential $\partial U(r, T)/\partial T$ is shown in Fig. 10.9, which is negative in the temperature range of interest. This indicates that the colloidal crystal will undergo compression upon heating, because the interaction potential in the heated region becomes less than that of the surrounding colder regions. Since the hot region is the illuminated region giving rise to the Kossel ring, compression of the region results in a decrease in the average interparticle spacing, which decreases the Kossel ring diameter.

As the sample photothermal response depends upon the temperature dependence of the force $[-\partial U(r, T)/\partial r]$, the situation is more complex. One can calculate this force by considering the linear array of particles shown in the inset of Fig. 10.7. Let the repulsive interaction potential between particles 1 and 2 be given by $U(T_1, R)$ and the potential between particles 2 and 3 by $U(T_2, R + \Delta R)$, where $T_1 > T_2$, R is the distance between the particles 1 and 2, and ΔR is the linear perturbation in R due to the increase in temperature. The net repulsive force between particles 1 and 2 is given by $-\partial U(T_1, R)/\partial r$ and the force between particles 2 and 3 by $-\partial U(T_2, R + \Delta R)/\partial r$. The net force on particle 2 is given by

$$\Delta F = -\frac{\partial}{\partial r} U(T_1, R) + \frac{\partial}{\partial r} U(T_2, R + \Delta R) \tag{10.8}$$

Because the temperature difference $(T_1 - T_2)$ and the change in interparticle spacing ΔR are small, $U(T_1, R)$ can be expanded in a Taylor series

$$U(T_1, R) = U(T_2, R + \Delta R) + \frac{\partial U(T_2, R + \Delta R)}{\partial T}(T_1 - T_2)$$
$$- \frac{\partial U(T_2, R + \Delta R)}{\partial r} \Delta R \tag{10.9}$$

Neglecting the higher-order terms,

$$\Delta F = -\frac{\partial}{\partial r} \left(\frac{\partial U(T_2, R + \Delta R)}{\partial T}(T_1 - T_2) - \frac{\partial U(T_2, R + \Delta R)}{\partial r} \Delta R \right) \tag{10.10}$$

Let $U(T_2, R + \Delta R) = U$ and $T_1 - T_2 = \delta T$. Because δT and ΔR are constant with respect to variations in r

$$\Delta F = \delta T \left[\frac{\partial}{\partial T} \left(-\frac{\partial U}{\partial r} \right) \right] + \Delta R \left(\frac{\partial^2 U}{\partial r^2} \right) \tag{10.11}$$

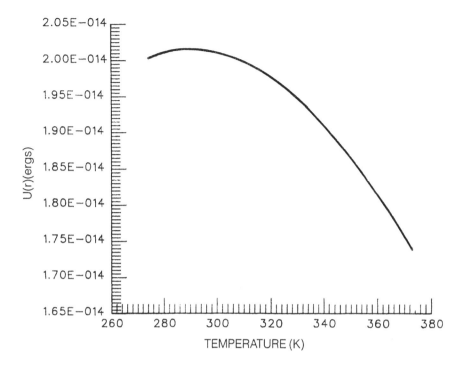

Figure 10.8   Plot of the interparticle interaction potential as a function of temperature. The interparticle separation is $r = 2.35 \times 10^{-5}$ cm, and the particle concentration is $n_p = 8.85 \times 10^{13}$ cm^{-3}.

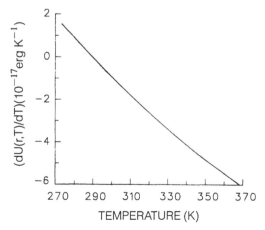

Figure 10.9   Plot of the temperature derivative of the interaction potential as a function of temperature.

Equation (10.11) can be rewritten as

$$\Delta F = \delta T \left[\left(\kappa + \frac{1}{R} \right) \frac{\partial U}{\partial T} + U \frac{\partial \kappa}{\partial T} \right] + U \, \Delta R \left(\kappa^2 + \frac{2\kappa}{R} + \frac{2}{R^2} \right) \quad (10.12)$$

In the present case δT, U, κ, and $\partial \kappa / \partial T$ are positive and $\partial U / \partial T$ is negative. Using Eq. (10.12) for the force acting on particle 2 due to the spatial temperature gradient, one can consider the situation close to $t = 0$, that is, when the pump is just turned on. At times slightly greater than $t = 0$, $\Delta R = 0$; the interparticle spacing is the same throughout the crystal. This means that ΔF is determined by the first term in Eq. (10.12), which is the temperature dependence of the repulsive force between particles. If this term is positive [i.e., $(\kappa + 1/R)|\partial U/\partial T| < U \, \partial \kappa / \partial T$], then ΔF is positive, the hot region expands, and particle 2 moves toward particle 3 (ΔR becomes negative) to balance this force. If $(\kappa + 1/R) \mid \partial U / \partial T \mid >$ $U \, \partial \kappa / \partial T$, ΔF becomes negative, the hot region contracts, and particle 2 moves toward particle 1 (making ΔR positive) to balance this force. It is really the sign of the product $[-\delta T (\partial^2 U / \partial r \, \partial T)]$ that determines whether the heated region in the crystal expands or contracts.

When the crystal is heated continuously, δT increases monotonically and ΔR increases or decreases to balance the resultant driving force (ΔF). If the product $[-\delta T (\partial^2 U / \partial r \, \partial T)]$ initially increases with temperature and then attains a maximum value and then decreases, the hot region will initially expand. Thus ΔR will initially be negative and attain a minimum value and then will increase in a positive direction as the crystal contracts. In such a case the crystal strain driving force is zero when ΔR is a minimum. Alternatively, the product $[-\delta T (\partial^2 U / \partial r \, \partial T)]$ may go through a minimum with temperature, and thus ΔR will go through a maximum. In such case, the driving force is zero at maximum ΔR. It will also be zero at steady state when the particles cease to move any farther due to the attainment of thermal equilibrium where the heat input is equal to the rate of heat lost to the surroundings.

In course of the present experiments the system continues to evolve. Upon heating an expansion is observed that then changes into contraction. This indicates that ΔR decreases initially, reaches a minimum value and then increases. This can be seen in Fig. 10.10, which shows the transmitted probe intensity (incident at an angle slightly greater than the Bragg angle)*versus* time when the crystal was heated by a 50 mW pump beam for 1 s. It starts expanding again about 200 ms after turning off the pump. $\Delta F = 0$ at the extremes of the transmittance curve.

One can calculate the value of ΔR that makes $\Delta F = 0$ for various values of ΔT_{max}

$$\delta T \left[\left(\kappa + \frac{1}{R} \right) \frac{\partial U}{\partial T} + U \frac{\partial \kappa}{\partial T} \right] + \Delta R \, U \left(\kappa^2 + \frac{2\kappa}{R} + \frac{2}{R^2} \right) = 0 \quad (10.13)$$

If ΔT_{max} is the difference in temperature between the center of the hot region

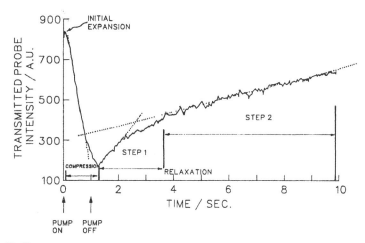

Figure 10.10 A transmission curve from a photothermal study of an absorbing colloidal crystal (see Fig. 10.5 for details). The angle of incidence of the probe beam is 1.5^o greater than the Bragg angle. Pump power = 50 mW and pump duration is 1 s.

and the edge and the radius of the beam W_b is ~ 0.1 mm, then

$$\delta T = \frac{\Delta T_{max}}{W_b} R \tag{10.14}$$

The values of $\partial U/\partial T$, κ, U, and $\partial \kappa/\partial T$ depend on temperature and hence the position of particle 2. Let particle 2 be midway between the center and edge of the hot region. Then these quantities can be calculated at $T = 291.5 + \Delta T_{max}/2$. Because R can be obtained from diffraction measurements, the value of ΔR for various values of ΔT_{max} can be calculated using Eqs. (10.13) and (10.14). Figure 10.11 shows the calculated dependence of $\Delta R/R$ on ΔT_{max}. From this plot and the experimental data (the maximum values of $\Delta R/R$ for various pump powers and pump duration) one can estimate the maximum temperature rise (ΔT_{max}) in the crystal for various pump powers and pump durations.

From Fig. 10.10 it may be seen that the hot region of the crystal continues to contract for about 200 ms after turning off the pump, before starting to expand toward the preheated value. This relaxation phenomenon appears to be a two-step process. When the pump is turned off, heat input ceases, but the hot region possesses a temperature gradient. The time constant for the thermal equilibrium of the hot region is given [49] by $\tau = W_b^2/4D \sim 18$ ms (where D is the thermal diffusivity of water, 1.46×10^{-3} cm^2 s^{-1}). The mean temperature and the temperature gradient of the hot region decrease with this ~ 18 ms time constant; however, the value of ΔR cannot change significantly during this short cooling time period because the collective diffusion coefficient (D_c), which determines the crystal

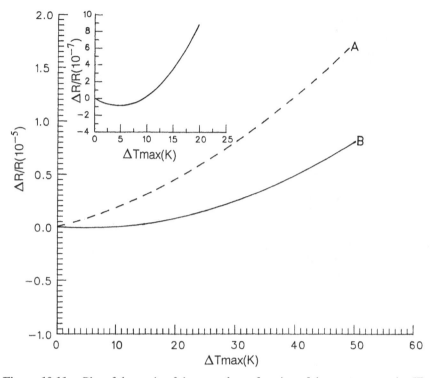

Figure 10.11 Plot of the strain of the crystal as a function of the temperature rise [Eq. (10.13)]. Curve A was computed using the measured surface charge of 2370e, and curve B utilized a renormalized surface charge of 1150e. The inset shows an expanded version of curve B.

strain relaxation rate, is much smaller. The collective diffusion coefficient of the particles can be determined from the following expression [50, 51]

$$D_c = \frac{D_0}{n_p k_B T} \left(B + \frac{4C}{3} \right),$$

(10.15)

where D_0 $(=k_B T / 6\pi \eta a)$ is the free particle diffusion coefficient, C $[= 4n_p (\kappa R)^2 U / 9]$ is the shear modulus, and $B[= 4n_p (\kappa R + 2)^2 U / 9]$ is the bulk modulus of the crystal [52]. For these crystals $D_c = 3.5 \times 10^{-6}$ cm^2 s^{-1}, which gives a resulting time constant (τ_p) for the response of the lattice of $\sim$7.2 s. This slow response time of the particle results in a large driving force [because the second term in Eq. (10.12) becomes very large], and the region previously heated expands rapidly. This rapid expansion is the first step of the relaxation process. During this first step, ΔR decreases rapidly, causing the driving force to decrease. In the second step of the relaxation process, the driving force is sufficiently small that ΔR follows the cooling. Hence the rate of cooling can be estimated from the slope of the second step of the relaxation curve.

Discussion

Figure 10.11 shows the values of $\Delta R/R$ that makes the driving force (ΔF) zero for various values of $\Delta T_{\max}$ [obtained from Eqs. (10.13) and (10.14)]. Curve A was obtained for a colloidal crystal having a particle concentration of 8.85×10^{13} cm^{-3}, a measured surface charge of $2370e$ per sphere, $a = 83$ nm, and for n_i of 1.018×10^{17} cm^{-3} (=0.169 mM; assumed to be equal to $n_p Z$). Curve A shows that only compression occurs on heating. Curve B was calculated for parameters the same as those for curve A except for the surface charge (Z is taken to be 1150). Curve B indicates that the heated region expands for low values of $\Delta T_{\max}$; $\Delta R/R$ becomes increasingly more negative up to $\Delta T_{\max} \sim 5$ K, after which it increases monotonically. The value of $\Delta R/R$ crosses zero at $\Delta T_{\max}$ $\sim$10 K. This curve indicates that upon heating the crystal initially expands. The expansion is maximum when the central temperature is elevated by ~ 5 K. For higher $\Delta T_{\max}$ values, the crystal only contracts. The inset of Fig. 10.11 shows the curve in the low $\Delta T_{\max}$ region on an expanded scale for clarity. It is important to note that the crystal expands on heating even when the pair potential decreases. This occurs because the positive temperature dependence of κ dominates the negative temperature dependence of U [Eq. (10.12)] for small values of $\Delta T_{\max}$ ($\sim$10 K); on the other hand, the temperature dependence of U dominates at larger values of $\Delta T_{\max}$.

The temperature difference at which expansion changes to contraction $(\Delta T)_r$ was calculated for various surface charge values and is shown in Fig. 10.12. From this figure it is clear that for charge $Z > 1340$, no initial expansion is predicted, while for $Z < 1340$ an initial expansion occurs and the $(\Delta T)_r$ increases with decreasing charge. The inset in Fig. 10.12 is a plot of the maximum expansion as a function of surface charge. This plot shows that the maximum expansion decreases with increasing charge and there is no expansion for $Z > 1340$. Obviously, if either the extent of the linear deformation or the temperature rise in the heated region is measured, Fig. 10.12 can be used to calculate the renormalized charge. For the colloid used here one observes an expansion during the first 100 ms, after which compression sets in. This indicates that the effective charge on the particles is less than $1340e$. Because the value of charge on the particle was measured to be $2370e$ it is clear that renormalization is required to use the screened Coulomb pair potential. Using the renormalization procedure of Alexander et al. [8], the renormalized charge is calculated to be $1150e$ and this value is used in all subsequent calculations. For this value of renormalized charge, $(\Delta T)_r$ turns out to be 5 K, and the calculated maximum expansion $(\Delta R/R)$ from Fig. 10.12 is -9.6×10^{-8}.

One can determine the maximum compression of the crystal and the corresponding $\Delta T_{\max}$ for various pump powers and pump durations using the estimated values of the initial expansion of the crystal. The observed maximum transmittance I_{T1} in Fig. 10.10 occurs when the crystal expansion is a maximum ($\Delta R/R = -9.6 \times 10^{-8}$). Similarly, the observed minimum transmittance

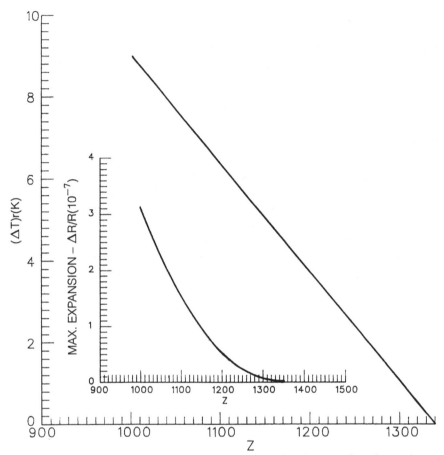

Figure 10.12 Plot of the temperature rise required for the crystal strain to change from expansion to compression as a function of the sphere charge. The inset shows the maximum expansion possible for various values of surface charge.

I_{T2} occurs when the crystal is maximally compressed. Hence the maximum compression can be related to the transmittance by

$$\left(\frac{\Delta R}{R}\right)_{max} = -9.6 \times 10^{-8} \left(\frac{I_{T0} - I_{T2}}{I_{T0} - I_{T1}}\right) \tag{10.16}$$

where I_{T0} is the observed transmittance at $t = 0$. The net crystal strain is extremely small (Fig. 10.11) and occurs on an essentially linear portion of the diffraction peak, as is evident from Fig. 10.9(c) of Rundquist et al. [12]. Given this value of $(\Delta R/R)_{max}$, we can use curve B of Fig. 10.11 to determine ΔT_{max}. I_{T1}/I_{T0} is observed to be the same for all pump powers and pump durations because the crystal expands to its maximum at $\Delta T_{max} \sim 5$ K. I_{T2}/I_{T0} depends on the pump power and pump duration, because the ΔT_{max} in the crystal increases with increasing pump power and duration.

Figure 10.13 shows the calculated value of ΔT_{max} as a function of pump power for pump durations of 1 s and 0.5 s. The value ΔT_{max} maximizes at about 30 K for the 0.5 s pulse. This saturation occurs because the thermal loss from the hot region to the surrounding cold region increases with temperature. The temperature increase of the hot region will also be a nonlinear function of the pump duration (for a given pump power). This is clearly illustrated in Fig. 10.13, where the difference in the slopes of the 1 and 0.5 s data is less than the expected factor of 2. The ΔT_{max} that occurs for a 1 s pump duration with 50 mW of laser power can be independently estimated to check the value derived from the model. The absorptivity of the dyed and undyed colloidal crystals of thickness $b_t (= 0.05$ cm) was measured to be 0.187 and 0.720, respectively. Assuming that the absorptivity of the undyed crystal comes from scattering from defects and phonons, the increased absorptivity of the dyed crystal comes from the dyed molecular absorption. The absorbed power P_w is calculated to be 27 mW. The maximum temperature rise in the crystal due to the absorption occurs at the center of the beam and can be determined from Eq. (16) of Dovichi [49]

$$\Delta T_{max} = \frac{P_w}{2\pi k_c b_t} \int_0^1 \frac{dt'}{(\tau + 2t')} \tag{10.17}$$

where k_c is the thermal conductivity of water and τ is the thermal equilibration time in water. Equation (10.17) gives a value of 33.7 K for ΔT_{max}, whereas the present model gives 25.7 K, as seen from Fig. 10.13. Equation (10.13) overestimates ΔT_{max} because it ignores heat loss through the walls of the container.

The saturation of ΔT_{max} is also evident in the data of Fig. 10.10. The rate of decrease in the transmittance with time slows during the 1 s interval of crystal

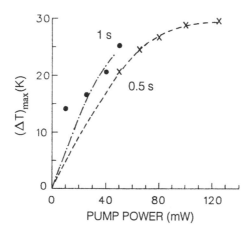

Figure 10.13 Plot of the maximum temperature rise in the crystal as a function of the pump laser beam. Pump duration of 0.5 s ($\times$) or 1 s ($\bullet$).

contraction that occurs while the pump beam is on. This decrease is most clearly evident for higher pump powers. During compression, ΔT_{max} and ΔR increase. If ΔT_{max} increases linearly with time, the rate of decrease of transmittance will be constant only if ΔR linearly follows ΔT_{max}. If ΔR lags behind ΔT_{max} because of a low collective diffusion rate, the rate of decrease of transmittance will initially have a smaller slope than ΔT_{max}, but this slope will increase with time because the driving force for compression increases with time. For high pump powers (>25 mW) the crystal continues to compress even after the pump is turned off. This behavior is not seen for low pump powers. For high pump powers, the rate of increase of ΔT_{max} is high and the response of the particles is too slow to follow the changing ΔT_{max}; that is, ΔR lags behind. When the pump is turned off, the driving force for compression is still large, and the particles continue to compress. For low pump powers the rate of increase of ΔT_{max} is sufficiently small that the change in ΔR does not significantly lag behind the change in temperature.

The linear deformation model also accounts for the relaxation behavior. When the pump is turned off, the compressed colloidal crystal must eventually expand. The initial driving force for expansion is zero, but δT and ΔR are positive and differ from zero. The rapid thermal equilibrium of the hot region in ~ 18 ms (the thermal equilibration time) causes ΔT_{max} and δT to decrease to $\Delta T'_{max}$ and $\delta T'$. However, ΔR cannot change significantly within an 18 ms time period because of the small collective particle diffusion coefficient. The strain driving force (ΔF), which is zero at the transmission minimum (Fig. 10.10) when the compressed colloid is about to expand, becomes increasingly positive with thermal equilibration. The increasing force causes the crystal to expand rapidly. As the value of ΔR decreases, ΔF becomes small, and the rate of particle motion can begin to follow the cooling rate. Hence the relaxation rate shown in Fig. 10.10 will show two different rates, which are evident at different time periods. These are illustrated in Fig. 10.10 by two straight lines, one with a larger slope (immediately after the pump is turned off) and another with a much smaller slope, at longer times. In the longer-time region ΔF is very small, and because ΔR follows $\Delta T'_{max}$, $\partial(\Delta F)/\partial T \sim 0$. Assuming that only ΔR and $\Delta T'_{max}$ depend on time and using Eq. (10.13), one gets $\partial \Delta F/\partial t = 0$; that is,

$$\frac{\partial \Delta F}{\partial t} = 0 = \frac{\partial \delta T'}{\partial t}\left[\left(k + \frac{1}{R}\right)\frac{\partial U}{\partial T} + U\frac{\partial K}{\partial T}\right]$$
$$+ \frac{\partial \Delta R}{\partial t}U\left(K^2 + \frac{2K}{R} + \frac{2}{R^2}\right) \tag{10.18}$$

One can determine $\partial \Delta R/\partial t$ by using Eq. (10.16) and the rate of increase of transmittance in step 2 of the relaxation process (Fig. 10.10). The rate of decrease of the temperature, $\partial \delta T'/\partial t$ can be calculated using Eq. (10.18). Figure 10.14 shows the calculated rate of cooling for the colloidal crystal heated to different values of ΔT_{max}. The rate of cooling increases nonlinearly with ΔT_{max}. This cooling rate differs from that for the illuminated region, which was earlier calculated to be $\tau = 18$ ms. τ monitors the decay of the thermal gradient over the hot

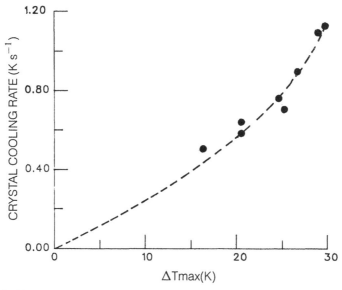

Figure 10.14 Plot of the rate of cooling of the colloidal crystal as a function of the temperature rise in the crystal.

region, whereas the cooling here monitors the thermal relaxation over the entire crystal. There are two thermal relaxation time processes involved. The fast 18 ms process involves relaxation of the thermal gradient within the probed portion of the sample while the slower process ($\sim$5 s) involves a relaxation of the thermal gradient over the entire sample. The two different thermal relaxation times come from two different length scales in the crystal. The 18 ms thermal relaxation time is associated with the thermal relaxation of the Gaussian-like thermal temperature distribution (which is approximated to a triangular distribution) within the probed region. This relaxes over the 200 μm beam width in the sample. This represents an experimentally defined relaxation time. In contrast, the 5 s relaxation is defined by the macroscopic thermal properties of the crystal.

The decrease in ΔT_{max} during the 18 ms of thermal equilibration ΔT_{th} (just after turning off the pump) can be obtained by extrapolating step 2 of the relaxation process of Fig. 10.10 back to $t = 1$ s. Let the extrapolated value of the transmittance at $t = 1$ s in the extrapolated line be I_{th}. The corresponding value of $\Delta R/R$ and $\Delta T'_{max}$ can be obtained from Eqs.(10.18), (10.13), and (10.14). Hence, $\Delta T_{th} = \Delta T_{max} - \Delta T'_{max}$ can be calculated. Figure 10.15 shows ΔT_{th} for various ΔT_{max}. It is observed that ΔT_{th} increases almost linearly with ΔT_{max}. For small pump powers, ΔT_{th} is insignificant, and the relaxation curve is a single straight line from the time the pump is turned off.

In the preceding discussion contributions to the change in lattice constant resulting from radiation pressure, photophoresis, and thermophoresis were neglected [53-57]. One can clearly neglect the radiation pressure since the colloidal

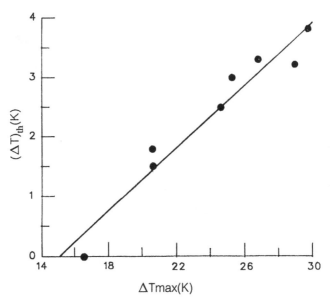

Figure 10.15   Plot of the initial drop in temperature due to the thermal equilibration (after the pump is turned off) as a function of the temperature rise in the crystal.

crystal of nonabsorbing spheres shows no lattice changes. The magnitudes of the lattice constant change due to radiation pressure, photophoresis, and thermophoresis are found to be ~4 orders of magnitude less than that due to the temperature-dependent Coulombic interaction for the incident intensities used in these studies (< 200 W/cm^2).

10.3.4 Measurement of Collective Diffusion Coefficient Based on Photothermal Compression of Colloidal Crystal

Figure 10.3 shows the schematic experimental setup to measure the collective diffusion coefficient in colloidal crystals based on photothermal compression. A particle concentration gradient is created via a local compression of the colloidal crystals in the high pump beam intensity (high-temperature) regions. This compression derives from the decrease of the interparticle repulsive interaction in these hot regions relative to the surrounding colder regions that are not illuminated. The pump beam light is incident on the crystal as a one-dimensional periodic array of bright fringes resulting in a periodic temperature grating. When the temperature increase is localized, particles will migrate from the adjacent lower-temperature regions to the locally hotter regions due to the force created by a gradient in the interaction potential, resulting in a particle concentration pattern that mimics the pump intensity profile. Particle motion depends on the particle concentration, the interaction potential, its derivative, and on the Debye layer thickness, and its temperature derivative [43]. Since the pump intensity is a one-dimensional periodic grating of bright fringes, the particle concentration profile will also be a one-dimensional array of high and low particle concentration. The periodic array of bright fringes extends across a large portion of the crystal; the height of the fringes is much greater than the diameter of the probe beam. When the pump intensity pattern is turned off, the magnitude of the interaction potential within the heated (illuminated) regions increases, achieving thermal equilibration within 0.1 s. The particle concentration grating relaxes by particle motion predominantly in one dimension along the grating axis toward a uniform particle concentration. As will be discussed, the relaxation time constant of this one-dimensional concentration grating is simply related to the collective diffusion coefficient. The inset of Fig. 10.11 shows that the colloidal crystal initially expands on local heating through ~5 K before it starts compressing when the ambient temperature is kept at 18.5 K. It indicates that the crystal on local heating will not expand if the ambient temperature is kept at 23.5 K or more. In order to avoid the hot region expanding before the compression sets in, the ambient temperature for these experiments is kept at 23.5 K.

The rate of relaxation of the grating is given by Fick's law, which can be written as [58]

$$\frac{\partial n}{\partial t} = D_c \nabla^2 n \tag{10.19}$$

Here the pump intensity distribution is assumed to be sinusoidal in the x direction and n is considered to be

$$n = n_t \sin\left(\frac{2\pi x}{x_0}\right) + n_p \tag{10.20}$$

where x is defined as the direction along the grating axis (see Fig. 10.3), x_0 is the spatial period of the intensity and concentration gratings, D_c is the collective diffusion coefficient, and n_p is the average particle concentration (9.28×10^{13} cm^{-3}). A solution to this differential equation is

$$n_t = n_m \exp(-t/\tau_p) \tag{10.21}$$

where n_m is the excess particle concentration over n_p in the center of the heated region (at the position of the probe beam) when the pump beam is turned off and τ_p is the crystal relaxation time constant, which is related to the collective diffusion coefficient by [28, 59, 60]

$$\tau_p = \frac{x_0^2}{4\pi^2 D_c} \tag{10.22}$$

Thus we expect the particle concentration grating to relax as a single exponential.

The change in the transmitted intensity is linearly proportional to the change in particle concentration close to the diffraction condition [43]. Therefore, the transmitted intensity should also change as a single exponential with the same time constant τ_p, and since the initial condition is such that the angle of incidence of the probe beam is slightly greater than the Bragg angle, the transmitted intensity should increase upon crystal relaxation. The collective diffusion coefficient is related to the long-wavelength limit of the static structure factor $S(0)$ by [50, 51]

$$D_c = \frac{D_0(1 - 6\phi)}{S(0)} \tag{10.23}$$

where D_0 is the free-particle self-diffusion coefficient. The factor $(1 - 6\phi)$ accounts for hydrodynamic interactions [61]. In the limit $Q \to 0$, the static structure factor is related to the elastic modulus of the crystal [50]

$$S(0) = \frac{n_p k_B T}{M} \tag{10.24}$$

The elastic modulus M is given by [52]

$$M = B + 4C/3 \tag{10.25}$$

where the bulk modulus B and shear modulus C for a BCC crystal are given by

$$B = 4n_p(\kappa R + 2)^2 U(R)/9 \tag{10.26}$$

$$C = 4n_p(\kappa R)^2 U(R)/9 \tag{10.27}$$

Therefore, a measurement of the crystal relaxation time constant, under conditions of one-dimensional diffusion, unambiguously characterizes the collective diffusion coefficient.

Figure 10.16 shows the relaxation data collected for the dyed crystal after illumination by the pump intensity pattern for 60 s. The pump power density per bright fringe was 2.6×10^{-2} W/cm^2. The background is also included in this plot to show that the transmitted intensity does not change unless the crystal is first illuminated with the pump beam. The data are reproducible within experimental error. The solid line through the data is a two-parameter nonlinear least-squares fit using the model

$$I = f_{\text{fit}} - g_{\text{fit}} \exp(-t/\tau_p) \tag{10.28}$$

with fitting parameters f_{fit} and g_{fit}. τ_p for this crystal was calculated to be 16.5 s using Eqs. (10.22)-(10.27) from the Coulomb pair potential given by Eqs. (10.5) and (10.6). For this calculation, the impurity concentration was assumed to be $n_p Z$, and a renormalized charge of $1150e$ was used [8]. A value of 16.5 s for τ_p has been used in Eq. (10.28). The agreement with the data is quite good. A three-parameter fit [where $1/\tau_p$ in Eq. (10.28) is replaced by h_{fit}, the third parameter] is also performed on the data. From this three-parameter fit, the relaxation time constant was determined to be 16.3 s, which is in good agreement with the calculated value. The best estimates of the parameters of the two- and three-parameter fits for this data set are listed in Table 10.1.

Figure 10.17 shows relaxation data collected at a higher fringe power density $(4.5 \times 10^{-2}$ W/cm^2) for different pump durations corresponding to (a) 30 and (b)

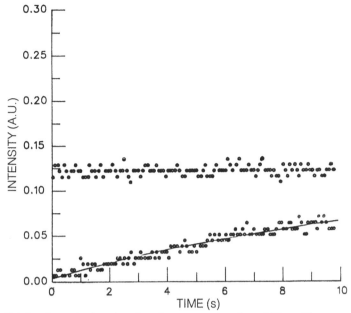

Figure 10.16 Crystal relaxation data for a pump duration of 60 s with a power density of 2.6×10^{-2} W/cm^2/fringe. The solid line through the data is the two-parameter fit with a relaxation time constant of $\tau_p = 16.5$ s. The data above the relaxation curve are the background, collected under the same experimental conditions, except the pump beam is blocked prior to the sample.

Table 10.1 Fitting parameters for a pump intensity of 2.6×10^{-2} W/cm^2/fringe and pump duration of 60 s

Number of paramaters	f_{fit}	g_{fit}	h_{fit} $(=1/\tau_p)$
2	0.14 ±0.0026	0.14 ±0.0034	—
3	0.14 ±0.0033	0.14 ±0.0032	0.061±0.0019

60 s. These sets overlap within experimental error, suggesting that the crystal is reaching the steady state in response to the temperature perturbation within 30 s. The solid lines through these data sets are again the two-parameter best fits with $\tau_p = 16.5$ s. The best-fit estimates of the parameters and the errors are listed in Table 10.2.

Figure 10.18 shows data collected at even higher power densities (6.4×10^{-2} W/cm^2) and for pump durations of 15 and 30 s, respectively. These data do not overlap because the first 15 s pump annealed the heated regions of the crystal, resulting in an improved signal for the 30 s pump data, which were collected next. In addition, using a relaxation time constant of 16.5 s results in a poor fit to

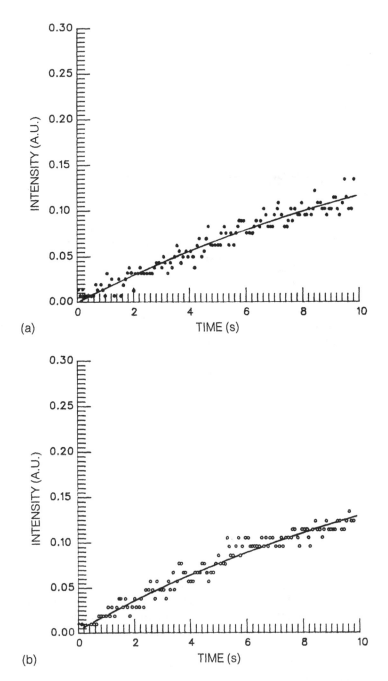

Figure 10.17 Crystal relaxation data with a pump power density of 4.5×10^{-2} W/cm^2/fringe for pump durations of (a) 30 and (b) 60 s, respectively. The solid lines through the data are the two-parameter fits with the relaxation time constant τ_p of 16.5 s.

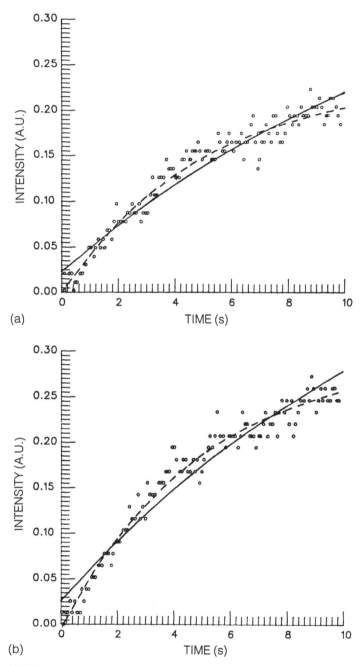

Figure 10.18   Crystal relaxation data with a pump power density of 6.4×10^{-2} W/cm^2/fringe for pump durations of (a) 15 and (b) 30 s, respectively. The solid lines through the data are the two-parameter fits with $\tau_p = 16.5$ s, and the dashed lines correspond to $\tau_p = 5$ s.

Table 10.2 Fitting parameters for a pump intensity of 4.5×10^{-2} W/cm^2/fringe and $\tau_p = 16.5$ s

Pump duration (s)	f_{fit}	g_{fit}
30	0.26 ±0.0045	0.26±0.0059
60	0.29 ±0.0043	0.28±0.0056

the data. If τ_p is decreased to 5 s, however, the fit improves. The parameters and errors for these data are listed in Table 10.3.

The relaxation time constant for the highest-intensity data is obtained to be less from Fig. 10.18, and, consequently, the collective diffusion coefficient is high [Eq. (10.22)]. The decrease in relaxation time constant with increasing incident pump intensity is probably due to a breakdown in the one-dimensional diffusion condition. If the spheres are not constrained to move only in one direction, then the relaxation of the concentration gradient will be faster due to the additional diffusion in other directions.

The signal-to-noise ratio for these measurements is a function of the crystal quality. A well-ordered crystal gives rise to a sharp, well-defined Kossel ring, and small changes in the ring diameter when the diffraction conditions are almost satisfied gives rise to a large change in the transmitted probe intensity. On the other hand, as the crystal defect density increases, the Kossel ring becomes more diffuse, and the contrast between the ring and the surroundings decreases. Changes in the ring diameter in this case do not result in a large change in the transmitted intensity. Therefore, the colloidal crystal must be well ordered for this method to be useful for determining the collective diffusion coefficient. The low-intensity pump probably results in a small temperature increase, which might not be sufficient to anneal the crystal. Hence the low pump intensity relaxation data is reproducible. At higher pump intensities (e.g., 6.5×10^{-2} W/cm^2/fringe), the temperature increase may be sufficient to anneal the crystal. Hence the signal-to-noise ratio improves at this higher pump intensity. At much higher pump intensities (> 0.15 W/cm^2/fringe), the temperature increase is sufficient to disorder the crystal considerably, resulting in a diffuse Kossel ring; this significantly decreases the signal-to-noise ratio for subsequent measurements.

Table 10.3 Fitting parameters for a pump intensity of 6.4×10^{-2} W/cm^2/fringe

Pump duration (s)	τ_p(s)	f_{fit}	g_{fit}
15	16.5	0.46 ±0.0082	0.44 ±0.011
15	5	0.24 ±0.0022	0.24 ±0.0045
30	16.5	0.58 ±0.01	0.55 ±0.013
30	5	0.30 ±0.0023	0.30 ±0.0047

10.4 Conclusion

It is shown that the average distance between particles in an electrostatically stabilized dyed colloidal crystal decreases upon local heating of the crystal by the laser beam. This phenomenon derives from the negative temperature dependence of the interparticle interaction potential. A linear deformation model is developed to explain this photothermal phenomenon in colloidal crystals using a screened Coulomb pair potential. Using this model, the temperature increase in an absorbing crystal upon high-intensity laser radiation is estimated. The photothermal response of these crystals can be used to examine the interparticle potential function. This study clearly indicates that the potential function only correctly predicts the interparticle potential if the colloid charge is renormalized. A new photothermal technique is described that allows one to monitor the interparticle potential function and determine the required charge renormalization. Further, a novel technique to determine the collective diffusion coefficient in colloidal crystals based on photothermal compression phenomena is presented. By illuminating an absorbing crystal, prepared from an aqueous suspension of dyed polystyrene spheres with a spatially periodic intensity profile, a particle concentration grating is created that follows the intensity pattern of the pump light. Since diffusion is constrained to primarily one dimension under these conditions, the collective diffusion coefficient is determined easily from the relaxation time constant of this concentration grating. The measured collective diffusion coefficient agrees well with the calculated value, assuming a screened Coulomb pair potential.

Acknowledgements

The author gratefully acknowledges Prof. Sanford A. Asher, S. Jagannathan, Paul A. Rundquist, Charles Brnardic, S. Xu, and Shalabh Tandon, who were collaborators of this work. This work has been carried out at the Department of Chemistry at the University of Pittsburgh, Pennsylvania, under the guidance and support of Prof. Sanford A. Asher.

References

1. Russel, W. B., Saville, D. A., and Schowalter, W. R., *Colloid Dispersions* (Cambridge University, New York, 1987).
2. Sood, A. K., in *Solid State Physics*, Eds. H. Ehrenreich, and D. Turnbull, (Academic, New York, 1991), **45**, p.1
3. Thirumalai, D., *J. Phys. Chem.* **93**, 5637 (1989).
4. Robbins, M, O., Kremer, K., and Grest, G. S., *J. Chem. Phys.* **88**, 3286 (1988).
5. Monovoukas, Y., and Gast, A. P., *J. Colloid. Interface Sci.* **128**, 533 (1989).
6. Derjaguin, B. V., and Landau, L., *Acta Physicochem. URSS* **14**, 633 (1941).
7. Vervey, E. J. W., and Overbeek, J. Th. G., *Theory of the Stability of Lyophobic Colloids* (Elsevier, Amsterdam, 1948).
8. Alexander, S., Chaikin, P. M., Grant, P., Morales, G. J., Pincus, P., and Hone, D., *J. Chem. Phys.* **80**, 5776 (1984).
9. Hiltner, P. A., and Krieger, I. M., *J. Chem. Phys.* **73**, 2386 (1969).
10. Asher, S. A., Flaugh, P. L., and Washinger, G., *Spectrocopy* **1**, 26 (1986).

11. Flaugh, P. L., O'Donnell, S. E., and Asher, S. A., *Appl. Spectroscopy* **38**, 847 (1984).
12. Rundquist, P. A., Photinos, P., Jagannathan, S., and Asher, S. A., *J. Chem. Phys.* **91**, 4932 (1989).
13. Spry, R. J., and Kosan, D. J., *Appl. Spectroscopy* **40**, 782 (1986).
14. Ackerson, B. J., in *Physics of Complex and Supermolecular Fluids*, Eds. N. A. Clark, and S. Safran, (Wiley-Interscience, New York, 1987).
15. Kesavamoorthy, R., and Arora, A, K., *J. Phys. A: Math. Gen.* **18**, 3389 (1985).
16. Carlson, R. J., and Asher, S. A., *Appl. Spectrosc.* **38**, 297 (1984).
17. Crandall, R. S., and Williams, R., *Science* **198**, 293 (1977).
18. Kesavamoorthy, R., Tandon, S., Xu, S., Jagannathan, S., and Asher, S. A., *J. Colloid Interface Sci.* **153**, 188 (1992).
19. Sogami, I. S., and Yoshiyama, T., *Phase Transitions* **21**, 171 (1990).
20. Richetti, F., Prost, J., and Clark, N, A., in *Physics of Complex and Supermolecular Fluids*, Eds. N. A. Clark, and S. Safran, (Wiley-Interscience, New York, 1987).
21. Williams, R., Crandall, R. S., and Wojtowicz, P. J., *Phys. Rev. Lett.* **37**, 348 (1976).
22. Daly, J. G., and Hastings, R., *J. Phys. Chem.* **85**, 294 (1981).
23. Okubo, T., *J. Chem. Soc. Faraday Trans. 1*, **82**, 3163 (1986).
24. Okubo, T., *J. Chem. Soc. Faraday Trans. 1*, **82**, 3185 (1986).
25. Chaikin, P. M., di Meglio, J. M., Dozier, W. D., Linsay, H. M., and Weitz, D. A., in *Physics of Complex and Supermolecular Fluids*, Eds. N. A. Clark, and S. Safran, (Wiley-Interscience, New York, 1987).
26. Berne, B. J., and Pecora, R., *Dynamic Light Scattering* (Wiley, New York, 1976).
27. Venkatesan, M., Hirtzel, C. S., and Rajagopalan, R., *J. Chem. Phys.* **82**, 5685 (1985).
28. Dozier, W. D., Lindsay, H. M., and Chaikin, P. M., *J. Phys. (Paris)* **C3**, 165 (1985).
29. Trotter, C. M., and Pinder, D. N., *J. Chem. Phys.* **75**, 118 (1981).
30. Marqusee, J. A., and Deutch, J. M., *J. Chem. Phys.* **73**, 5396 (1980).
31. Qiu, X., Ou-Yang, H. D., Pine, D. J., and Chaikin, P. M., *Phys. Rev. Lett.* **61**, 2554 (1988).
32. Qiu, X., Ou-Yang, H. D., and Chaikin, P. M., *J. Phys. (Paris)* **49**, 1043 (1988).
33. Ackerson, B. J., and Chawdhury, A., *Faraday Discuss. Chem. Soc.* **83**, 309 (1987).
34. Fraden, S., Hurd, A. J., and Meyer, R. B., *Phys. Rev. Lett.* **63**, 2373 (1989).
35. Okubo, T., *J. Chem. Soc. Faraday Trans. 1*, **83**, 2487 (1987).
36. Tomita, M., and van de Ven, T. G. M., *J. Phys. Chem.* **89**, 1291 (1985).
37. Okubo, T., *J. Am. Chem. Soc.* **112**, 5420 (1990).
38. Mathis, C., Bossis, G., and Brady, J. F., *J. Colloid. Interface Sci.* **126.**, 16 (1988).
39. Tomita, M., and van de Ven, T. G. M., *J. Colloid. Interface Sci.* **99**, 374 (1984).
40. Clark, N. A., Hurd, A. J., and Ackerson, B. J., *Nature* **281**, 57 (1979).
41. Van den Hul, H. J., and Vanderhoff, J. W., *J. Electroanal. Chem.* **37**, 161 (1972).
42. Schaefer, D. W., *J. Chem. Phys.* **66**, 3980 (1977).
43. Kesavamoorthy, R., Jagannathan, S., Rundquist, P. A., and Asher, S, A., *J. Chem. Phys.* **94**, 5172 (1991).
44. Rundquist, P. A., Jagannathan, S., Kesavamoorthy, R., Brnardic, C., Xu, S., and Asher, S. A., *J. Chem. Phys.* **94**, 711 (1991).
45. Hiltner, P. A., Papir, Y. S., and Krieger, I. M., *J. Phys. Chem.* **75**, 1881 (1971).
46. Bateman, J. B., Weneck, E. J., and Eshler, D. C., *J. Colloid Interface Sci.* **14**, 308 (1959).
47. Pieranski, P., *Contemp. Phys.* **24**, 25 (1983).
48. Rundquist, P. A., Kesavamoorthy, R., Jagannathan, S., and Asher, S. A., *J. Chem. Phys.* **95**, 1249 (1991).
49. Dovichi, N., *Crit. Rev. Anal. Chem.* **17**, 357 (1987).
50. Gruner, F., and Lehmann, W. P., *J. Phys. A: Math. Gen.* **15**, 2847 (1982).
51. Pusey, P. N., and Tough, R. J. A., in *Dynamic Light Scattering: Application of Photon Correlation Spectroscopy*, Ed. R. Pecora, (Plenum, New York, 1985).
52. Kesavamoorthy, R., and Arora, A. K., *J. Phys. C: Condens. Matter.* **19**, 2833 (1986).
53. Chawdhury, A., Ackerson, B. J., and Clark, N. A., *Phys. Rev. Lett.* **55**, 833 (1985).

54. Ashkin, A., *Phys. Rev. Lett.* **24**, 156 (1970).
55. Kerker, M., *Am. Sci.* **62**, 92 (1974).
56. Preining, O., in *Aerosol Science*, Ed. Davies, C. N. (Academic, New York, 1966).
57. Waldmann, L., and Schmitt, K. H., in *Aerosol Science*, Ed. C. N. Davies (Academic, New York, 1966).
58. Heimenz, P. C., *Principles of Colloid and Surface Chemistry*, 2nd ed. (Dekker, New York, 1986).
59. Smith, B. A., and McConnell, H. M., *Proc. Natl. Acad. Sci. USA* **75**, 2759 (1978).
60. Eichler, H., Salje, G., and Stahl, H., *J. Appl. Phys.* **44**, 5383 (1973).
61. Ackerson, B. J., *J. Chem. Phys.* **64**, 242 (1976).

11

Theory of Interactions in Charged Colloids

Bo Jönsson, Torbjörn Åkesson, and Clifford E. Woodward

The interaction between charged particles has successfully been described within the dielectric continuum model using the Poisson-Boltzmann equation. This so-called DLVO theory has been extensively applied to charged aqueous colloidal suspensions containing monovalent salt. Surprisingly enough the theory fails completely whenever divalent (counter)ions are present. The origin of the breakdown of the mean-field approach is its neglect of ionic correlations, which become dominant for divalent species. The electrostatic correlations contribute an attractive component to the osmotic pressure, which will eventually make the net osmotic pressure negative. In order to describe these correlations, one generally needs to perform Monte Carlo simulations or solve integral equations; both require substantial numerical effort. There exists, however, a simple model system consisting of two spherical macroions and counterions, which can be more simply solved. For this system, the analogy with quantum-mechanical dispersion forces is immediate, providing an intuitive feeling for the attraction. At sufficiently high salt concentration, additional hard-core correlations will contribute to the attractive forces. Further, attractive colloidal forces due to a completely different origin were proposed by Sogami and Ise a decade ago. This has led to some controversy over the nature of the effective pair potential in charged colloidal suspensions. In this review we also attempt to state the problem in a clear and concise fashion in an effort to highlight this controversy.

11.1 Introduction

The stability of macromolecular aggregates in solution is determined by the force acting between them. If the particles are charged, then the force has a strong and long-range component, which often has a profound influence on the macroscopic properties displayed by the solution. The typical picture is that a charged surface in its immediate neighborhood has an accumulation of charges of opposite sign - a phenomenon usually referred to as an *electric double layer* [1, 2]. There are many examples of the importance of electrostatic interactions between charged macromolecules in biological systems [3, 4, 5], and examples can also be found in technical applications. The famous theory of colloidal stability due to Derjaguin-Landau-Verwey-Overbeek (the DLVO theory) [6, 7] has as one of its main ingredients the electrostatic force.

The traditional theoretical approach to the electric double layer is to treat the water (the only solvent we will consider here) as a structureless continuum and to apply a mean-field approximation to the explicit ionic interactions. The mean-field approach, in terms of the Poisson-Boltzmann (PB) equation, has been very successful in describing a wide range of electric double-layer phenomena. The number and the diversity of applications where the PB equation has been successfully applied is so large that we are led to the conclusion that both the primitive model and the mean field approach are legitimate approximations. It seems as if the molecular nature of the solvent can be neglected down to surprisingly small dimensions [8, 5]. The numerically most striking examples of the accuracy of the PB equation have come from direct force measurements using a surface force apparatus [9, 10]. Despite the almost universal applicability of the PB equation, there still exists a number of situations where it apparently fails, not only quantitatively but also qualitatively, to describe experimental observations. One common feature of these exceptional circumstances is the presence of multivalent ions.

The breakdown of the PB approximation can be ascribed to ionic correlation effects left out in the mean-field theory but naturally included in, for example, Monte Carlo simulations. The neglect of (ionic) correlations is standard in approximate many-body theories in science, and it is well known that in sufficiently strongly coupled systems it becomes an invalid approach. The PB equation assumes that the charge distribution is given by the mean potential, and we should expect an increasingly inferior description at high electrostatic coupling. This increased coupling can be brought about in different ways, for example, by increasing the charge of the aggregates, lowering the temperature, lowering the relative permittivity, or, more abruptly, by exchanging monovalent with divalent (counter)ions.

In order to investigate the PB approximation, one can choose two alternative paths: (1) use a more refined theory, which takes ionic correlations into account, or (2) perform Monte Carlo (MC) simulations, which will, within statistical uncertainties, give the exact answer. We will discuss both approaches, although with an emphasis on the latter. MC simulations became applicable to complex systems at the end of the 1970s, and, to our knowledge, the first simulation of an electric double layer was published 1979 [11]. Since then hundreds of simulations of electrical double layers in varying geometries have appeared in the literature, and it is beyond the scope of this review to try to cover this field exhaustively. Our objectives will be limited to the description of ion correlation effects on colloidal stability.

In the next section, we review MC simulations for a planar electrical double layer containing only counterions. An intuitive understanding of the attractive correlation force can be obtained by considering a classical analogue of the quantum-mechanical dispersion or London force [12], which is presented in Section 11.3. The addition of salt and very concentrated double layers introduce additional hard-core correlations, which also give rise to attractive forces, but via a completely different mechanism as is described in Section 11.4. Section 11.5 discusses the hypernetted-chain (HNC) equation, which has been shown to be a very accurate integral equation theory for charged systems, and it gives under most circumstances virtually identical results when compared to MC simulations. Here we also discuss some recent very promising density-functional approaches. A special section is devoted to the much debated theory of Sogami and Ise [13].

11.2 Ion Correlations in Planar Double Layers

Consider an electrical double layer consisting of two infinite and uniformly charged walls confining their neutralizing counterions. The separation between the walls will be denoted by b, and assuming the z axis perpendicular to the walls, we will place them at the z coordinates $\pm b/2$ (see Fig. 11.1). The relative permittivity is assumed to be that of water at room temperature. Further technical details can be found in the original publications [14, 15].

Figure 11.2(a) shows a comparison of the counterion profile obtained from the PB equation and from MC simulations. The figure deceptively leads one to believe that the PB equation is virtually exact, while a closer look at Fig. 11.2(b) reveals that the divalent counterion concentration differs qualitatively at the midplane in the two calculations. To recognize the importance of this discrepancy, we have to consider the expression for the osmotic pressure, which is given by the contact theorem [16]. The theorem can be derived by considering the interactions across a fictitious plane parallel to the walls. In fact any plane can be chosen, but the simplest relation is obtained if the plane coincides with one of the charged walls [17], for example, at $z = b/2$

$$p = k_B T \sum_i c_i \left(\frac{b}{2}\right) - \frac{\sigma^2}{2\epsilon_0 \epsilon} \tag{11.1}$$

The summation over the concentrations c_i runs over all species i and σ is the surface charge, $\epsilon_0 \epsilon$ is the dielectric permittivity, k_B is Boltzmann's constant, and

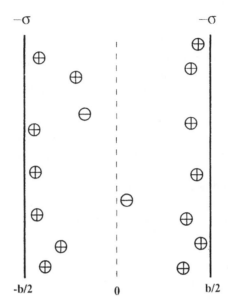

Figure 11.1   Schematic representation of an electric double layer. The two uniformly charged walls confine the mobile co- and counterions.

T is the absolute temperature. An alternative is to use the midplane, and then the expression for the osmotic pressure takes the form

$$p = k_B T \sum_i c_i(0) + p^{\text{corr}} \tag{11.2}$$

where the second term on the right-hand side (rhs) in the general case contains contributions from both electrostatic and hard-core correlations. The first term on the rhs of Eq. (11.2) is obviously positive (repulsive), but the correlation pressure is negative (attractive) except in very dense systems, where hard-core correlations can cause a sign change. For a single double layer with one charged wall at $z = b/2$ and one neutral wall at $z = 0$, the correlation term will disappear, and the osmotic pressure is given by

$$p = k_B T \sum_i c_i(0) \tag{11.3}$$

By combining Eqs. (11.1) and (11.3), one obtains an exact expression for the concentration drop over the double layer

$$\sum_i c_i\left(\frac{b}{2}\right) - \sum_i c_i(0) = \frac{\sigma^2}{2\epsilon_0 \epsilon \, k_B T} \tag{11.4}$$

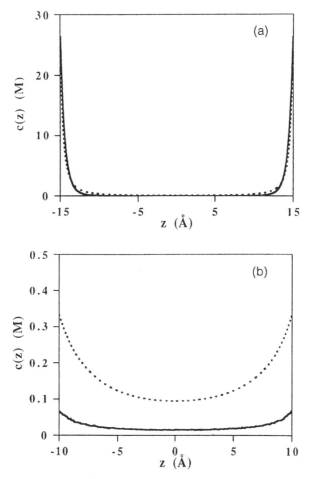

Figure 11.2 Counterion profiles in an electrical double layer. The solid line is from the MC simulation, and the dashed line is from the PB equation. The wall-wall separation is 30 Å, and σ is 0.317 C/m^2. The ionic hard-core radius of the divalent counterions is zero. (a) The full width and (b) the midpart of the double layer.

The pressure given by Eqs.(11.1)–(11.3) is the internal osmotic pressure. The net osmotic pressure, which is the physically relevant quantity, is given by

$$P_{osm} = p - p_{bulk} \tag{11.5}$$

where p_{bulk} is the osmotic pressure of the bulk, with which the double layer is in equilibrium. In this section, we restrict the discussion to salt-free systems; hence p_{bulk} is zero and the net pressure is equal to the internal pressure.

The contributions to the correlation pressure can be formulated as appropriate integrals over the two-particle densities, but the explicit expressions become rather involved, and we will not to repeat them here [18]. Several important conclusions can be drawn from Eqs. (11.1)–(11.4). The PB equation obeys Eq. (11.4); hence for large separations it will give an exact representation of the ion distribution near the charged wall, since p_{osm} will go to zero. This has been demonstrated for a number of aggregates of varying geometries [11, 19-22]. It is also clear that the PB equation will, for most situations of experimental interest, overestimate the osmotic pressure, as the correlation contribution is usually negative. Finally, if the PB equation fails to predict the ionic concentration at the midplane, then it will fail to predict a correct osmotic pressure. Figure 11.2(b) shows that this is the case for divalent counterions. Figure 11.3 displays the same phenomenon, with the osmotic pressure plotted as

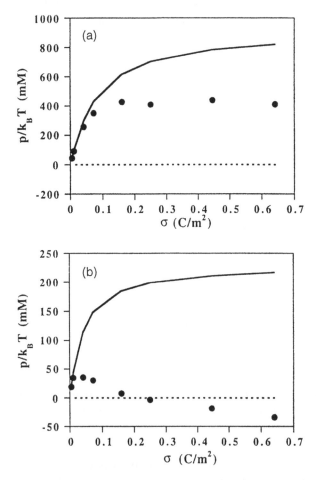

Figure 11.3 The total osmotic pressure as a function of surface charge density from the PB equation (solid line) and from Monte Carlo simulations (symbols); (a) monovalent counterions and (b) divalent counterions. The ionic hard-core radius is zero, and the wall-wall separation is 20 Å.

a function of surface charge density. The PB curve monotonically approaches a finite limiting value equal to $[(\pi k_B T)/(z_i eb)]^2 \epsilon_0 \epsilon$, while simulations predict $p/k_B T$ to go through a maximum before it eventually becomes negative (z_i is the counterion valency). Figure 11.3 clearly demonstrates that the common procedure [23] of determining the surface charge density from a measurement of the force between two charged surfaces is fraught with numerical difficulties.

It is illustrative to consider separately the two components of the pressure in Eq.(11.2). Such a division of the total pressure for double layers with monovalent or divalent counterions is made in Fig. 11.4. The correlation pressure is approximately the same for the two systems (as a matter of fact, it is more negative with monovalent than divalent counterions), and it is in the entropic term, $k_B T \sum_i c_i(0)$, that the large difference emerges [24]. Monte Carlo simulations predict a dramatic decrease in the midplane concentration when monovalent ions are exchanged with divalent ones. The PB equation does also show a decreased concentration, but not to the same extent as MC simulations. The MC result is remarkable in the sense that $c(0)$ goes to zero when at the same time the average ionic concentration increases. Literally, it means lower concentration (at the midplane) with more ions in the double layer! One way to rationalize these results is to examine the energy and entropy contributions to the total free energy of the double layer. The energy minimum is one with all counterions sitting at the walls perfectly correlated. This situation is of course counteracted by entropy, and the compromise is the diffuse double layer. From these considerations it

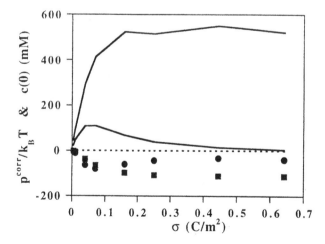

Figure 11.4 The correlation contribution $p^{\mathrm{corr}}/k_B T$ and the entropic contribution $c(0)$ to the osmotic pressure as a function of surface charge density. The upper curve and squares represent a double layer with monovalent counterions, and the lower curve and circles represent divalent counterions. The ionic hard-core radius is zero, and the wall-wall separation is 20 Å.

can be understood that the energy cost in the mean-field approach prevents the (uncorrelated) ions from accumulating near the charged walls as much as they do in MC simulations.

At this point it might be appropriate to pause and look for experimental verifications of the attractive interaction between equally charged particles. A direct comparison with experiment is difficult, except possibly with measurements using the osmotic stress technique or surface force apparatus. As a matter of fact, attractive forces have been observed with the latter technique in charged phospholipid systems in the presence of Ca^{2+} ions [25]. The nonswelling behavior of clays has long been a puzzling phenomenon [26], usually explained by specific nonelectrostatic interactions, but the attractive correlation force might be the proper explanation. Similarly, lamellar liquid crystalline systems with monovalent counterions usually swell in excess water, but with divalent counterions one finds a substantially reduced water uptake [27]. Biologically, the ordering of DNA upon addition of divalent ions [28, 29] and the restricted swelling of different plant membranes, which do not follow simple PB behavior, might also partly be explained by an attractive force. More indirectly, anomalous coagulation kinetics found in charged latex systems may be understood in terms of an additional attractive component in the colloid-colloid interaction. In order to explain the ordering of these highly charged latex particles, taking place at low salt content [30], one also needs to invoke an attractive force.

11.3 Ion Correlation in Spherical Double Layers

Let us first make clear that the derivation to follow is purely classical, but the use of statistical-mechanical perturbation theory leads to an expression very similar to the one obtained for quantum-mechanical dispersion forces [12]. It is of course possible to perform a similar analysis in any geometry, but none will give as neat algebraic expressions as the spherical one (for an analysis of the planar system see Attard et al. [31]).

Consider two spherical cells, depicted in Fig. 11.5, each of radius R_c and each containing one spherical macroion of charge Z and radius a. We will only consider salt-free cells; thus each cell also contains the necessary amount of neutralizing counterions of valency z_i. A distance r separates the macroion centers. The excess free energy due to the interaction of the two subsystems can be written as

$$\Delta F = k_B T \ln\langle \exp[-\Delta U(r)/k_B T]\rangle_0 \tag{11.6}$$

$\Delta U(r)$ is the interaction energy between the cells for a particular configuration of counterions. The angular brackets $<>_0$ represent an average over the counterions in the individual noninteracting cells. The interaction energy $\Delta U(r)$ can be written as a two-center multipole expansion [32], where the different terms represent charge-charge, charge-dipole, dipole-dipole interactions, etc. It is obvious from symmetry considerations that the two first terms will disappear and that the first

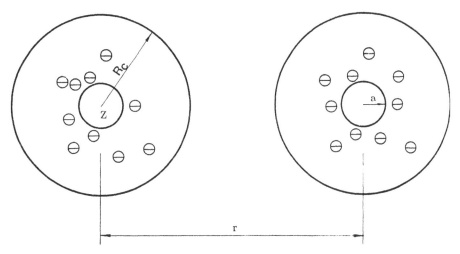

Figure 11.5 Schematic picture of a model system with two macroions of charge Z and radius a with neutralizing counterions in their respective cells of radius R_c.

nonvanishing term is the dipole-dipole interaction. By expanding the logarithm in Eq. (11.4) and keeping only second-order terms in the interaction and including the first nonzero term in the multipole expansion, one obtains

$$\Delta F = \text{const.} \frac{Q_A^{(p)} Q_B^{(p)}}{r^6} \qquad (11.7)$$

where $Q^{(p)}$ is a generalized polarizability for the counterions in their respective cells defined as

$$Q^{(p)} = \left\langle \left(\sum_j z_i e r_j \right)^2 \right\rangle_0 \qquad (11.8)$$

Higher-order polarizabilities will appear for higher multipole interactions. The analogy with quantum-mechanical dispersion forces is direct, where in the present case the $Q^{(p)}$s have taken the place of atomic polarizabilities. The classical polarizabilities vary with aggregate charge and cell radius, but the difference between a system with monovalent or divalent counterions is small (see Table 11.1 and Fig. 11.6). The asymptotic result from the second-order theory, Eqs. (11.7) and (11.8), is surprisingly accurate (see Table 11.1 and Fig. 11.6).

This is an important observation, since the generalized polarizabilities can be evaluated quite accurately using simple Poisson-Boltzmann theory [31, 32]. Finally, we note that the correlation force, obtained as the derivative of Eq. (11.6) or (11.7), is orders of magnitude larger than the usual van der Waals force between comparable hydrocarbon spheres in water.

Table 11.1   Classical polarizabilities for different cell radii and charge

	Classical polarizabilities $Q^{(p)}$ ($e^2\,\text{Å}^2$), see Eq. (11.8)			
R_c (Å)	25	40	60	100
$z_i=1$	2.0	5.5	13	44
$z_i=2$	2.0	5.1	13	42
$z_i=1$	4.0	11	23	81
$z_i=2$	4.1	7.0	16	55

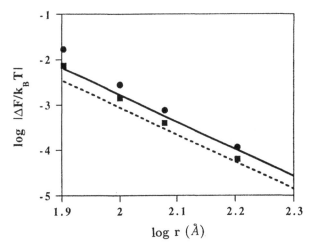

Figure 11.6   The interaction free energy for two spherical cells (see Fig. 11.5) in units of $k_B T$ as a function of their separation. The symbols are MC results using Eq. (11.6), and the curves are asymptotic results from Eq. (11.7). The circles and solid line are for monovalent counterions, and the squares and dashed line are for divalent counterions. $R_c = 40$ Å, $a = 18$ Å, and $Z = 60e$.

Thus the spherical system not only provides a pedagogic analogy with quantum-mechanical dispersion forces demonstrating the origin of the attractive force component, but it also makes possible a simple yet accurate evaluation of the force via the asymptotic second-order expression, Eq. (11.7). It is important to notice, however, that this approach gives an estimate only of the direct correlation force. There is also an indirect effect, since the one-particle distribution will change as a consequence of ionic correlations, which in turns affects the entropic contribution to the force [cf. Eq. (11.2)]. This second effect is significantly larger and unfortunately less amenable to a simple theoretical analysis. It will, however, not give rise to an attractive force, but it will substantially reduce the repulsive component, as seen in Figs. 11.3 and 11.4.

11.4 Salt Effects on Ionic Correlations

In systems with counterions only, the electrostatic repulsion between the ions creates a so-called correlation hole in the ionic atmosphere around each ion. Physical properties of the system are therefore unaffected by the ionic size, provided that the ionic diameter is small compared to the correlation hole. Significant contributions due to hard-core correlations are thus observed only at high volume fractions (i.e., large ions or dense systems). However, in the presence of an electrolyte, collisions mainly between positive and negative ions most be considered even at low volume fractions. As a matter of fact, the understanding of salt effects on the double-layer interaction was for a long time limited by the absence of simulation methods to determine hard-core contribution to p^{corr} [18, 33] or accurate contact densities [15]. In the presence of salt it was therefore possible to determine the osmotic pressure only in systems with one charged and one neutral wall [19, 34, 35]. For such systems the simulation results show that the PB equation works rather well as long as the counterions are monovalent. The discrepancies between PB and MC results increase, however, with increasing salt concentration, that is, at a higher electrostatic coupling.

Recently, improved simulation techniques have been developed, making it possible to study double-layer forces in systems with *two* charged walls [15, 18, 33]. As seen in Fig. 11.7 the PB equation is a reasonable accurate approximation for

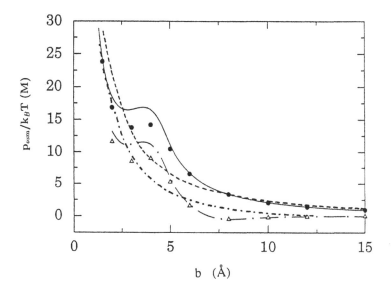

Figure 11.7 The osmotic pressure between two charged surfaces ($\sigma = 0.267 C/m^2$) in a 1:1 salt. The ionic diameter is 4.25 Å. (Full line) anisotropic HNC, (- - -) PB, and (filled sphere) MC results in a salt-free system. (- . -) anisotropic HNC, (-.-) PB and (triangle) MC results in a 2 M 1:1 salt solution (from Kjellander et al. [18].)

systems with monovalent ions at low and intermediate salt concentrations. The shoulder at 2-4 Å is to a large extent caused by the hard-core interactions. At larger separations the electrostatic correlations dominate, and the PB pressure is more repulsive than the simulated one. Note that the pressure may even become attractive at high salt concentrations. This could be a consequence of different ionic correlations in the double layer and in the bulk, affecting both the pressure and the chemical potential in the two phases. In Fig. 11.8 the net osmotic pressure as a function of the separation is given for double layers immersed in a 2:1 electrolyte (divalent counterions). At high concentrations the pressure is strongly oscillatory, which is caused by the formation of layers with alternating excess amount of counterions or coions. This layering is due to a complex interplay between the electrostatic and hard-core interactions. The PB approximation cannot reproduce these results, and the theory becomes qualitatively wrong for divalent ions, even in the presence of salt.

It is also found that interacting double layers containing 2:1 or 2:2 salt show a similar behavior as long as the divalent ion is the counterion. This has also been observed for a single charged surface, while such a system with divalent coions follows the PB predictions [34]. Thus multivalent counterions can cause

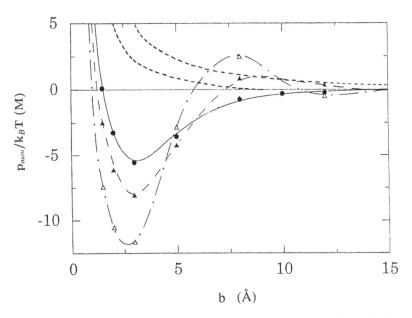

Figure 11.8 The osmotic pressure for systems with divalent counterions. (Full line) anisotropic HNC and (filled sphere) MC results with only counterions, (- - -) AHNC and (filled triangle) MC in an 1 M 1:2 salt solution, (- . -) AHNC and (delta) MC in 2 M salt. The broken curves (- - -) show the PB results; upper curve salt-free system, lower curve 2 M salt. σ is 0.267 C/m^2 and the ionic diameter 4.25 Å(from Kjellander et al. [18].)

an unexpected behavior, but until recently only little attention has been paid to the effect of multivalent coions. In the work by Bolhuis et al. [36] it is found that the net osmotic pressure, even at a low salt content in the double layer, may become strongly attractive at short separations (Fig. 11.9). This attraction is caused by a hard-core mechanism, though in an indirect way. The correlation pressures in the double layer and in the bulk are roughly the same at equilibrium, while the entropic pressure in the bulk exceeds the entropic double-layer pressure, giving rise to the attraction. The high bulk density is a consequence of the rapidly increasing chemical potential of the salt in the double layer at small separations. The change in the chemical potential is surprisingly not accompanied by a large increase of the internal pressure. Increasing the size or the valency of the coion is found to strengthen the attraction. The surface charge density also has a large effect on the osmotic pressure being repulsive at low densities and strongly attractive at high values. Also the osmotic pressure obtained at constant bulk concentration (constant chemical potential) shows a significant attraction at resonable high electrolyte concentrations, resembling the double-layer forces in a 1:1 electrolyte (cf. Fig. 11.7).

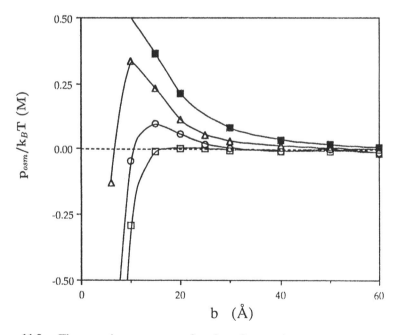

Figure 11.9   The osmotic pressure as a function of separation at constant salt concentration in the double layer. From top to bottom the curves represent 0.01 M, 0.025 M, 0.05 M, and 0.11 M salt solution. 1:2 salt with divalent coions. σ is 0.224 C/m^2 and the ionic diameter is 4.25 Å (from Bolhuis et al. [36]).

11.5 Correlated Theories

The lack of ionic correlations in the PB theory has been known for many years, and the possibility of attractive double-layer interactions has been mentioned in the literature [37], but the theoretical means to perform the necessary calculations were not available at that time. Attractive colloid-colloid interactions were seen in an early study of colloidal interactions using the HNC approximation [38]. The attraction appeared at very short separations, but the calculations have been criticized [39], and the attraction is now considered to be due to an artefact in the particular HNC approximation used. Several other integral equation approaches have been attempted, but the most successful so far is the anisotropic HNC procedure (AHNC) [40]; a similar theory is found to be equally accurate for single charged surfaces [41].

The AHNC results for double layers containing 1:1, 2:1 (divalent counterions), and 2:2 electrolytes [42, 43] have been compared by simulations [18, 33]. The agreement between AHNC and MC results is almost perfect, except for a slightly overestimated hard-core contribution to the pressure, which could be large in 1:1 electrolytes at high volume fractions (Fig. 11.10). In the case of divalent counterions, excellent agreement is obtained for all separations (cf. Fig. 11.8).

In double layers with important hard-core interactions one can, however, use a modified HNC theory adapted to anisotropic systems, called reference HNC (RHNC) [44]. This approximation is based on the observation that the bridge function, which is set equal to zero in the HNC closure, belongs to the same family of curves for many pair potentials. This means that the bridge function obtained for hard spheres in a reference system can be used in the integral equations to obtain pair correlation functions. Such functions have been calculated by anisotropic RHNC in a salt-free double layer with monovalent counterions. Compared to the pair correlation functions obtained from simulations, the agreement is within statistical error perfect [45], and consequently the RHNC theory predicts a correct net osmotic pressure even at small separations (cf. Fig. 11.7).

Both Monte Carlo simulations and solutions to the AHNC equation require substantial numerical calculations, and a simple yet accurate theory has been long sought. Nonlocal density- functional theories have recently been tried out by several groups [46, 47], and the results so far seem very promising. Tang et al. [46] were able to reproduce attractive double-layer forces found in planar systems with divalent counterions both at high surface charge density and high salt concentration. They also found that hard-core correlations give rise to attractive interactions at high particle densities, as discussed in one of the previous sections. Penfold et al. [47] have studied spherical macroions with similar success. To arrive at the final equations in the nonlocal density-functional theory requires some lengthy algebra. Once this price is paid, then the computational efforts are negligible.

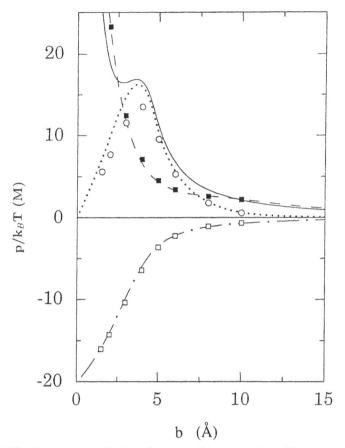

Figure 11.10 Pressure contributions for monovalent counterions. The curves represent anisotropic HNC results and the symbols, MC results. (- - -) and (■) entropic component, (...) and (o) collision (ion-ion core contact) contribution, (-.-) and (square) electrostatic correlation component. For reference, the total pressure obtained from AHNC is shown (full line). σ is 0.267 C/m^2; the ionic diameter is 4.25 Å, no salt. The RHNC results coincide with MC (from Kjellander et al. [18]).

11.6 Sogami-Ise Theory

A decade ago Sogami and Ise (SI) proposed a new theory for charged colloid interactions based on the use of the Gibbs free energy [13]. The traditional DLVO potential is based on the Helmholtz free energy. The use of the latter follows from the approximation that, even at constant atmospheric pressure, once all species are in solution, the total solution volume remains essentially constant, irrespective of the configurations of the charged particles. Treating the solvent as a dielectric medium, as in the primitive model, then leads to a picture of the charged solute species as an effective fluid confined to a fixed volume equal to the solution volume. As the number of particles, temperature, and volume are fixed, the natural thermodynamic potential is the Helmholtz free energy.

Sogami and Ise made the observation that if the amount of salt in addition to the counterions is small, then the local concentration of the mobile electrolyte will be affected by the colloid configurations. Indeed the mobile electrolyte would be largely confined to the region about the colloidal particles. This led them to introduce a local "solute" volume V, which is quite distinct from and generally less than the solution volume V^*. Sogami and Ise chose to model the behavior of V by assuming that the mobile ions were constrained within a semipermeable membrane (permeable to solvent only) that surrounded the fixed macroions. The volume V contained by the membrane was supposedly determined by a constant osmotic pressure p_0 acting at the membrane surface. Of course p_0 is a fictitious construct; the containment of most of the mobile electrolyte (no additional salt) is due to the electrostatic potential in the region about the fixed colloid particles. Nevertheless, the constant-pressure system does become more exact as p_0 approaches zero.

The natural thermodynamic potential at constant pressure is the Gibbs free energy G. For the system as defined by Sogami and Ise, G is given by

$$G\,[\mathbf{R},\,V(\mathbf{R})] = F\,[\mathbf{R},\,V(\mathbf{R})] + p_0 V(\mathbf{R}) \qquad (11.9)$$

where $F\,[\mathbf{R},\,V(\mathbf{R})]$ is the Helmholtz free energy and we have included the dependences on the macroion configuration $\mathbf{R}$. The linearized Poisson-Boltzmann expression for F is [13]

$$F\big[\mathbf{R},\,V(\mathbf{R})\big] = \sum_{i=1}^{N} f^{(1)}\big[\mathbf{r}_i,\,V(\mathbf{R})\big] + \sum_{i<j}^{N,N} f^{(2)}\big[|\mathbf{r}_i - \mathbf{r}_j|,\,V(\mathbf{R})\big] \qquad (11.10)$$

where the positions of the N macroions is denoted by $\mathbf{R} = (\mathbf{r}_1, \ldots, \mathbf{r}_N)$. The quantity $f^{(1)}$ is a one-particle potential and $f^{(2)}$ is the usual pair potential that appears in the DLVO theory. The equilibrium value for $V(\mathbf{R})$ is given by

$$\left(\frac{\partial G}{\partial V}\right)_{\mathbf{R}} = \left(\frac{\partial F}{\partial V}\right)_{\mathbf{R}} + p_0$$
$$= 0 \qquad (11.11)$$

Sogami and Ise avoided the need to determine $V(\mathbf{R})$ by using the following expression for G

$$G = \sum_{k=1}^{N_{sp}} \mu_k\,[\mathbf{R},\,V(\mathbf{R})]\,N_k \qquad (11.12)$$

where the sum is over the number of mobile ions and surface species, μ_k is the chemical potential of the kth species, and N_k the number. The chemical potentials were obtained by direct differentiation of F Eq. [11.10]. This leads to explict expressions for G in terms of single and pair potentials, much like the result for F. The remarkable feature of the Sogami and Ise expression is that the pair potential contains an attractive region for similarly charged colloid particles, a result not observed in F. This result has recently been supported by Smalley [48], who attempted to rederive the Sogami and Ise potential using different thermodynamic arguments.

Overbeek [49] has criticized the use of the Gibbs free energy on the grounds more or less stated above; that is, the solution volume remains essentially constant even at atmospheric pressure. That criticism, however, seems not to have been properly aimed at the osmotic system envisaged by Sogami and Ise. That is, the constant applied pressure is not due to the atmosphere, but is a fictitious construct to validate the use of a Gibbs free energy for the solute particles. More recently, however, Woodward [50] has shown that the attractive interaction obtained by Sogami and Ise is spurious and is due to the incorrect application of Eq. (11.12) to a nonuniform system. That is, Eq. (11.12) follows from the fact that a uniform system is extensive with respect to the number of particles and the volume. This property is intrinsically linked to the system's symmetry and will not be true in the general case, where the total electrostatic potential due to the fixed colloid particles has no symmetry at all. Woodward has shown that, in a simple planar system, a proper evaluation of the Gibbs free energy as envisaged by Sogami and Ise leads to repulsive interactions. Furthermore, the Gibbs free energy result approaches the usual DLVO potential as p_0 approaches zero.

A simple argument presented here shows that the Gibbs free energy can only predict attractive pair interactions if the corresponding Helmoholtz free energy does

$$\frac{dG}{d\mathbf{R}} = \left(\frac{\partial G}{\partial \mathbf{R}}\right)_V + \left(\frac{\partial G}{\partial V}\right)_{\mathbf{R}} \frac{dV}{d\mathbf{R}}$$

$$= \left(\frac{\partial F}{\partial \mathbf{R}}\right)_V \qquad (11.13)$$

The forces between particles, as determined by the Gibbs free energy, are equal to the forces obtained from the Helmholtz free energy, but at a varying mobile ion density, determined by $V(\mathbf{R})$. As the Poisson-Boltzmann free energy predicts repulsive forces at all densities, the attractive forces obtained by Sogami and Ise appear to be erroneous.

11.7 Conclusions

In this review we have attempted to give an overview of the current understanding of interactions between charged colloids. The fundamental concepts of Derjaguin, Landau, Verwey, and Overbeek remain the foundations upon which the modern theories are built. This notwithstanding, the surprising impact of correlation effects in strongly coupled systems cannot be understated. Indeed, the discovery of *attractive* interactions between like charged particles, due to ionic correlations, attests to their importance. Computer simulations and sophisticated integral equation methods have been indispensible in unravelling these correlation mechanisms. Nevertheless, it is possible to understand attractive correlations as being analogous to quantum dispersion forces. The example we present of two interacting spherical cells using perturbation theory is very similar to the treatment of dispersion forces between two atoms. In the latter case attractions are due to correlated quantum fluctuations of the electron density; in the former it is due to thermal fluctuations of the mobile electrolyte. This example can lead one to believe that correlations leading to attractive interactions are due mainly to the counterions to the colloid particles. However, when the salt concentration becomes high, repulsive correlations become important. They can provide a mechanism for attractive interactions between colloid particles at small separations. This attraction appears to be driven by the depletion of coions in the region between the walls.

The electrostatic attractions proposed in the theory by Sogami and Ise are of a source entirely different from any discussed here. Indeed the Sogami-Ise potential is derived from the mean-field Poisson-Boltzmann theory, which neglects ionic correlations altogether. It is our opinion that the theoretical results by those authors are erroneous. If some experiments cannot be explained by the classical DLVO theory, the Sogami-Ise model does not provide legitimate physical reasons for the discrepancies - the answers must be sought elsewhere. Unfortunately, for the theoretician, many colloidal systems are far from ideal charged spheres or planes, this oftenmakes interpreting experiments difficult and inconclusive. However, it is this richness in the behavior of colloidal systems that ultimately leads to their usefulness.

References

1. Gouy, G., *J. Phys.* **9**, 457 (1910).
2. Chapman, D. L., *Philos. Mag.* **25**, 475 (1913).
3. Tanford, C., and Kirkwood, J. G., *J. Am. Chem. Soc.* **79**, 5333 (1957).
4. Bratko, D., Luzar, A., and Chen, S -H., *Bioelectrochem. Bioenerg.* **20**, 291 (1988).
5. Svensson, B., Jönsson Bo, Woodward, C. E., and Linse, S., *Biochemistry* **30**, 5209 (1991).
6. Derjaguin, B. V., and Landau, L., *Acta Phys. Chim. URSS* **14**, 633 (1941).
7. Verwey, E. J. W., and Overbeek, J. Th. G., *Theory of the Stability of Lyophobic Colloids* (Elsevier, Amsterdam, 1948).
8. Bader, J. S., and Chandler, D., *J. Phys. Chem.* **96**, 6423 (1992).
9. Pashley, R. M., *J. Coll. Interface Sci.* **83**, 531 (1981).
10. Israelachvili, J. N., and Adams, G. E., *J. Chem. Soc. Faraday Trans. I* **74**, 975 (1978).
11. Torrie, G. M., and Valleau, J. P., *Chem. Phys. Lett.* **65**, 343 (1979).
12. London, F., *Z. Phys. Chem. B.* **11**, 222 (1930).
13. Sogami, I., and Ise, N., *J. Chem. Phys.* **81**, 6320 (1984).
14. Jönsson, Bo, Wennerström, H., and Halle, B., *J. Phys. Chem.* **84**, 2179 (1980).
15. Åkesson, T., and Jönsson, Bo, *Electrochim. Acta* **36**, 1723 (1991).
16. Henderson, D., Blum, L., and Lebowitz, J. L., *J. Electraanal. Chem.* **102**, 315 (1979).
17. Wennerström, H., Jönsson, Bo, and Linse, P., *J. Chem. Phys.* **76**, 4665 (1982).
18. Kjellander, R., Åkesson, T., Jönsson, Bo, and Marcelja, S., *J. Chem. Phys.* **97**, 1424 (1992).

19. Torrie, G. M., and Valleau, J. P., *J. Chem. Phys.* **73**, 5807 (1980).
20. Bratko, D., and Vlachy, V., *Chem. Phys. Lett.* **90**, 434 (1982).
21. Linse, P., Gunnarsson, G., and Jönsson, Bo, *J. Phys. Chem.* **86**, 413 (1982).
22. Murthy, C. S., Bacquet, R. J., and Rossky, P., *J. Phys. Chem.* **89**, 701 (1985).
23. Pashley, R. M., McGuiggan, P. M., and Ninham, B. W., *J. Phys. Chem.* **90**, 5841 (1986).
24. Guldbrand, L., Jönsson, Bo, Wennerström, H., and Linse, P., *J. Chem. Phys.* **80**, 2221 (1984).
25. Marra, J., *Biophys. J.* **50**, 815 (1986).
26. Norrish, K., and Quirke, J. P., *Nature* **18**, 120 (1954).
27. Khan, A., Fontell, K., and Lindman, B., *J. Colloid Interface Sci.* **101**, 193 (1984).
28. Rau, D. C., Lee, B., and Parsegian, V. A., *Proc. Natl. Acad. Sci. USA* **81**, 2621 (1984).
29. Guldbrand, L., Nilsson, L., and Nordenskiöld, L., *J. Chem. Phys.* **85**, 6686 (1986).
30. Ise, N., Okubo, T., Sugimura, M., Ito, K., and Nolte, H. J., *J. Chem. Phys.* **78**, 536 (1983).
31. Attard, P., Kjellander, R., Mitchell, D. J., and Jönsson, Bo, *J. Chem. Phys.* **89**, 1664 (1988).
32. Woodward, C. E., Jönsson, Bo, and Åkesson, T., *J. Chem. Phys.* **89**, 5145 (1988).
33. Valleau, J. P., Ivkov, R., and Torrier, G. M., *J. Chem. Phys.* **95**, 520 (1991).
34. Torrie, G. M., and Valleau, J. P., *J. Phys. Chem.* **86**, 3251 (1982).
35. Svensson, B., Woodward, C. E., and Jönsson, Bo, *J. Phys. Chem.* **94**, 2105 (1990).
36. Bolhuis, P. G., Åkesson, T., and Jönsson, Bo, *J. Chem. Phys.* **98**, 8096 (1993).
37. Oosawa, F., *Polyelectrolytes* (Marcel Dekker, New York, 1971).
38. Patey, G. N., *J. Chem. Phys.* **72**, 5763 (1980).
39. Teubner, M., *J. Chem. Phys.* **75**, 1907 (1981).
40. Kjellander, R., and Marcelja, S., *J. Chem. Phys.* **82**, 2122 (1985).
41. Plischke, M., and Henderson, D., *J. Chem. Phys.* **90**, 5738 (1989).
42. Kjellander, R., and Marcelja, S., *J. de Physique* **49**, 1009 (1988).
43. Kjellander, R., and Marcelja, S., *Chem. Phys. Lett.* **127**, 402 (1986).
44. Kjellander, R., and Sarman, S., *Molec. Phys.* **70**, 215 (1990).
45. Greberg, H., Kjellander, R., and Åkesson, T. (to be published).
46. Tang, Z., Scriven, L, E., and Davis, H. T., *J. Chem. Phys.* **97**, 9258 (1992).
47. Penfold, R., Jönsson, Bo, and Nordholm, S., *J. Chem. Phys.* **99**, 497 (1993).
48. Smalley, M. V., *Molec. Phys.* **6**, 1251 (1990).
49. Overbeek, J. Th. G., *J. Chem. Phys.* **87**, 4406 (1987).
50. Woodward, C. E., *J. Chem. Phys.* **89**, 5140 (1988).

12

Long-Range Attraction in Charged Colloids

M. V. Smalley

The effective interaction between highly charged plates immersed in an electrolyte solution has been treated exactly within mean-field theory by Sogami, Shinohara, and Smalley. The adiabatic potentials, expressed in terms of elliptic integrals, have long-range weak attractive parts as well as medium-range strong repulsive parts. The same features are reproduced by the counterion dominance solution, which allows the adiabatic pair potential to be expressed in terms of elementary functions. The former solution has been criticized by Overbeek, and the latter solution has been criticized by Ettelaie and by Levine and Hall. These criticisms and replies to them are reviewed. It is shown that the Levine-Hall criticism is based on a mathematical error and that the terms that Ettelaie and Overbeek append to the SSS theory lead to the system not being at thermal equilibrium. The earlier debate concerning the Gibbs free energy formalism of the Sogami theory is briefly reviewed. The latter formalism is adapted to the properties of the two-phase (gel) region of colloid stability, which is observed in clay swelling. It yields the predictions that the interplate separation be a constant number of Debye screening lengths and that, for a constant surface potential, the ratio of the electrolyte concentration in the supernatant fluid to the average electrolyte concentration in the gel phase be constant. These predictions are in agreement with the experimental results on the n-butylammonium vermiculite system, which is an excellent model system for one-dimensional colloids and which clearly manifests the long-range attraction in charged colloids.

12.1 Introduction

"The principle of science, the definition, almost, is the following: *The test of all knowledge is experiment.* Experiment is the *sole judge* of scientific truth." [1] (Feynman's italics).

The long-range attraction in charged colloids is an established experimental fact. The most extensive experimental evidence has been obtained from systems of charged polystyrene sulfonate latex spheres [2] and the original formalism of mean-field theory addressing the fact was for the effective potential between charged spherical particles in dilute suspension [3, 4]. More recently, the positive adsorption of ionic micelles near a like-charged surface monolayer [5] and the positive adsorption of latex particles near a like-charged cover glass [6] have demonstrated the existence of the attraction for the case of the interaction of a sphere with a flat plate, but the best evidence comes from the plate-plate interaction.

The plate-plate interaction has been studied in detail in the n-butylammonium vermiculite system, which was first reported by Walker [7]. As long ago as 1962, Garrett and Walker [8] suggested that the homogeneous looking green-yellow gels formed when n-butylammonium vermiculite crystals are soaked in water or dilute aqueous solutions of n-butylammonium salts could be used to provide a sensitive test of theoretical models of colloid stability, and in 1963, reporting a preliminary X-ray study of the gels, Norrish and Rausel-Colom [9] noted that the cohesive force in the system at interplate separations of the order of 200 Å was far too great to be explained by van der Waals forces.

In recent years the system has been studied as a function of five variables: the volume fraction ϕ of the crystals in the condensed matter system, the electrolyte concentration c of the soaking solution, the temperature T, hydrostatic pressure p_h, and uniaxial stress p_u along the swelling axis. Smalley et al. [10] studied the p_h, T behavior at $\phi < 0.01$, $c = 0.1$ M, $p_u = 0$; Braganza et al. [11] studied the c,T behavior at $\phi < 0.01$, $p_h = 1$ atm, $p_u = 0$; Crawford et al. [12] studied the p_u, c behavior at $\phi < 0.01$, $T = 10^0$C, $p_h = 1$ atm; and most recently, Williams et al. [13] studied the ϕ, c, T behavior at $p_h = 1$ atm, $p_u = 0$. The most important facts are as follows.

1. The gels show no tendency to disperse into the solvent medium when the latter is held in gross excess ($\phi < 0.01$) with respect to the macroions [7-12].
2. There is a *reversible* phase transition, both with respect to temperature [11, 13] and hydrostatic pressure [10] between the crystalline (primary minimum) and gel (secondary minimum) states of the system.
3. Throughout the concentration range $0.001M < c < 0.1$ M, the d spacing between the parallel plates (denoted by l_c, the lattice parameter along the c axis, the swelling axis of the mineral) varies approximately as the inverse square root of the concentration of the soaking solution. In other words, the plate separation is nearly equal to a constant number of Debye screening lengths κ^{-1} [7, 11–13].

4. Throughout the concentration range $0.001M < c < 0.1$ M, the d spacing between the parallel plates decreases linearly with the logarithm of the applied uniaxial stress along the swelling axis, the quantitative relation showing that the surface potential is equal to 70 mV and constant with respect to the salt concentration [12].

5. Throughout the concentration range $0.001M < c < 0.1$ M, the n-butyl-ammonium chloride concentration in the supernatant fluid is 2.6 ± 0.4 times the average concentration in the gel phase [13].

Fact 1 shows that there is a long-range (100 to 1000 Å) attractive force between the charged plates. Fact 2 shows that the colloidal (gel) state is a true thermodynamic phase of the system. This already invalidates DLVO theory [14, 15], which is based on the assumption that lyophobic sols are thermodynamically unstable. Fact 3 is very strong evidence for the Sogami theory [3, 4] because its adaptation to the plate macroion case [16] yields the prediction that l_c is proportional to $c^{-1/2}$. By contrast, the DLVO theory predicts a very rapid, $l_c^4 \exp(-\kappa l_c)$, variation that cannot be fitted to the experimental results [11]. Facts 4 and 5 will be addressed later in the chapter.

Such observations on a well-defined system of plate macroions are very interesting from a theoretical point of view. This is because, unlike the general three-dimensional case, the one-dimensional problem of the interaction of two parallel flat plates immersed in an electrolyte solution lends itself to an exact mathematical solution of the appropriate Poisson-Boltzmann equation. The importance of a rigorous analysis of the one-dimensional problem was stressed by Verwey and Overbeek in their *Theory of the Stability of Lyophobic Colloids* [14]. According to Verwey and Overbeek [14], this "exact" solution leads to a purely repulsive pair potential between the charged particles, a conclusion that cannot be maintained *on experimental grounds* in the light of the results on the n-butylammonium vermiculite gels. The "exact" solution [14], which has been a cornerstone of colloid science for nearly half a century, is wrong. The exact mean-field theory solution to the one-dimensional colloid problem was given by Sogami et al. [17].

12.2 The Exact Mean-Field Theory Solution

In the calculations of Sogami et al. [17] two infinite plates immersed symmetrically in an electrolyte solution divided the solution into two disconnected regions, the inner region R_i between the plates and the outer region R_o extending to infinity from the outer surfaces of the plates. In spite of this apparent division of the solution due to the one-dimensional idealization, an essential feature of the model was that *for a real charged plate system thermal equilibrium is achieved throughout the solution both inside and outside the plates*. The mean electric potential and ion distributions in R_i and R_o were derived from the postulate of global thermal equilibrium throughout the whole system. The geometry of the problem and example results for the electric potential and ion distributions, which were recently verified by Overbeek [18], are shown in Fig. 12.1.

The Helmholtz free energy was formulated as a functional of the mean electric potential and exact solutions of the Poisson-Boltzmann equation under Dirichlet and Neumann boundary conditions enabled Sogami et al. [17] to express the free energy analytically in terms of Legendre's elliptic integrals of the first and second kinds. The essential feature of the results is that there are three terms that contribute to the adiabatic (Helmholtz) pair potential between the plates, which is expressed in the following form

$$U(r) = U^{\text{el}}(r) + U_i^o(r) + U_o^o(r) \tag{12.1}$$

where r is the plate separation. The purely repulsive part $U_i^o(r)$ represents the osmotic contribution of the ions that are trapped in the inner region R_i and the purely attractive part $U_o^o(r)$ originates in the osmotic pressure of the small ions in R_o exerted against the surfaces of the plates from outside. The other component $U^{\text{el}}(r)$ results from the electric field energy E_i in the inner region R_i through the Debye-Hückel charging-up procedure. An example calculation of these three terms, which were recently verified by Overbeek [18], is shown in Fig. 12.2.

The adiabatic potentials of the charged plates, which were determined as the parts of the free energy that depend on the interplate distance, turn out to have long-range weak attractive parts as well as medium-range strong repulsive parts irrespective of the type of boundary conditions. While the repulsion originates mainly in the osmotic pressure of the excess ions trapped between the plates by the large surface charges, the attraction arises from an electric pull from the intermediate cloud of excess counterions between the plates.

The purely repulsive DLVO potential

$$U_{\text{DLVO}}(r) \propto e^{-\kappa r} \tag{12.2}$$

where κ is the inverse Debye length, corresponds, approximately but essentially, to the sum of the osmotic components $U_i^o(r) + U_o^o(r)$ in Eq. (12.1). It is the other component $U^{\text{el}}(r)$ of the pair potential that leads to the long-range attraction. The DLVO theory ignored the component $U^{\text{el}}(r)$ and, as a result, predicted a pure repulsion.

Langmuir [19] was the first to recognize that the van der Waals force is not able to act at the considerable distances necessary to explain colloid stability, and he postulated that the source of the attractive forces was to be found in the electric forces acting between the small ions and the charged macroionic particles. Langmuir derived a formula for the interplate force caused solely by the osmotic pressure exerted by the excess of small ions between the plates by the following argument. The fixed surface charges of the plates work to increase the local concentration of counterions at the midpoint between the plates by an amount $n_+(0) - n_+(\infty)$ and decrease that of coions by $n_-(\infty) - n_-(0)$, where $n_\pm(0)$ and $n_\pm(\infty)$ are the simple ion number densities at the midplane and at an infinite distance from the plates, respectively. Hence the net

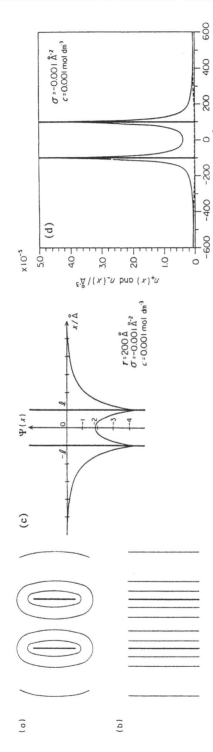

Figure 12.1   The geometry of the problem. (a) Schematic illustration of the equipotential lines around two charged plates with finite width immersed in an electrolyte solution, and (b) representation of the one-dimensional idealization of this situation for two infinite plates. (c) Variation of the dimensionless electric potential $\Psi(x) = e\psi(x)/k_BT$ against the coordinate x, whose origin is taken at the midpoint between the plates with separation $r = 200$ Å, and (d) distributions of small ions $n_i(x) = n_\infty \exp[-z_i\Psi(x)]$ with valency z_i ($z_\pm = \pm1$) around the plates with separation $r = 200$ Å. In (c) and (d), the concentration of the external soaking solution is $c = 0.001$ mol dm^{-3} and the surface charge density is taken to be $\sigma = -0.001\text{Å}^{-2}$ in the Neumann model, which corresponds to the surface potential $\Psi_s = -4.35$ in the Dirichlet model.

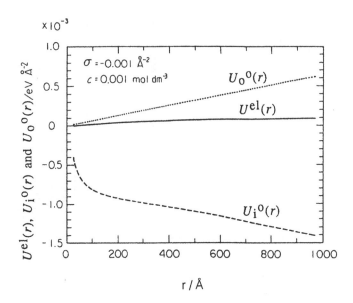

Figure 12.2 Behavior of the electric component $U^{el}(r)$ (solid line) and the osmotic components $U_i^o(r)$ (dashed line) and $U_o^o(r)$ (dotted line) of the adiabatic pair potential as a function of the interplate distance r for the surface charge density $Z_0 = -0.001$ Å^{-2} in the Neumann model (the surface potential $\Psi_s = -4.35$ in the Dirichlet model). The concentration of the external soaking solution is taken to be $c = 0.001$ mol dm^{-3}.

deviation of the concentration of small ions at $x = 0$ from the homogeneous background is given by $n_+(0) + n_-(0) - 2n_\infty$, where $n_+(\infty) = n_-(\infty) = n_\infty$. This deviation results in an osmotic pressure p pushing the plates apart expressed in the form

$$p = n_\infty k_B T \left(e^{-\psi(0)} + e^{\psi(0)} - 2 \right) \qquad (12.3)$$

Since $e^{-\psi(o)} + e^{\psi(0)} \geq 2$ for any value of $\psi(0)$, this expression gives rise to a purely repulsive pair potential for the charged plates. Derjaguin, and also Verwey and Overbeek, arrived at Eq. (12.3) by pursuing different ways of reasoning, and the DLVO repulsive potential given in Eq. (12.2) was obtained when the solution of the linearized Poisson-Boltzmann equation was used for $\psi(0)$. Where Langmuir's work differs from that of all other authors in the famous controversy concerning the interaction of ionic colloidal particles in the 1940s [20] lies in his recognition that in the complete expression for the interaction there is an additional term of purely electric nature. The Sogami et al. calculations [17] bear out his brilliant chemical intuition in a particularly clear way.

12.3 The Counterion Dominance Solution

An obvious feature of the exact ion distributions shown in Fig. 12.1(d) is that counterions dominate coions in the inner region R_i between the charged macroionic plates. This immediately suggests that, in R_i, the full nonlinear Poisson-Boltzmann equation

$$\frac{d^2\Psi}{dx^2} = \kappa^2 \sinh \Psi \tag{12.4}$$

is better approximated by the neglect of coions, as

$$\frac{d^2\Psi}{dx^2} = -\frac{\kappa^2}{2} e^{-\Psi} \tag{12.5}$$

than it is by the linearized Poisson-Boltzmann equation

$$\frac{d^2\Psi}{dx^2} = \kappa^2 \Psi \tag{12.6}$$

where $\Psi(x)$ is a dimensionless electric potential, equal to $e\psi(x)/k_BT$. Sogami et al. [21] used Eq. (12.5) to obtain the adiabatic pair potential, calling the approximation the counterion dominance approximation.

Note that the variable x in Eqs. (12.4)–(12.6) must be distinguished from the interplate separation r. In order to calculate the electric potential $\Psi(x)$ and the ion distributions $n_+(x)$ and $n_-(x)$ throughout the solution, it is necessary to fix the interplate separation r. The Helmholtz free energy of the system $F(r)$ is calculated from an x-dependent integration at a fixed value of r. Then r is changed to a new value, say r', and $\Psi(x)$, $n_+(x)$, and $n_-(x)$ are recalculated at this new interplate separation, and the solutions are used to recalculate the Helmholtz free energy of the system, $F(r')$. This is the nature of the adiabatic approximation used by Sogami et al. [17, 21], who obtained the pair potential $U(r)$ in Eq. (12.1) by taking the difference $F(r) - F(\infty)$.

It turns out that the expressions for $U^{el}(r)$, $U_i^o(r)$, and $U_o^o(r)$ obtained from the counterion dominance approximation are very similar to those obtained from the exact solution allowing for the small, but finite, number of coions in R_i. A major advantage of the counterion dominance approximation is that it allows the adiabatic pair potential to be expressed in terms of elementary functions. The field energy term, $U^{el}(r)$, which is missing from DLVO theory, is expressed within the counterion dominance approximation in the following simple form

$$U^{el}(r) = 2|Z_i|k_BT[-1 + u^2/q + \ln(q + u^2/q)] \tag{12.7}$$

where Z_i is the charge on the *inner* surfaces of the plates and the dimensionless quantities u and q are defined by the equations

$$q = \pi\lambda_B|Z_i|r \tag{12.8}$$
$$a_d = |Z_i|/n_\infty l \tag{12.9}$$
$$u = [(q/a_d)\exp(-e\psi_0/k_BT)]^{1/2} \tag{12.10}$$

where λ_B is the Bjerrum length, $r = 2l$ is the interplate separation, n_∞ is the number density of simple ions far from the plates, and ψ_0 is the electric potential at the midplane between the plates. The parameters q and u are related by

$$q = u \tan u \tag{12.11}$$

and the relation

$$(\kappa^2 r/4u) \cos u = -2\pi \lambda_B |\sigma| + [(2\pi \lambda_B \sigma)^2 + \kappa^2]^{1/2} \tag{12.12}$$

together with Eqs. (12.8) and (12.11) determines the valency on the inner surface Z_i as a function of the valency of an isolated plate σ and the interplate distance r.

Both the counterion dominance solution of the one-dimensional colloid problem [21] and the exact mean-field theory solution [17] have been criticized as incorrect. Levine and Hall [22] and Ettelaie [23] criticized the counterion dominance solution, and Overbeek [18] criticized the exact solution. These criticisms have been refuted [24, 25], and the refutations are summarized in the following.

12.3.1 The Levine-Hall Criticism

Equations (12.7)–(12.12) enable the Levine and Hall criticism [22] to be quickly refuted, because it is based on a mathematical error in taking the r derivative of $U^{el}(r)$. Equations (12.8), (12.11), and (12.12) lead to the following relations for the r derivatives

$$\frac{\partial u}{\partial r} = \frac{1}{r} \frac{u}{1+q} \tag{12.13}$$

$$\frac{\partial q}{\partial r} = \frac{1}{r} \frac{q + q^2 + u^2}{1+q} \tag{12.14}$$

$$\frac{\partial |Z_i|}{\partial r} = \frac{|Z_i|}{r} \frac{u^2}{q + q^2} \tag{12.15}$$

Since U^{el} is a function of u, q, and $|Z_i|$, the r derivative of the field energy term was calculated as [21]

$$\frac{\partial U^{el}}{\partial r} = \frac{\partial U^{el}}{\partial u} \frac{\partial u}{\partial r} + \frac{\partial U^{el}}{\partial q} \frac{\partial q}{\partial r} + \frac{\partial U^{el}}{\partial |Z_i|} \frac{\partial |Z_i|}{\partial r} \tag{12.16}$$

from which

$$\frac{\partial U^{el}}{\partial r} = \frac{2k_B T |Z_i|}{qr} \left(\frac{u^2}{1+q} \ln \frac{2u}{\sin 2u} + u(\tan u - u) \right) \tag{12.17}$$

follows readily.

Levine and Hall used two alternative derivations of the electrical free energy, and the form of $U^{el}(r)$ is not in doubt. However, they failed to notice that $|Z_i|$ is the valency on the inner surface of the plates and that it varies depending on the interplate separation r. They form the derivative of U^{el} with respect to r, holding $|Z_i|$ constant, and compound this error with a further mathematical error. Repeating their equation (37) as

$$\frac{\partial U^{el}}{\partial r} = 2k_B T |Z_i| \frac{(\tan u - u)(\tan u + u \sec^2 u)}{u \tan^2 u} \left(\frac{\partial u}{\partial r}\right)_{Z_i = const} \quad (12.18)$$

it is straightforward to prove that this equation is indeed the same as that obtained by taking the first two terms of Eq. (12.16). It can be written as

$$\frac{\partial U^{el}}{\partial u}\frac{\partial u}{\partial r} + \frac{\partial U^{el}}{\partial q}\frac{\partial q}{\partial r} = \frac{2k_B T |Z_i|}{qr}\left(\frac{(q + q^2 + u^2)(q - u^2)}{q(1+q)}\right) \quad (12.19)$$

However, Eq. (12.19) [or (12.18)] does *not* reduce to

$$\frac{\partial U^{el}}{\partial u}\frac{\partial u}{\partial r} + \frac{\partial U^{el}}{\partial q}\frac{\partial q}{\partial r} = \frac{2k_B T |Z_i|}{qr}(q - u^2) \quad (12.20)$$

as claimed by Levine and Hall in their equation (39). This is shown most clearly by writing the third term in Eq. (12.16) as

$$\frac{\partial U^{el}}{\partial |Z_i|}\frac{\partial |Z_i|}{\partial r} = \frac{2k_B T |Z_i|}{qr}\left(\frac{(-u^2)(q - u^2)}{q(1+q)} + \frac{u^2}{1+q}\ln\frac{2u}{\sin 2u}\right) \quad (12.21)$$

Of course, the sum of Eqs. (12.19) and (12.21) is Eq. (12.17), the correct form for the r derivative. Equation (12.20) is false. It is neither the correct form for the r derivative, nor the form that would be obtained from the incorrect procedure of taking the r derivative with $|Z_i|$ held constant. The pleasing cancellation of terms that reduces the term in large parentheses in Eq. (12.19) and the first term in large parentheses in Eq. (12.21) to the simple form $(q - u^2)$ only occurs when the necessary dependence of $|Z_i|$ on r is taken into account. It does not occur as a result of reducing Eq. (12.18) as claimed by Levine and Hall. As "equation" (12.20) is the sole basis for their claim [22] that the attractive branch in $U(r)$ disappears, their criticism of the counterion dominance solution [21] is quite meaningless.

12.3.2 The Ettelaie Criticism

The Ettelaie criticism [23] is more interesting. It is based on the idea that, although their individual expressions for $U^{el}(r)$, $U_i^o(r)$, and $U_o^o(r)$ are correct, Sogami et al. [21] "do not correctly deal with a simple but important term in the Helmholtz free energy." Ettelaie works under the assumption that the adsorption of an ion from the solution onto the solid surface involves a change in the environment of that ion. He refers to μ_g as "the chemical potential of an ion in the solution" (the subscript g indicates a gaslike phase), μ_s as the "chemical potential of the same ion on the surface" (the subscript s indicates a solidlike phase) and ψ_s as the "electric potential difference between the surface and the solution." He writes

$\Delta \mu_c = \mu_s - \mu_g$ and states that "When equilibrium between the plate and the solution is finally established, the chemical potential difference between the two phases becomes zero, leading to the important relation

$$\psi_s = \Delta \mu_c / e \qquad (12.22)$$

He refers to Eq. (12.22) (equation (2) in his paper) as the Nernst equation and states his approach explicitly. "The only important point is that the Nernst equation (2) should correctly be obeyed, if one is considering a constant surface potential problem." His criticism is clearly based on the idea that "there exists a chemical potential difference $\Delta \mu_c$ between the two phases."

It is a misinterpretation of the concept of a phase to consider these two possible sites for a counterion as separate phases. This confuses the macroscopic property of a well-defined phase (a solid) in equilibrium with another well-defined phase (a homogeneous electrolyte solution) with an internal property of an inhomogeneous single phase, the macroionic (or colloidal) phase. It makes a division of the macroionic phase into two regions, which has no physical counterpart. Even if we do make this arbitrary division, in the manner of Ettelaie, his method is still wrong.

First, the two "phases" considered by Ettelaie do not have the same chemical composition. One phase (the surface adsorbed phase) contains the macroion, and the other phase (the solution) does not contain the macroion. The condition for thermal equilibrium in these circumstances has been clearly stated by Guggenheim [26]: "For the distribution of the ionic species i between two phases α, β of different chemical composition the equilibrium condition is equality of the electrochemical potential μ_i, that is to say

$$\mu_i^\alpha = \mu_i^\beta \qquad (12.23)$$

Any splitting of $\mu_i^\beta - \mu_i^\alpha$ into a chemical part and an electrical part is in general arbitrary and without physical significance."

There is therefore no basis for Ettelaie's version of this equation, expressing the (electro)chemical potential difference (the brackets are used because Ettelaie does not make any distinction between these two quantities) between the two phases as $\mu_s = \mu_g + e\psi_s$. The mathematical equation corresponding to the statement that "the chemical potential difference between the two phases becomes zero" is in fact

$$\Delta \mu_c = 0 \qquad (12.24)$$

and *not* Eq. (12.22). The statement corresponding to Eq. (12.22) is that the chemical potential difference between the two phases becomes equal to a finite potential energy term. In Ettelaie's two-state model for the simple ion system the boundary between the "solidlike" and "gaslike" regions is a plane where the chemical potential gradient is infinite due to a discontinuous change in the μ_i.

In the SSS approach [17, 21], the simple ion system is treated as a one-state system, and $n_+(x)$ and $n_-(x)$ are obtained from the Boltzmann distribution

$$n_i(x) = n_\infty \exp[-z_i e\psi(x)/k_BT] \tag{12.25}$$

where n_∞ is the number density far from the plates and z_i is the valency. *Equation (12.25) expresses the condition of thermal equilibrium over all regions.* It does not matter whether a counterion is adsorbed onto the surface of the plates, whether it is in the double layers trapped between the plates, whether it is close to the external surfaces of the plates, or whether it is in the infinite reservoir of electrolyte solution far from the plates: If the system is at thermal equilibrium, the electrochemical potential of the simple ions is everywhere constant.

It is easy to prove that the SSS system is at thermal equilibrium by taking logarithms of the Boltzmann equation (12.25) and rearranging as follows:

$$k_BT \ln n_i(x) + e\psi(x) = k_BT \ln n_\infty \tag{12.26}$$

The left-hand side of Eq. (12.26) is the electrochemical potential of the ion i and can be written as $\mu_i(x)$. In this form

$$\mu_i(x) = k_BT \ln n_\infty \tag{12.27}$$

Equation (12.27) shows that the electrochemical potential of the ion is constant (x-independent and r-independent) and equal to its value at an infinite distance from the plates, as it must be at thermal equilibrium in the infinite dilution limit.

To prove that the system treated by Ettelaie is not at thermal equilibrium, consider the consequences of adding on the $e\psi_s$ term to Eq. (12.27). Since the "gas" of small ions has been treated by the Poisson-Boltzmann equation, Eq. (12.27) can be written as

$$\mu_g = k_BT \ln n_\infty \tag{12.28}$$

There is nothing exceptional in this equation: since the electrochemical potential is constant everywhere within the "gaslike phase," $\mu(x)$ can be replaced by the constant μ_g. Inserting the Nernst equation (12.22) to obtain the electrochemical potential in the "solidlike" phase

$$\mu_s = k_BT \ln n_\infty + e\psi_s = k_BT \ln \left[n_\infty \exp \left(\frac{e\psi_s}{k_BT} \right) \right] \tag{12.29}$$

now begs the question: "What is the surface-adsorbed phase in equilibrium with?" The immediately apparent answer is that the surface-adsorbed phase is in equilibrium with a simple ion gas of number density $n_\infty \exp(e\psi_s/k_BT)$. In other words, it is not in equilibrium with the experimentally measurable and controllable number density of ions n_∞: The application of the Nernst equation (12.22) leads to the system not being in thermal equilibrium.

12.3.3 The Overbeek Criticism

In [18], Overbeek verifies the mean potential function, the ion distributions, and the analytic forms of $U^{el}(r)$, $U_i^o(r)$, and $U_o^o(r)$, which were derived in [17], but the main point of his paper is to try to prove the conclusions of SSS theory wrong. In effect, Overbeek introduces a three-state model for the simple ion system, as follows.

1. Ions in the outer region, far from the plates, where $\psi = 0$ and therefore $n_+ = n_- = n_\infty$. This region is labeled R_∞, as an abbreviation for R_o, $r \to \infty$.
2. Ions in the inner region, in the diffuse double layers, where ψ is of the order of tens of millivolts and therefore $n_+ \gg n_-$. This region is labeled R_i, as before.
3. Ions adsorbed on to the surface of the macroions in the inner region, where $\psi = \psi_s$ is so large that $n_+ \to \infty$, $n_- \to 0$. This region is labeled R_s.

Overbeek notices that when the plates are brought closer together, ions leave the region R_i by two processes.

1. Counterions adsorb onto the surfaces of the plates, accounting for a decrease in $|Z_i|$. This is represented by $R_i \to R_s$.
2. Pairs of counterions and coions migrate from the region between the plates into the external solution. This is represented by $R_i \to R_\infty$.

However, Overbeek fails to notice that the type (1) and type (2) processes are *consequences* of maintaining thermal equilibrium throughout the system, and he appends *extra* terms to the electrochemical potential of the simple ion system to allow for them. For each ion adsorbed onto the surface ($R_i \to R_s$) he adds $e\psi_s$ to the Helmholtz free energy of the system and for each ion that migrates into the external solution ($R_i \to R_\infty$) he adds $k_B T \ln n_\infty$ to the Helmholtz free energy of the system. He calls these terms "corrections" to Eq. (12.1).

The type (1) "correction" is the same term as that introduced by Ettelaie and has been treated above. The type (2) "correction" is described in the abstract of Ref. [18] as follows : " . . . in the calculation of the free energy of the system, the ions are considered to make a nonelectric contribution to the free energy of the order of $-20k_B T$ per ion in a 0.001 M solution. Unfortunately, the authors count fewer ions when the plates are close together, than when they are far apart . . . ." It is clear that 0.05 g of salt in 1000 g of water (in the text Overbeek considers the example of a 0.001 molar solution of a salt of molecular weight approximately equal to 50 g) represents a mass fraction of 5×10^{-5} and that the natural logarithm of this number is -9.9, which is approximately -10. It is clear that simple salts dissolve in water as ion pairs and that the $k_B T \ln(5 \times 10^{-5}) = -10 k_B T$ term should be doubled to $-20k_B T$ to represent the ideal entropy of mixing of salt "molecules" and water *relative to the standard states of the solid salt and pure solvent*. If a mole of salt is added to a cubic meter of water this is surely an important term to consider in the dissolution process.

This, however, is not the case at issue. In the SSS model, and likewise in DLVO theory, two parallel plates are immersed in an electrolyte solution, for example, a 0.001 M solution, and the simple ions redistribute themselves around the macroionic plates according to the surface potential and separation of the plates, relative to their homogeneous distribution as a simple electrolyte solution. In other words, the 0.001 M solution is the standard state in this case. In the type (2) process it is obvious that no ions are really lost from the system, but are simply taken from the inner region to the infinite amount of solution far from the double layers. Adding an *extra* free energy term of $-20k_B T$ per migrating ion pair does not make sense. Only if they were removed as solid salt would we have to make allowance for the $-20k_B T$ contribution, but this addition to the theoretical model treated in Refs.[17, 21], and likewise [14, 15], is against thermodynamics. The differences in the two approaches can be summarized as follows:

- Sogami et al. [17]: $\mu_i(R_i) - \mu_i(R_s) = 0$, expressing the necessary condition for thermodynamic equilibrium between the surface-adsorbed layer and the diffuse double layer, and $\mu_i(R_i) - \mu_i(R_\infty) = 0$, expressing the necessary condition for thermodynamic equilibrium between the inner and outer regions.
- Overbeek [18]: $\mu_i(R_i) - \mu_i(R_s) = |e\psi_s|$, which destroys thermodynamic equilibrium between the surface-adsorbed layer and the diffuse double layer and $\mu_i(R_i) - \mu_i(R_\infty) = k_B T \ln n_i$, which destroys thermodynamic equilibrium between the inner and outer regions.

12.4 The Gibbs Free Energy Formalism

Reference [18] is the second attempt Overbeek has made to try to prove the Sogami theory wrong, the first [27] being leveled at the original formulation using the linearized Poisson-Boltzmann equation and the Gibbs free energy formalism. Overbeek [27] suggested that the Gibbs free energy expression employed by Sogami and Ise [4] was incomplete because the electrical contribution of the solvent was omitted. The equation he used to justify this was

$$N_i \mu_i + N_{solv} \mu_{solv} = 0 \qquad (12.30)$$

which states an independent relation between the chemical potentials of the solvent molecules and the small ions in a region that also contains the macroions. Ise et al. [28] first pointed out that Overbeek violated the Gibbs-Duhem equation by omitting the contribution of the macroions in Eq. (12.30), the correct Gibbs-Duhem equation being

$$N_i \mu_i + N_{solv} \mu_{solv} + N_{part} \mu_{part} = 0 \qquad (12.31)$$

and this rebuttal was amplified [16] by showing that Eq. (12.30) leads to the inescapable and implausible conclusion that "there is *no* free energy associated with the electrical double layers." As pointed out by Schmitz [29] in an independent review of the debate: "Since this conclusion is clearly wrong, one must seek a different argument from inclusion of the solvent to invalidate the theory proposed by Sogami and Ise." Woodward [30] did attempt some different arguments, whose shortcomings have also been pointed out by Schmitz [29]. Of greater interest are predictions of the Sogami theory for the behavior of plate macroions that have been obtained from the Gibbs free energy formalism.

The crucial feature of the Sogami potential for the interaction of plate macroions [16] is that it has a minimum at the position r_{min} defined by

$$r_{min} = \frac{1}{\kappa}\left(4 + 2a\kappa\frac{\sinh(2a\kappa)}{1 + \cosh(2a\kappa)}\right) \tag{12.32}$$

where $2a$ is the plate thickness. For monovalent ions in water at 25°C the inverse Debye length κ is defined by

$$\kappa^2 = 0.107c \tag{12.33}$$

with κ expressed in units of Å^{-1} and the concentration of the electrolyte solution c expressed in moles/liter. In [16], theory was connected to the experimental results on the n-butylammonium vermiculite gels by assuming that r_{min} could be related directly to the observed c axis d spacing and that c was given directly by the concentration of the soaking solution. With $2a = 19.4$ Å, the observed lattice constant for the n-butylammonium vermiculite crystals [11], the two terms in Eq. (12.32) are as given in the second and third columns of Table 12.1, where the second term has been labeled $f(a, \kappa)$. Column 4 of Table 12.1 gives the predictions for r_{min} as a function of c and the fifth column expresses this as the appropriate number of Debye screening lengths N_κ. This shows clearly that in dilute ($c < 0.01$ M) electrolyte solutions the position of the minimum in the pair potential between macroionic plates is given by the remarkably simple relation

$$\kappa r_{min} = 4 \tag{12.34}$$

This localizes the plates at a distance of four Debye screening lengths and so provides a qualitative explanation for Fact 3.

Table 12.1 The Sogami theory prediction for the plate separation, r_{min}, as a function of the electrolyte concentration c. The second column gives $4/\kappa$, the leading term in the prediction, and the third column gives $f(a, \kappa)$, the second term on the right-hand side in Eq. (12.32). The fifth column gives the number of Debye screening lengths N_κ at the secondary minimum

c (M)	$4/\kappa$ (Å)	$f(a, \kappa)$ (Å)	r_{min} (Å)	N_κ
0.001	400	2	402	4.0
0.003	231	3	234	4.1
0.01	126	6	132	4.2
0.03	73	9	82	4.5
0.1	40	15	55	5.5

The interaction between the charged plates is mediated by the electrolyte concentration between the plates and the approximation $c_{gel} = c_{ex}$, namely, that the average electrolyte concentration in the gel phase is equal to the electrolyte concentration in the supernatant fluid, is no longer tenable in view of the calculations in [17]. The Gibbs free energy formalism has recently been re-examined in this light and used to calculate the electrolyte distribution between the two phases in the two-phase region of colloid stability [31]. The calculations take Fact 4 as given (vide infra), namely, that the macroionic plates maintain a constant surface potential irrespective of the concentration of the soaking solution. The salt distribution between the gel phase and the supernatant fluid was calculated by describing the ion exclusion from the gel phase in the manner of Klaarenbeek [32]. The distribution of ions was described as the expulsion of a certain number of ions of the same sign as the colloid. For a single flat double layer on a negatively charged wall, the ratio g_r of the deficit of negative ions to the total double-layer charge is given by the integral [32]

$$g_r = \frac{\{-\int_0^{\psi_s} \left[\exp\left(\frac{e\psi}{k_B T}\right) - 1\right] \frac{dr}{d\psi}\, d\psi\}}{\{\int_0^{\psi_s} \left[\exp\left(-\frac{e\psi}{k_B T}\right) - \exp\left(\frac{e\psi}{k_B T}\right)\right] \frac{dr}{d\psi}\, d\psi\}} \qquad (12.35)$$

which is solved by

$$g_r = -\frac{\exp(e\psi_s/2k_B T) - 1}{\exp(-e\psi_s/2k_B T) - \exp(e\psi_s/2k_B T)} \qquad (12.36)$$

where ψ_s is the surface potential. For $|\psi_s| = 70$ mV, $g_r = 0.20$ for a single flat double layer, and it was shown [31] that when allowance is made for the finite value of the midplane potential [see Fig. 12.1(c)], $g_r = 0.17$ for this surface potential in the two-plate problem.

Noting that the relationship between the (net) surface charge σ and the (effective) surface potential ψ_s is given by [17] (this relation is common to Sogami theory and DLVO theory)

$$\sigma = (\kappa/2\pi\lambda_B) \sinh(e\psi_s/2k_B T) \qquad (12.37)$$

the calculation of the salt fractionation effect is extremely straightforward. If N_{de} is the total number of coions per unit area expelled from the half-region between the midplane and the surface, this deficit of negative ions is given by

$$N_{de} = g_r \sigma \qquad (12.38)$$

and the average number density of the deficit n_{de} is given by

$$n_{de} = g_r \sigma / l \qquad (12.39)$$

where l is the half-separation of the plates. Since the surface charge is proportional to κ [Eq. (12.37)] and l is inversely proportional to κ [Eq. (12.34)], n_{de} is proportional to κ^2. The term on the left-hand side of Eq. (12.39) can be converted into a deficit molarity c_{de} using

$$n_{de} = 6.02 \times 10^{-4} c_{de} \tag{12.40}$$

where n is expressed in Å^{-3} and c in moles/liter, and the right-hand side can be converted into a molarity using Eq. (12.33). This yields

$$c_{de} = 0.64c \tag{12.41}$$

for ψ_s equal to 70 mV, and generally that the deficit molarity is proportional to the molarity of the salt solution. The simple linear relation (12.41) leaves a certain ambiguity in the definition of c, the salt concentration. However, in the limit of low sol concentration $c = c_{ex}$, the salt concentration in the supernatant fluid. Since

$$c_{gel} = c_{ex} - c_{de} \tag{12.42}$$

the ratio of the salt concentrations in the supernatant fluid and the gel phase is seen to be constant (for a given constant surface potential). This is expressed by

$$c_{ex}/c_{gel} = s \tag{12.43}$$

where the salt fractionation factor has been labeled by s, which is consistent with the notation used in [13]. The results for the Sogami theory prediction for s are given in Table 12.2. An interesting feature of Table 12.2 is that s has a turning point as a function of ψ_s. This may be an artefact of using the nonlinear relation of the surface potential to the surface charge, Eq. (12.37), in an otherwise fully linearized theory, but it may also be a genuine effect. It is noteworthy that for the n-butylammonium vermiculite system s lies close to the maximum possible ratio. For $\psi_s = 70$ mV, $s = 2.8$, which renders it straightforward to measure experimentally [13]. The experimental results on the n-butylammonium vermiculite gels are in good agreement with this prediction; see Fact 5. The ratio is approximately constant across the range of salt concentations between 0.001 and 0.1 M, its average value being equal to 2.6 ± 0.4 [13]. The accuracy

Table 12.2 Some illustrative values of the salt fractionation factor s as a function of the surface potential (ψ_s)

ψ_s (mV)	s
0	1.0
25	1.6
50	2.3
70	2.8
75	2.9
100	3.1
200	2.3

of this prediction is strong evidence for the Sogami theory. This prediction also led to a quantitatively accurate prediction of the observed d values as a function of electrolyte concentration c in the n-butylammonium vermiculite system, as shown in the following.

The Sogami theory result $r_{min} = 4/\kappa$ only yields an experimentally testable prediction of r_{min} versus c if the relation between κ and c is known. In [16] it was assumed that

$$\kappa_{gel} = \kappa_{ex} = \kappa_b \tag{12.44}$$

where the symbols represent the inverse Debye screening lengths in the gel, supernatant fluid, and globally, respectively. This gives an unambiguous experimental definition of κ because the supernatant fluid is for all practical purposes a simple electrolyte solution. Its experimentally controlled, known concentration then determines κ_{ex} via Eq. (12.33). However, in view of the relation (12.43), Eq. (12.44) is no longer tenable.

The simplest possible procedure was adopted [31], that of assuming that neither the macroions nor the counterions contribute to the screening length. This approximation is discussed in [29] and [31]. Since κ is then determined solely by the electrolyte concentration, it was logical to choose

$$\kappa_{gel}^2 = 0.107c_{gel} \tag{12.45}$$

in accordance with Eq. (12.33). An immediate benefit of this choice is that it gives a new prediction for the d spacings, l_c, as a function of c in the limit of low sol concentrations ($\phi < 0.01$). In this limit $c_b = c_{ex}$ because the salt that is excluded from the gel phase has a negligible effect in the large volume that remains as the supernatant fluid. For the n-butylammonium vermiculite gels, $c_b = c_{ex} = 2.8c_{gel}$. Together with Eqs. (12.33) and (12.45), this implies $\kappa_{ex} = 1.7\kappa_{gel}$ and substitution of κ_{gel} into Eq. (12.34) gives the new prediction: $\kappa_{ex}r_{min} = 6.7$. This gives *quantitative* agreement with the observed interlayer spacings in the low-ϕ, low-c limit. For example, for $\phi < 0.01$, $c_{ex} = 0.001$ M ($\kappa_{ex} = 0.0103$ Å^{-1}) the prediction is $r_{min} = l_c = 650$ Å, and the observations lie between 550 and 680 Å [11, 12]. Note that equating r_{min} with l_c, the d spacing in the gel phase, is consistent with the approach used in [16] and the preceding section.

When the salt fractionation effect is taken into account, the Sogami theory predicts that the secondary minimum between plate macroions should lie at approximately 7 Debye screening lengths, if the latter is calculated from the added electrolyte concentration. It has recently been shown [33] that the Sogami minimum between spherical macroions also lies at approximately 7 Debye screening lengths. This is a very interesting result. To quote Hunter [34], summarizing a vast body of colloid science data, "The secondary minimum that occurs at larger distances ($\approx 7\kappa^{-1}$) is responsible for a number of important effects in colloidal suspensions . . ." By contrast, the DLVO theory, which balances a κ-independent attractive force (van der Waals force) with a repulsive force that decays exponentially with κ (double layer force) has no explanation to offer for this global fact.

12.5 The Stress Field

One global fact for which DLVO theory does offer a superficially plausible explanation is that the force between charged colloidal particles decays approximately exponentially with their separation. This was established by the famous mica force balance experiments in the late 1970s [35] and early 1980s [36-38] and had been observed in clay swelling as early as 1963 [9]. The n-butylammonium vermiculite system is no exception [12, 39]: see Fact 4. Note that the radius of curvature of the mica cylinders used in the surface force apparatus [35] is so large that the Derjaguin approximation [40] can be used to map the interaction onto that of flat plates, so that the following applies to both cases, except that the salt fractionation effect has not been solved in the crossed cylinder geometry.

The interplate spacing r is given by Eq. (12.46) as a function of the applied uniaxial stress p_u

$$\ln p_u \propto -\kappa r \tag{12.46}$$

or $\ln(dU/dr) \propto -\kappa r$, where U is the relevant thermodynamic function. This apparently means $(dU/dr) \propto e^{-\kappa r}$, so that $U \propto e^{-\kappa r}$, which is the DLVO potential, given by Eq. (12.2).

In the Sogami theory, the full Gibbs pair potential between parallel charged plates in an electrolyte solution [16] is expressed by

$$U_{mn}^G = \frac{2\pi e^2}{\epsilon} Z_m Z_n \frac{\exp(-\kappa r_{mn})}{\kappa} \\ \left\{ [1 + \cosh(2a\kappa)](3 - \kappa r_{mn}) + 2a \sinh(2a\kappa) \right\} \tag{12.47}$$

where Z_m and Z_n are the charges on the mth and nth macroion and r_{mn} is their separation. In the dilute electrolyte limit, for $a\kappa \ll 1$, this function can be expressed approximately as

$$U(r) = Be^{-\kappa r} \left(3 - \kappa_{gel} r \right) \tag{12.48}$$

where the thermodynamic potential has been abbreviated by $U(r)$ and B is an electrostatic constant. Differentiating this function with respect to r gives the following approximate expression for the force $F_p(r)$ acting between the plates

$$F_p(r) = -dU(r)/-dr \propto e^{-\kappa r}(4 - \kappa_{gel} r) \tag{12.49}$$

Of course, $F_p(r) = 0$ at $\kappa_{gel}r = 4$, the equilibrium separation of the plates expressed by Eq. (12.34). The natural way to adapt this function to allow for the
salt fractionation effect is to write

$$F_p(r) \propto e^{-\kappa r}(7 - \kappa_{ex}r) \tag{12.50}$$

fixing the equilibrium separation of the plates at its observed value of 7 Debye
screening lengths. Since experimental data are invariably plotted in the form $\ln p_u$
versus r, Eq. (12.50) is plotted in logarithmic form in Fig. 12.3.

The Sogami prediction is given by the curvilinear line in Fig. 12.3. "Error bars"
have been added to it at intervals of half Debye screening lengths to represent
the uncertainty inherent in any experimental determination of the force between
colloidal particles. A straight line has been drawn through these simulated "experimental" points. It departs noticeably from the true curvilinear plot at separations
greater than 6 Debye screening lengths. This shows that in order to make a direct
measurement of the attractive tail in the Sogami potential, it will be necessary to
collect accurate force-distance data at separations greater than 6 Debye screening
lengths, which corresponds approximately to 2000 Å in a 10^{-4} M solution, 600
Å in a 10^{-3} M solution, and 200 Å in a 10^{-2} M solution. At closer separations

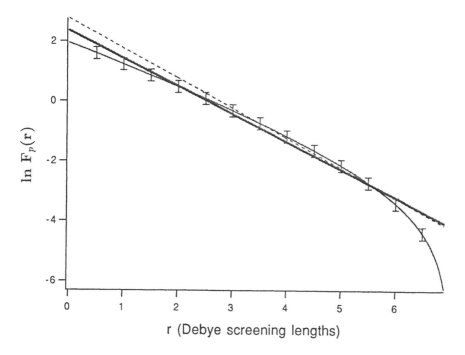

Figure 12.3   Plots of the logarithm of the uniaxial stress against the interplate separation.
The curved line is the Sogami prediction. The "error bars" marked on it simulate the errors
in a typical experiment. The solid straight line is the apparently linear Sogami prediction.
The dotted line is the DLVO prediction.

the Sogami potential predicts an exponential decay of the interparticle force, to the limit of experimental error, in just the same way as the DLVO potential. The mica force balance experiments and clay squeezing experiments to date have probed only the medium-range strong repulsion between the plates; obviously the DLVO theory would not have survived so long if this osmotic repulsion were not an important feature of the interaction.

The DLVO potential has been represented by the dotted line in Fig. 12.3, adjusted to reproduce the same force at half the equilibrium separation (of course, with a purely repulsive potential there is no equilibrium separation). This shows that the gradients of the effectively linear Sogami plot and the actually linear DLVO plot are different, and so give rise to a different interpretation of κ. They also have different intercepts and so give rise to a different interpretation of the surface potential from the same data. Unfortunately, there are not as yet any sufficiently accurate independent measurements of κ and ψ_s to make use of these subtle differences between the two potentials. Simply considering the qualitative fact expressed by Eq. (12.46) in isolation (Fact 4) does not distinguish between the two potentials. It is only by taking a global view of the system (Facts 1-5) that the advantage of the Sogami potential becomes apparent.

12.6 Discussion

Within the framework of mean-field theory, the solvent is treated as a uniform dielectric medium, and the mobile ions are treated as point charges. It is therefore a description of the medium-range and long-range interactions between colloidal particles (100 to 1000 Å in the case of colloidal plates, the observed range in the n-butylammonium vermiculite gels). In order to extend this mean-field description so as to describe short-range behavior, it will be necessary to take into account many effects such as dispersion forces, the size of the small ions, and the molecular degrees of freedom of the solvent. The short-range interaction in the parallel-plate problem has been studied using Monte Carlo simulations [41-43], but it is difficult to compare the results with those of Sogami theory because the simulations consider a maximum interplate separation of 60 Å, somewhat below the lower limit of applicability of mean-field theory [31] and somewhat below the range of experimental interest.

The mean-field theory remains the central theory for understanding and predicting colloidal phenomena, and it is therefore essential that the correct mean-field theory framework be established. Sogami et al. strongly stressed [17, 21] that their results are in fundamental disagreement with DLVO theory *within* the mean-field formalism, encapsulated in the Poisson-Boltzmann equation, and qualitatively in agreement with experiment. Experiment is the sole judge of scientific truth and alone decides that the Sogami theory is the true theory of colloid stability, until someone finds a counterexample that invalidates it. Nevertheless, the purely theoretical questions raised by the other school, all of which have attempted to disprove the new theory *within* the mean-field formalism it employs

[18, 23, 24, 27, 30], have to be answered on theoretical grounds, and a substantial proportion of this article has been devoted to summarizing the recent debate.

Fortunately, there is a system of plate macroions that approximates well the theoretical ideal of a stack of evenly spaced, perfectly parallel plates, the n-butylammonium vermiculite system. The reason why the electrical phenomena are seen so clearly in this model system is because the n-butylammonium ion approximates as closely as any ion does to ideal behavior: The enthalpy of solution of simple n-butylammonium salts is nearly equal to zero [44], and their partial molar volumes are nearly independent of concentration [45], corresponding to a density of 1.00 g/cm^3, implying that there are no special ion-solvent effects between n-butylammonium ions and water.

Sogami theory provides a quantitative explanation for Facts 1 and 3 and provides a qualitative explanation for Fact 2 [16, 46], which is a definite counterexample to DLVO theory. Furthermore, if Fact 4 is taken as given, Sogami theory also provides a quantitative explanation of Fact 5. In short, all of the experimental evidence points to the n-butylammonium vermiculite swelling being governed by the Sogami potential with the Dirichlet boundary condition (constant surface potential). There is no possibility that the experimental results presented in the first section could be explained by DLVO theory. Furthermore, it is highly unlikely that both the d spacings and salt fractionation effect in the two-phase region could be accurately predicted from electrical theory if any of the other forces that are commonly introduced into colloid theory, such as hydrophobic forces or van der Waals forces, played any significant role.

The most interesting experimental question which remains is "How general is the Dirichlet boundary condition in colloid science?" The surface charge and surface potential of charged colloidal particles are difficult to measure experimentally; so data are scarce. Apart from the n-butylammonium vermiculite results [12, 13], Low [47] has reported that the surface potential remains constant at about 60 mV as a function of electrolyte concentration in the montmorillonite clays, but this is another "one-dimensional" system. It would be very interesting to devise experiments to test whether or not the Dirichlet boundary condition applied to well-characterized three-dimensional systems, such as the polystyrene sulfonate latex spheres, another experimental system that proves the existence of the long-range attraction [2, 28, 48, 49].

The most interesting theoretical question which remains is "How do we build on the insight given by the exact mean-field theory solution to the one-dimensional colloid problem in treating the three-dimensional case?" Of course, the full nonlinear Poisson-Boltzmann equation has not yet been solved even for the simplest three-dimensional problem, that of interacting spherical particles [29], and in these circumstances appropriate methods of calculating the Helmholtz free energy have to be developed. A possible candidate for such a method has been outlined by Sogami [50]. In this approach extra parameters have to be introduced, namely, the effective surface charge and effective diameter of a dressed macroion. However, the one-dimensional case has been solved without these difficulties.

Acknowledgements

I would like to thank Professors I. S. Sogami and N. Ise for a critical reading of the manuscript. I also thank the ERATO Project of the Japan Research Development Corporation (JRDC) for their support.

References

1. Feynman, R. P., *The Feynman Lectures on Physics* (Addison-Wesley, Menlo Park, California, 1963), Vol. 1, p. 1.
2. Dosho, S., Ise, N., Ito, K., Iwai, S., Kitano, H., Matsuoka, H., Nakamura, H., Okumura, H., Ono, T., Sogami, I. S., Ueno, Y., Yoshida, H., and Yoshiyama, T., *Langmuir* **9**, 394 (1993).
3. Sogami, I., *Phys. Lett.* **96A**, 199 (1983).
4. Sogami, I., and Ise, N., *J. Chem. Phys.* **81**, 6320 (1984).
5. Lu, J. R., Simister, E. A., Thomas, R. K., and Penfold, J. J., *J. Phys. Chem.* **97**, 13907 (1993).
6. Ito, K., Muramoto, T., and Kitano, H., *J. Am. Chem. Soc.* **117**, 5005 (1995).
7. Walker, G. F., *Nature* **187**, 312 (1960).
8. Garrett, W. G., and Walker, G. F., *Clays Clay Min.* **9**, 557 (1962).
9. Norrish, K., and Rausel-Colom, J. A., *Clays Clay Min.* **10**, 123 (1963)
10. Smalley, M. V., Thomas, R. K., Braganza, L. F., and Matsuo, T., *Clays Clay Min.* **37**, 474 (1989).
11. Braganza, L. F., Crawford, R. J., Smalley, M. V., and Thomas, R. K., *Clays Clay Min.* **38**, 90 (1990).
12. Crawford, R. J., Smalley, M. V., and Thomas, R. K., *Adv. Colloid Interface Sci.* **34**, 537 (1991).
13. Williams, G. D., Moody, K. R., Smalley, M. V., and King, S. M., *Clays Clay Min.* **42**, 614 (1994).
14. Derjaguin, B. V., and Landau, L., *Acta Physicochim. URSS* **14**, 633 (1941).
15. Verwey, E. J. W., and Overbeek, J. Th. G., *Theory of the Stability of Lyophobic Colloids* (Elsevier, Amsterdam, 1948).
16. Smalley, M. V., *Molec. Phys.* **71**, 1251 (1990).
17. Sogami, I. S., Shinohara, T., and Smalley, M. V., *Molec. Phys.* **76**, 1 (1992).
18. Overbeek, J. Th. G., *Molec. Phys.* **80**, 685 (1993).
19. Langmuir, I., *J. Chem. Phys*, **6**, 873 (1938).
20. Verwey, E. J. W., and Overbeek, J. Th. G., *Discuss. Faraday Soc.* **B42**, 117 (1946).
21. Sogami, I. S., Shinohara, T., and Smalley, M. V., *Molec. Phys.* **74**, 599 (1991).
22. Levine, S., and Hall, D. G., *Langmuir* **8**, 1090 (1992).
23. Ettelaie, R., *Langmuir* **9**, 1888 (1993).
24. Smalley, M. V., *Langmuir* **11**, 1813 (1995).
25. Smalley, M. V., and Sogami, I. S., *Molec. Phys.* **85**, 869 (1995).
26. Guggenheim, E. A., *Thermodynamics* (North Holland, Amsterdam, 1957).
27. Overbeek, J. Th. G., *J. Chem. Phys.* **87**, 4406 (1987).
28. Ise, N., Matsuoka, H., Ito, K., Yoshida, H., and Yamanaka, J., *Langmuir* **6**, 296 (1990).
29. Schmitz, K. S., *Macroions in Solution and Colloidal Suspension* (VCH, New York, 1993).
30. Woodward, C. E., *J. Chem. Phys.* **89**, 5140 (1988).
31. Smalley, M. V., *Langmuir* **10**, 2884 (1994).
32. Klaarenbeek, F. W., *Ph.D. Thesis* (Utrecht University, 1946).
33. Ise, N., and Smalley. M. V., *Phys. Rev.* **B 50**, 16722 (1994).
34. Hunter, R. J., *Foundations of Colloid Science* (Clarendon Press, Oxford, 1987).
35. Israelachvili, J. N., and Adams, G. E., *J. Chem. Soc. Faraday Trans. 1* **74**, 975 (1978).
36. Pashley, R. M., *J. Colloid Interface Sci.* **80**, 153 (1981).
37. Pashley, R. M., *J. Colloid Interface Sci.* **83**, 531 (1981).
38. Pashley, R. M., and Israelachvili, J. N., *Colloids Surfaces* **2**, 169 (1981).

39. Rausel-Colom, J. A., *Trans. Faraday Soc.* **60**, 190 (1964).
40. Derjaguin, B. V., *Kolloid-Z* **69**, 155 (1934).
41. Valleau, J. P., Ivkov, R., and Torrie, G. M., *J. Chem. Phys.* **95**, 520 (1991).
42. Kjellander, R., Akesson, T., Jonsson, B., and Marcelja, S., *J. Chem. Phys.* **97**, 1424 (1992).
43. Bolhuis, P. G., Akesson, T., and Jonsson, B., *J. Chem. Phys.* **98**, 8096 (1993).
44. Krishnan, C. V., and Friedman, H. L., *J. Phys. Chem.* **74**, 3900 (1979).
45. Desnoyers, J. E., and Arel, M., *Can. J. Chem.* **45**, 359 (1967).
46. Smalley, M. V., *Prog. Colloid Polymer Sci.* **97**, 59 (1994).
47. Low, P. F., *Langmuir* **3**, 18 (1987).
48. Ito, K., Yoshida, H., and Ise, N., *Science* **263**, 66 (1994).
49. Kamenetzky, E. A., Magliocco, L. G., and Panzer, H. P., *Science* **263**, 207 (1994).
50. Sogami, I., in *Ordering and Organisation in Ionic Solutions*, Eds. N. Ise and I. Sogami (World Scientific, NJ, 1988), p. 624.

13

Determination of Interparticle Potentials from Static Structure Factors

Raj Rajagopalan

There is considerable interest in probing the nature of interaction forces in colloidal dispersions, as evident from the discussions presented in earlier chapters of this volume. Numerous attempts have been made in the past to extract information on the effective interparticle interaction forces by fitting experimentally observed static structure factors with structure factors calculated theoretically or by computer simulations. As is well known, such a procedure has serious drawbacks, two of the most important being: (1) the relative insensitivity of the forward theories to the details of the potentials in predicting the structure, and (2) the bias introduced in the procedure by a priori assumptions concerning the nature and the form of the interaction potentials. In view of these, the inverse problem of extracting effective potentials from the structure factor has received some attention in the literature. Solving the inverse problem, however, has its own advantages as well as disadvantages. In this chapter, we present a brief discussion of the inverse problem and follow it with a discussion of a predictor/corrector inversion method, based on separating out the effects of excluded volume due to the repulsive core of the potential from the effects of any attractive "tail" of the potential. The advantages and difficulties of extracting information on the effective potential from structural data are also discussed.

13.1 Introduction

The nature of the interaction forces in colloidal dispersions has received much attention in the literature, as is evident from the emphasis given to this subject in numerous textbooks and monographs dating back to the early work of Verwey and Overbeek [1-4]. Traditionally, colloid stability (particularly in dilute dispersions), electrokinetic phenomena, and particle deposition, adhesion, and resuspension phenomena have been the motivation for this attention to colloidal forces [5]. However, as should be evident from the various contributions to this volume, the nature of the interaction forces in colloids has become a topic of considerable interest in recent years for at least three additional reasons: (1) The structure and properties of colloids at large concentrations (or when the interaction forces are strong) can be predicted and manipulated with the help of theories of liquid-state physics [5-8]; (2) as reviewed in Chapter 1 of this volume [9] and illustrated in others, model colloids can be used as excellent model many-body systems to study phenomena in condensed matter physics that are otherwise inaccessible experimentally at convenient length and time scales [9,10]; and (3) furthermore, as pointed out in Chapter 1 [9], recent experiments using charged colloids have raised some interesting questions about the nature of interaction forces between charged particles in a solution of coions and counterions. Therefore, a systematic investigation of long-range interactions in systems of charged particles is desirable.

These and related issues form the motivation for studying the relation between the microstructure and the interparticle potential in model colloids and for developing reliable methods to extract information about the potential from the observed structure and properties.

13.1.1 Direct Experimental Measurement of Interaction Forces

As a results of the important role played by colloidal forces in a number of applications and in a wide class of interfacial phenomena, numerous experimental innovations have appeared over the years for determining colloidal forces directly or indirectly. Notable among these is the so-called "surface force apparatus," which can measure directly the force of interaction between two coated or uncoated mica surfaces immersed in liquids [4]. Although this technique corresponds to measuring the forces between two flat surfaces (i.e., not between two *particles*), it has contributed to an enhanced understanding of electrostatic and steric forces in colloids as well as to the interpretation of the so-called structural forces [4] and the potentials of mean forces between surfaces in the presence of added polymer or micellar entities [11,12]. It appears that a direct measurement of forces between a colloidal particle and a planar surface may also be possible through the use of total internal reflection microscopy and evanescent wave scattering measurements [13]. Alternatively, evanescent wave scattering from a large number of particles near a surface can also be measured using static light scattering techniques in order to get statistically better estimates for the distribution of distances of the particles from the surface [14]. Although such measurements can also be used to deduce the particle/surface interaction potentials, they require one to assume a particular functional form of pair potential in advance of the calculations and, therefore, cannot be considered a "direct" method.

An important point to note about the above techniques, however, is that they measure forces between *two* objects (in vacuum or in a medium); that is, the measurements correspond to forces between particles at extreme dilution. It is, therefore, important to develop complementary methods that can provide *effective* pair-interaction forces from measurements that are obtained from dense or strongly interacting systems. An important case of the latter is offered by micellar systems (either ionic or nonionic); the micelles, being thermodynamic entities, cease to exist upon dilution. The measurement of effective (or "dressed") charges and interaction forces in such systems thus requires methods that can account for (and "unscramble" the effects of) collective interactions implicit in the measured properties that are used as the basis for inferring the interaction forces.

13.1.2 Obtaining Effective Interaction Forces from Static Structure Factors

In the present chapter, we discuss a procedure for obtaining effective interaction potentials from the static structure factor $S(Q)$, where Q is the magnitude of the scattering vector $\mathbf{Q}$, for fluids or dispersions with "liquidlike" structure. Although we shall consider only spherically symmetric interaction forces, no a priori restrictions concerning the functional form of the interaction potentials will be assumed. This problem falls under a broad category known as *inverse problems* and carries with it a number of attendant issues such as whether the problem is well-posed or ill-posed, the uniqueness and existence of the solution, and the stability of the solution to errors in experimental data. Although many of these issues can be addressed adequately and in a reasonably rigorous manner, we shall restrict our attention here to an outline of the inverse problem and a method of solution.

13.2 Interaction Potential from Structure Factors: Traditional Approaches

Traditionally, static structure factors have been used primarily to gain an understanding of the structure and ordering in the materials under study and to infer the global features of the interactions. In contrast, the inverse problem of extracting effective interaction potentials from observed structure has received very limited attention. The broader implications of the inverse problem, overviews of past attempts to address the inversion of static structural data, and the pros and cons of the available options are presented elsewhere (see, for example, [15]). It is, therefore, sufficient for our purpose here to recognize the following two common approaches that have been popular in the literature:

1. *Inversions based on integral-equation closures.* This class of methods, based on the closure relation for integral-equation theories such as the Percus-Yevick (PY) and the hypernetted-chain (HNC) equations [15], is, in principle, a formal inversion technique (albeit highly approximate and inadequate) and dates back to the pioneering work of March and co-workers [16-18] in their attempts to understand of interactions in rare-gas liquids and liquid metals. The integral-equation closures relate $S(Q)$ (or a related function such as the direct correlation function $c(r)$ [15]) to the pair potential $U(r)$ directly, and one can therefore obtain $U(r)$ in a straightforward manner from the given $S(Q)$. We shall call this approach *single-step* inversion to differentiate it from the iterative, *predictor-corrector* method of the type presented in this chapter. As has already been illustrated [15,19], the quality of the effective potentials obtained using the PY and HNC closures is far from satisfactory. Numerous examples of the failure of classical integral-equation closures for inversion are given by Ailawadi [15].

2. *Fitting the experimental static structure factors to those calculated from assumed potentials.* This approach is, in essence, a trial-and-error procedure but is, nevertheless, the more popular of the two, both in the area of classical liquids as well as in colloid science. It has been used extensively in the study of dense dispersions, ionic and nonionic micellar solutions, and microemulsions. For instance, this approach has been found to be very useful for determining parameters such as the aggregation numbers and the "dressed" charges of micelles and some biomolecules from small-angle neutron scattering data [20,21]. However, this method requires a priori information on the nature and the functional form of the interaction potentials and, therefore, could lead to serious errors since such information is not always available in advance. Any assumption concerning the details of the potential thus biases the results one obtains from the method. An example of such a situation may be seen from the results presented by Tata et al. [22] for a charged colloid. Tata et al. find that their experimental data can be "fitted," with about the same level of accuracy, using either a purely repulsive, monotonically decreasing Yukawa potential or a Sogami potential [23,9], which has a minimum and a long-range attractive tail (see Figs. 13.1 and 13.2). We shall return to this example in Section 5.2 for an explanation for the above conflicting results.

Other than the few comments presented in Section 5, we shall not concern ourselves with the reasons for the successes or failures of the above approaches. Our objective here is to explore an alternative to the above that seeks to avoid the inaccuracies implicit in the simpler integral-equation closures and the uncertainties introduced by a priori assumptions in the fitting methods. The following section outlines the philosophy and the essential details of the method, and this is followed by a few illustrations of its use. These and additional details are available in Refs. [24, 25].

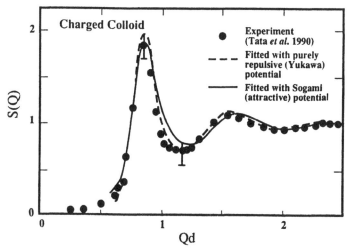

Figure 13.1   Experimental structure factor for a charged dispersion "fitted" with structure factors calculated using two different potentials shown in Fig. 13.2. The calculated $S(Q)$ for the Yukawa potential is based on the rescaled mean-spherical approximation. The $S(q)$ for the Sogami potential is based on Brownian dynamics computer experiments. The particle diameter is denoted by d [22]. [Reprinted with permission from B. V. R., Tata et al., *Pramana-J. Phys.* **34**, 23 (1990) Copyright 1990 The Indian Academy of Sciences]

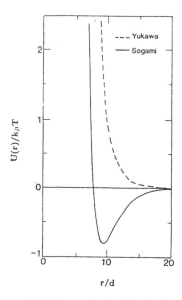

Figure 13.2   The Yukawa and the Sogami potentials for the dispersion whose structure factor is shown in Fig. 13.1. The experimental parameters are given in [22]. [Reprinted with permission from B. V. R., Tata et al., *Pramana-J. Phys.* **34**, 23 (1990) Copyright 1990 The Indian Academy of Sciences]

13.3 Unscrambling the Details of the Interaction Forces from Static Structure Factors

13.3.1 Separation of the Excluded-Volume Effect from the Effects of Attraction

As pointed out elsewhere [25], the key to unscrambling the information on interaction forces contained in the static structure is to identify a procedure that correctly sorts out the individual features of the potential that determine the coarse-grained and the fine-grained structures of the fluid. For this, one can use the so-called van der Waals (vdW) picture of "liquids," which posits that the core of the potential and the effective density of the fluid determine the coarser features of the structure while the rest can be attributed to the tail of the potential. As is well known, this thesis forms the backbone of perturbation theories in statistical thermodynamics and is satisfied by a large class of materials, including the ones of interest here [26,27]. Since excellent descriptions of various forms of perturbation formalisms (for the forward problem) are readily available in standard books [26], we shall restrict ourselves to the essential details of the inversion method in this chapter. Additional details and extensions to cases not presented here are given in Guérin [28].

We begin with the experimental static structure factor $S(Q)$ at known intervals of the scattering vector $\mathbf{Q}$. The radial distribution function $g(r)$ is related to $S(Q)$ through

$$g(r) = \frac{1}{n_p(2\pi)^3} \int dQ \; [S(Q) - 1] \exp(i\mathbf{Q} \cdot \mathbf{r}) \tag{13.1}$$

The magnitude Q of the scattering vector $\mathbf{Q}$ is defined as

$$Q = \frac{4\pi n}{\lambda} \sin\left(\frac{\theta}{2}\right)$$

where θ is the angle at which the intensity of the scattering is measured, λ is the wavelength of the radiation used in the experiment, and n is the refractive index of the carrier fluid in which the particles are suspended (equal to unity in the case of X-ray or neutron scattering).

13.3.2 Repulsive Core of the Interaction Potential

The vdW picture implies that the (as yet unknown) potential $U(r)$ can be written as

$$U(r) = U_o(r) + U_p(r) \tag{13.2}$$

where $U_o(r)$ is a suitable reference potential (usually an appropriate representation of the hard core) and $U_p(r)$ is the perturbation (usually the soft, slowly varying tail). The potential $U_o(r)$ can be replaced, provisionally, by an effective hard-sphere potential $U_d(r)$ of diameter d_n, determined using an appropriate prescription [see Eq. (13.4)], to define a trial potential $U_T(r)$ as

$$U_T(r) = U_d(r) + U_p(r) \tag{13.3}$$

Numerous options exist for the selection of d_n [for a given $U_o(r)$]; however, the following criterion is known to lead to better thermodynamic consistency [29]:

$$\int \frac{\partial y_d(r)}{\partial d} \, (e^{-\beta U_o(r)} - e^{-\beta U_d(r)}) \, d\mathbf{r} = 0 \tag{13.4}$$

where $y_d(r)$ is the so-called *cavity function* [26] corresponding to $U_d(r)$.

13.3.3 Attractive Segment of the Potential

The structure factor corresponding to $U_T(r)$ can be derived from the so-called optimized random-phase approximation and is given by

$$S_T(Q) = \frac{S_d(Q)}{[1 + S_d(Q) n_p \beta U_p(Q)]} \tag{13.5}$$

where n_p is the number density, β^{-1} is the product of the Boltzmann constant k_B and the temperature T, $S_d(Q)$ is the structure factor of a hard-sphere fluid of diameter d_n, and $U_p(Q)$ is the Fourier transform of $U_p(r)$. Associated with this $U_p(r)$ is what is often referred to as a "renormalized potential" $\phi(r)$ that makes the radial distribution function inside the core become zero, as demanded by the excluded-volume criterion. The Fourier transform of the function $\phi(r)$ is related to $S_d(Q)$ and $U_p(Q)$ by

$$\phi(Q) = -\frac{S_d(Q)^2 \beta U_p(Q)}{[1 + S_d(Q) n_p \beta U_p(Q)]} \tag{13.6}$$

The renormalized potential $\phi(r)$ can be used to derive a higher level of approximation for $g(r)$ than implicit in Eq. (13.5)

$$g(r) = y_d(r) \, e^{[\phi(r) - \beta U_o(r)]} \tag{13.7}$$

which takes into account the softness of the core of the potential.

13.3.4 Piecing Together the Full Potential

The inversion of a given structure factor proceeds as follows: One first needs an estimate of the effective hard-sphere diameter d_n that represents the first approximation to the excluded-volume effect of the core of the potential on the structure. This can be obtained from the well-known relation [26] between the isothermal compressibility and the known structure factor at $Q = 0$

$$\frac{1}{S(0)} = \left(\frac{\partial(\beta P_{osm})}{\partial n_p}\right)_T \tag{13.8}$$

where the pressure P_{osm} (the osmotic pressure in the case of colloids) can be substituted from an equation of state such as the Carnahan-Starling equation for hard-sphere fluids. Once d_n is estimated, $S_d(Q)$ can be obtained using the solution of the Percus-Yevick equation [26]. Then Eq. (13.5) and the given structure-factor data [substituted in place of $S_T(Q)$] lead to a first estimate of the perturbative part of the effective potential and subsequently to the renormalized potential through Eq. (13.6). Since $\phi(r)$ and $y_d(r)$ are now known, the "softened" core $U_o(r)$ can then be determined from Eq. (13.7) with the experimental radial distribution function, $g_{exp}(r)$, in place of $g(r)$. The first approximation to the effective potential is thus given by

$$\beta U(r) = \ln\left(\frac{y_d(r)}{g_{exp}(r)}\right) + \phi(r) + \beta U_p(r) \tag{13.9}$$

This first approximation may need further refinements since the effective hard-sphere diameter d_n is estimated originally rather crudely using Eq. (13.8). One can refine the result from Eq. (13.9) through successive iterations by returning to the above scheme with a new hard-sphere diameter obtained [using Eq. (13.4)] for the extracted potential. We call Eq. (13.9) the "predictor," and the refinement of the hard-sphere diameter the "corrector." This predictor-corrector method converges very rapidly (often within a few iterations).

13.4 Test of the Inversion Procedure

We shall begin with some examples of the results obtained using the above method for a few classes of fluids representative of a broad spectrum of interaction potentials found in colloidal systems and simple liquids. For this, we start with some known potentials and use computer simulations to generate $S(Q)$ and radial distribution functions $g(r)$. These are then used in the inversion scheme to investigate how faithfully the features of the known potentials are recovered by the inversion method. Monte Carlo (MC) simulations are used (with the number of particles ranging from about 3000 to 11,000) for generating $g(r)$ and $S(Q)$. The need for highly accurate computer simulations (with appropriate estimates for statistical errors in the simulations, e.g., confidence intervals) even for testing the accuracy of the various closures for the integral equations is illustrated in Salgi et al. [30], and some of the observations and details relevant in the present context may be found there.

13.4.1 Model Potentials

Three types of potentials have been selected here for testing the inversion procedure:

1. The first is a Lennard-Jones (LJ) potential, which is chosen in such a way that it mimics the essential features of a typical colloid potential in an electrostatically stabilized dispersion with a moderate "secondary" minimum. As is well known, the LJ potential has the following form:

$$\beta U_{LJ}(r) = 4\beta\epsilon_{LJ}\left[\left(\frac{d_c}{r}\right)^{12} - \left(\frac{d_c}{r}\right)^{6}\right] \tag{13.10}$$

where ϵ_{LJ} is the minimum in the potential and d_c is the collision diameter (defined as the value of r at which the potential is zero; see Fig. 13.3).

2. The next is an *exponential-6* potential, which has a much softer core than the LJ potential. This potential has the following form:

$$\beta U_{exp-6}(r) = A\left[\exp\left(-\frac{Br}{d_c}\right) - \left(\frac{d_c}{r}\right)^{6}\right] \tag{13.11}$$

where A and B are two parameters that are chosen such that the location and the magnitude of the minimum the exp-6 potential coincide with those of the LJ potential used. The reason for such a choice is discussed in the following.

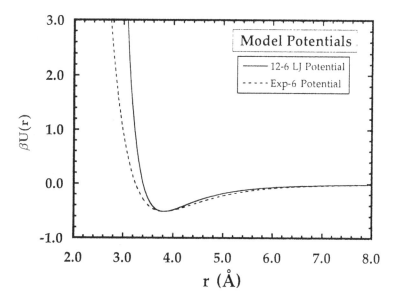

Figure 13.3   The model Lennard-Jones and exponential-6 potentials used for calculating the structure factors shown in Fig. 13.4.

3. Finally, a colloidal potential of the Derjaguin-Landau-Verwey-Overbeek (DLVO) form [3] that includes more complex features typical in colloidal systems is also considered. The model colloid potential used here has a hard core (of unit diameter), a strong London-van der Waals attraction U_A, a weak electrostatic repulsion U_E, and a steep Born-type repulsion near the core, U_B:

$$\beta U(r) = \beta U_A(r) + \beta U_E(r) + \beta U_B(r) \tag{13.12}$$

where r is the center-to-center distance of separation between the particles in units of particle diameter. The expression for the London-van der Waals interaction U_A between two identical spheres is given by [3, 27]:

$$\beta U_A(r) = -N_{Lo} \left[\frac{1}{(2r^2)} + \frac{1}{2(r^2-1)} + \ln\left(\frac{r^2-1}{r^2}\right) \right] \tag{13.13}$$

where $N_{Lo} = \beta A_H/6$. The Hamaker constant A_H is a material property. We represent the repulsive contribution U_E by the Yukawa form, which reproduces the essential features well for particles with constant surface-charge densities and thick electrical double layers [1,3], that is, small inverse screening length, $\kappa a < 2.5$:

$$\beta U_E(r) = N_E \frac{e^{-\kappa a(r-1)}}{r} \tag{13.14}$$

where a is the radius of the particles (equal to 0.5 in our dimensionless units) and the premultiplication factor N_E is related to the contact potential. The other repulsive contribution, U_B, despite its very short range, needs to be taken into consideration for dispersions with moderate or low electrostatic interactions in order to represent the steep repulsion adequately at near-contact distances. An expression of the following form has been used here for the Born repulsion U_B so that the appropriate shape of the pair potential could be generated:

$$\beta U_B(r) = N_B(r-1)^{-12} \tag{13.15}$$

with N_B being a scale factor that we call the Born parameter. The numerical values that have been used for the dimensionless parameters are $N_{Lo} = 1.025$, $N_E = 10.0$, $\kappa a = 3.75$, and $N_B = 2.275 \times 10^{-20}$. The resulting pair potential has both a primary minimum and a secondary minimum, with a moderate energy barrier in between. The attractive force immediately beyond the primary minimum is significant in magnitude. The potential resulting from Eqs. (13.12)-(13.15) is shown in Fig. 13.7, along with the potential inverted from the structure factor.

13.4.2 Sample Results

We begin with the results for the LJ and exp-6 potentials. The two pair potentials are shown in Fig. 13.3. The LJ potential corresponds to $d_c = 3.4\text{Å}$ and $\beta\epsilon_{LJ} =$

−0.51 in Eq. (13.10). As mentioned earlier, the parameters for the exp-6 potential have been chosen in such a way that the magnitude of the minimum in the potentials and the location of the minimum remain the same for both potentials (see Fig. 13.3) so that the tails of both potentials are practically identical and only the cores differ. This allows one to test the relative accuracy of the inversion method with respect to two qualitatively and quantitatively different cores while maintaining the rest of the potential the same. The structure factors for these two potentials are calculated for a range of densities using MC simulations. The MC results for a reduced density of $n_p r_{min}^3 = 0.8$, where r_{min} is the location of the minimum in the potential, are shown in Fig. 13.4 for both potentials. These are based on simulations in which about 11,000 particles are used, with a maximum cutoff distance of half the length of the simulation box. The averages are taken over at least 20 million configurations, after the effects of the initial configuration are eliminated.

The effect of the softness of the core on the long-range configurational structure (i.e., the low-Q range of the structure factor) is strikingly brought out in the figure. One sees a sharp increase in $S(Q)$ for low-Q values for the exp-6-potential, indicating the compressibility of the fluid interacting through the softer potential. In contrast, the dominance of the core of the potential in determining the long-range statistical ordering in the LJ fluid may be seen by the larger value of the first peak in the corresponding $S(Q)$ and the low value of the isothermal compressibility. Thus the structure of the LJ liquid under the given conditions of temperature and density is consistent with the assumptions implicit in the

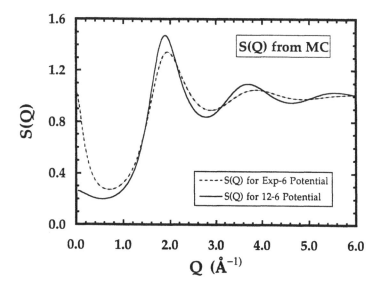

Figure 13.4 The structure factors calculated using Monte Carlo simulations for the two potentials shown in Fig. 13.3.

inversion scheme, and an excellent agreement is seen between the known potential and the inverted potential in this case (see Fig. 13.5). What is noteworthy is that almost a similar level of success is seen in the case of the exp-6 potential as well, as illustrated in Fig. 13.6. The inversion method recovers both the core and the tail of the potential extremely well. It is also worth noting that the calculations converge within about 5 to 10 iterations in each case, the convergence is uniform, and the intermediate values of the potential at any r "bracket" the final value.

Next the results obtained for the DLVO potential are shown in Fig. 13.7. As evident from the figure, the inversion method predicts the near-field interactions (i.e., close to the primary minimum) quite accurately, even with the structure-factor results for the highest volume fraction (0.4) tried here. It can also be seen from Fig. 13.7 that the inverted pair potential is in excellent agreement with the known potential. All the major features of the potential, including the shape, are reproduced very well. As noted earlier, the tail of the potential (i.e., the potential beyond the primary minimum) has a steeply varying attraction followed by a slowly varying repulsion as r increases. (The tail shows a very small, roughly $0.05\,k_B T$, secondary minimum in this case.) Nevertheless, the inversion method performs quite well. Overall, the inversion procedure appears capable of reproducing both the near-field and the far-field behavior of the potential very well, and in a physically consistent manner, as long as the given structure factor is sufficiently accurate.

13.5 Discussion

In this section, we review briefly how practical it is to use inversion schemes to extract the details of the interaction potentials from static structure factors. In order to appreciate the difficulties involved in using inversion methods, it is useful to consider first the role of the direct correlation function $c(r)$ in the inversion procedures. This will illustrate some of the problems related to the influence of the accuracy of the structure factor data and of the range of Qs over which they are available on the accuracy and the stability of inversion procedures (the one discussed here as well as others). Following this, we shall comment on some of the other practical difficulties (e.g., polydispersity effects and multibody interactions) that place severe restrictions on the potential success of the inversion.

13.5.1 The Role of Direct Correlation Functions

An essential feature of almost all inversion methods is that they depend, directly or indirectly, on successful retrieval of the direct correlation function $c(r)$ from the given structure factor. This function is related to $S(Q)$ by the relation

$$c(Q) = \frac{[S(Q) - 1]}{n_p S(Q)} \tag{13.16}$$

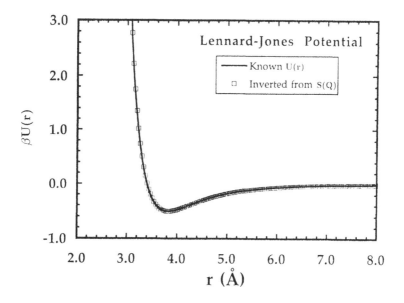

Figure 13.5 The known LJ potential is compared with the potential inverted from the corresponding structure factor shown in Fig. 13.4.

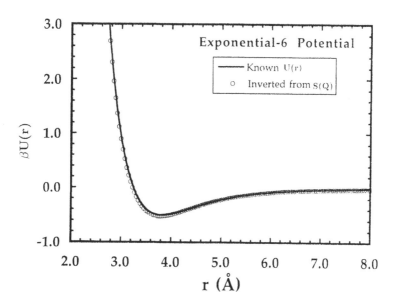

Figure 13.6 The known exp-6 potential is compared with the potential inverted from the corresponding structure factor shown in Fig. 13.4.

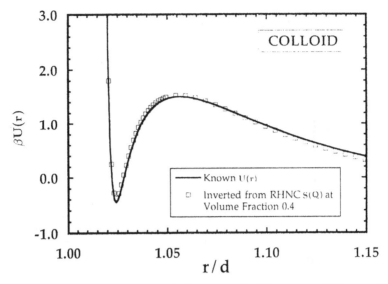

Figure 13.7   The known DLVO potential is compared with the potential inverted from the structure factor for a volume fraction of 0.4.

where $c(Q)$ is the Fourier transform of $c(r)$. The correlation function $c(r)$ is sensitive to errors in $S(Q)$ (particularly in the low-Q region), and this may be compounded further because of numerical errors introduced by Fourier transforms when the Q-space information is transformed to real space. This is one of the major reasons for the failure of inversion approaches based on simple integral-equation closures such as the PY and HNC approximations, which we discussed briefly in an earlier section. This statement also applies to a number of other approaches such as the ones based on the mean-spherical approximation (MSA) and generalized mean-spherical approximation (GMSA), which are outlined elsewhere [24]. The same is true in the case of the single-step RHNC (reference hypernetted-chain) method (i.e., without a "corrector" step as done in our method), although the latter is generally superior to the others mentioned because of the appearance of the reference hard-sphere bridge-function correction (to the HNC closure) in the predictor.

The direct correlation function does not appear explicitly in the equations presented earlier for the inversion scheme discussed in this chapter; however, it is implicitly present in the calculation of $U_p(r)$, although its effect is diminished to some extent since $U_p(r)$ is proportional to the difference between $c(r)$ of the entire potential and $c_d(r)$, the direct correlation function of the hard-sphere reference fluid. In fact, it can be shown that the method presented here is an iteration based on the RHNC closure. Generally the inaccuracies in $c(r)$ are at the source of what makes inversions difficult even when the predictor is reliable. For this reason, the corrector needs to be designed very carefully. We have chosen to correct only the influence of the effective excluded volume at each step. This allows one to have a relatively fast and easy-to-use scheme for the inversion, but its adequacy for dispersions at high concentrations requires further examination. An alternative is to use computer simulations as correctors [to generate $S(Q)$ from the potential inverted at each step] for subsequent steps, as was done by Reatto et al. [31]. While this is expected to be more accurate, its use is computationally demanding and time consuming, and the convergence of the procedure is not clear.

13.5.2 Some Practical Considerations

A number of practical issues usually determine the usefulness and the efficacy of any inversion technique. We shall discuss some of these briefly in this section. Three issues that are particularly important in this context are:

1. *The sensitivity of the method to errors in* $S(Q)$. This includes experimental errors in $S(Q)$ as well as the related issue of the extension of $S(Q)$ beyond the finite range over which they are usually accessible. The extension of experimental $S(Q)$ includes: (1) the zero-Q limit [i.e., Q values smaller than Q_{peak}, the location of the first peak in $S(Q)$] is not always accessible experimentally to a sufficient extent, and one may need to extend the data in this region by suitable approximations; and (2) a similar situation may also arise for large Q values (i.e., one may find that sufficiently large values of Q for which $S(Q)$ tends to unity are not reachable experimentally).

   The importance of accurate values of $S(Q)$ may be seen from the conflicting results illustrated in Fig. 13.1, in which the experimental data seem to be in agreement with the structure factors based on both the Yukawa and the Sogami potentials. In this particular case, it is the difference in the two computed structure factors in the low-Q region that distinguishes one potential from the other (see Table 13.1)—a region for which experimental data are not available and for which data are, in general, difficult to get accurately. The example illustrated in Figs. 13.1 and 13.2 and in Table 13.1 thus illustrates (1) the importance of $S(Q)$ in the low-Q region, (2) the need for highly accurate data, and (3) the potential implications in extending the data beyond the range of Qs for which they are available.

   The issue of experimental errors in $S(Q)$ has been addressed by Guérin [28] to a limited extent and our calculations show that a 10% error in the most important region of $S(Q)$, that is, the low-Q region (especially for $Q \leq 0.8Q_{peak}$), causes roughly a 5% error in the region of the minimum in the potential. This is in sharp contrast to previous methods (including the PY and HNC methods), where one sees a doubling of the error in the region of the minimum in the potential [24,32]. A similar analysis for errors in the region beyond the second peak in $S(Q)$ reveals that the inversion is insensitive to errors for large Qs. Thus the method seems to be very stable relative to numerical and experimental errors. The fact that the inversion is stable with respect to errors in the large-Q region does not imply that the truncation of

Table 13.1 Low-Q structure factors based on Monte Carlo simulations for the Yukawa and Sogami potentials shown in Fig. 13.2 for the conditions used in Fig. 13.1

	Yukawa		Sogami	
Qd	$S(Q)$	Error	$S(Q)$	Error
0.11956	0.0175	0.00207	0.0835	0.00599
0.23912	0.0228	0.00118	0.0716	0.00236
0.35867	0.0398	0.00157	0.0805	0.00243
0.47823	0.0796	0.00455	0.1235	0.00378
0.59799	0.2077	0.02149	0.2575	0.00884

The 95% confidence intervals are given by $S(Q)\pm$ Error.
Please see Salgi et al. [30] for additional details. (Due to a typographical error, the value of d in [30] is incorrect; the correct value is 1.09×10^{-7} nm.)

$S(Q)$ in the large-Q region is of no consequence. The truncation of $S(Q)$ before $S(Q)$ reaches the asymptotic limit of unity sufficiently closely (or the nonavailability of experimental data in this region) has a severe effect on the quality of the radial distribution function derived from the data. This is an important problem that remains essentially unaddressed adequately at present, although maximum-entropy formalisms and other techniques have been explored to some extent in this context.

2. *The effects of polydispersity of the dispersions.* It is well known that the effects of polydispersity are equally important. The structure of polydisperse colloids has not been studied in great detail so far, but a reasonable amount of information for hard-spheres, adhesive hard-spheres, and charged spheres with and without attraction is currently available [33]. These results show that the average structure factor deviates only slightly from the structure factor based on the average particles as long as the standard deviation of size polydispersity (or polydispersity in other parameters) is within about 10% of the average size (or the average value of the appropriate parameter). In such a case, the inversion will lead to an "average" *effective* potential. If decoupling approximations [33] can be used, then one can again obtain an average effective potential. For binary or multicomponent systems, the inversion method presented here can be extended appropriately (as shown by Guérin [28]), if the partial structure factors are known.

3. *The effects of multibody interactions.* Any inversion technique can provide only an *effective* pair potential if multibody interactions are significant. The inversion scheme presented here may be supplemented further by imposing other "consistency" criteria in such cases (such as forcing osmotic pressure or some other chosen property based on the inverted potential to be the same

as that of the actual system) so as to guarantee a useful effective potential in more complex cases.

13.6 Concluding Remarks

Scattering data (preferably in combination with suitable property measurements) provide the best opportunity for studying interaction forces in condensed matter. If there are sufficiently strong physical grounds for choosing the form of the potential, the "fitting" method described in Section 13.2 offers a straightforward and relatively easy route to extracting information on interaction forces. However, it has serious drawbacks, as illustrated in Figs. 13.1 and 13.2, when no independent arguments can be advanced for selecting the functional form of the potential.

The formal inversion methods are clearly the better option when a priori assumptions concerning the potential are not desirable or cannot be justified. However, such methods require data of very high quality and over the entire range of Q and are still untested. The method described in this chapter is still an approximate one since the corrector does not use the full inverted potential at each step of the iteration. However, the advantage is that it is much simpler to use than, for example, a simulation-based corrector. Moreover, it appears to be sufficient to discriminate between drastically different potentials like the ones shown in Fig. 13.2. More detailed and extensive investigations of the inverse problem posed in this chapter are, therefore, expected to be very useful. In the case of colloids, however, polydispersity of sizes, charges, etc. restrict the use of inversion methods to model colloids whose relevant parameters can be assumed to be highly monodisperse.

Acknowledgements

We thank the U.S. National Science Foundation and the Petroleum Research Funds administered by the American Chemical Society for partial support of some of the work presented here.

References

1. Verwey, E. J. W., and Overbeek, J. Th. G., *Theory of the Stability of Lyophobic Colloids* (Elsevier, Amsterdam, 1948).
2. Hunter, R. J., *Foundations of Colloid Science, Vols. I & II* (Oxford Univ. Press, Oxford, UK, 1987, 1989).
3. Hiemenz, P. C., and Rajagopalan R., *Principles of Colloid and Surface Chemistry*, 3rd Edition (Marcel Dekker, New York, 1996).
4. Israelachvili, J. N., *Intermolecular and Surface Forces*, 2nd Edition (Academic, London, 1991).
5. Hirtzel, C. S., and Rajagopalan R., *Advanced Topics in Colloidal Phenomena* (Noyes, Park Ridge, NJ, 1985).
6. Safran, S. A., and Clark, N. A., Eds., *Physics of Complex and Supermolecular Fluids* (Wiley, New York, 1987).

7. Ise, N., and Sogami, I., Eds., *Ordering and Organisation in Ionic Solutions* (World Scientific, Singapore, 1988).
8. Chen, S.-H., and Rajagopalan, R., Eds., *Micellar Solutions and Microemulsions: Structure, Dynamics, and Statistical Thermodynamics* (Springer-Verlag, New York, 1990).
9. Arora, A. K., and Rajagopalan, R. Chapter 1, this volume.
10. Murray, C. A., and Wenk, R. A., *Phys. Rev. Lett.* **62**, 1643 (1989).
11. Richetti, P. and Kékicheff, P., *Phys. Rev. Lett.* **68**, 1951 (1992).
12. Parker, J. L., Richetti, P., Kékicheff, P., and Sarman, S., *Phys. Rev. Lett.* **68**, 1955 (1992).
13. Prieve, D. C., Luo, F., and Lanni. F., *Faraday Discuss. Chem. Soc.* **83**, 22 (1987).
14. Schumacher, G. A., and van de Ven, T. G. M., *Langmuir* **7**, 2028 (1991).
15. Ailawadi, N. K., *Phys. Rep.* **57**, 241 (1980).
16. Johnson, M. D., and March, N. H., *Phys. Lett.* **3**, 313 (1963).
17. Johnson, M. D., Hutchinson, P., and March, N. H., *Proc. Roy. Soc. London* **A282**, 283 (1964).
18. March, N. H., *Liquid Metals: Concepts and Theory* (Cambridge Univ. Press, Cambridge, 1990).
19. Rajagopalan R. and Hirtzel, C. S., *J. Colloid Interface Sci.* **118**, 422 (1987).
20. Degiorgio, V., and Corti, M., Eds., *Physics of Amphiphiles: Micelles, Vesicles and Microemulsions* (North-Holland, Amsterdam, The Netherlands, 1985).
21. Chen, S.-H., Sheu, E. Y., Klaus, J. and Hoffmann, H., *J. Appl. Cryst.* **21**, 751 (1988).
22. Tata, B. V. R., Sood, A. K., and Kesavamoorthy, R., *Pramana* **34**, 23 (1990).
23. Sogami, I., *Phys. Lett.* **A96**, 199 (1983).
24. Sen, A. K., and Rajagopalan, R., *Phys. Chem. Liq.* **22**, 195 (1991).
25. Rajagopalan, R., *Langmuir* **8**, 2898 (1992).
26. Hansen, J.-P., and McDonald, I. R., *Theory of Simple Liquids*, 2nd Edition (Academic, New York, 1986).
27. Castillo, C. A., Rajagopalan, R., and Hirtzel, C. S., *Rev. in Chem. Eng.* **2**, 237 (1984).
28. Guérin, J.-F., PhD Dissertation, University of Houston, 1992.
29. Lado, F., *Mol. Phys.* **52**, 871 (1984).
30. Salgi, P., Guérin J.-F., and Rajagopalan, R., *Coll. Polym. Sci.* **270**, 785 (1992).
31. Reatto, L., Levesque, D., and Weis, J. J., *Phys. Rev.* **A33**, 345 (1986).
32. Ballentine, L. E., and Jones, J. C., *Can. J. Phys.* **51**, 1831 (1973).
33. Salgi, P., and Rajagopalan, R., *Adv. Colloid Interface Sci.* **43**, 169 (1993).

Index

Lightning Source UK Ltd.
Milton Keynes UK
UKOW06n0321181115

262933UK00001B/41/P